Parasitology

Volume 124   Supplement 2002

Parasites in marine systems

EDITED BY
R. POULIN

CO-ORDINATING EDITOR
L. H. CHAPPELL

CAMBRIDGE
UNIVERSITY PRESS

CAMBRIDGE UNIVERSITY PRESS
Cambridge, New York, Melbourne, Madrid, Cape Town, Singapore, São Paulo

Cambridge University Press
The Edinburgh Building, Cambridge CB2 8RU, UK

Published in the United States of America by Cambridge University Press, New York

www.cambridge.org
Information on this title: www.cambridge.org/9780521534123
First published 2002

A catalogue record for this publication is available from the British Library

ISBN 978-0-521-53412-3 paperback

Transferred to digital printing 2007

Contents

Contents

List of contributions

Parasites in marine systems

EDITED BY R. POULIN

CO-ORDINATING EDITOR L. H. CHAPPELL

Preface

The sea is the last frontier for biologists. The oceans are not only a major source of food for much of humanity, they are also the repository of most of the Earth's biodiversity. We know relatively little about the interactions among their inhabitants, however, and this is especially true with respect to parasitic organisms. This supplement consists of 12 review articles, written by international experts, each summarising and synthesising the available information on key aspects of the biology of marine parasites.

The topics included cover the evolution and ecology of marine host-parasite associations, as well as applied aspects of marine parasitology with commercial repercussions. Hoberg & Klassen and Cribb *et al.* begin by looking at the co-evolution and biogeography of host-parasite associations in different marine systems. An evolutionary perspective is necessary to understand how assemblages of different parasite species have come to share the same hosts, but it is ecological processes that determine the structure of these assemblages. The next two papers examine the species interactions and structure of parasite assemblages in the two best-studied marine host-parasite systems: larval trematodes in gastropods (Curtis), and ectoparasites on fish (Morand *et al.*). Fish ectoparasites are unusual in that they are subjected to predation by cleaner organisms, and thus are at the basis of one of the best-known marine symbioses; Grutter takes a look at these cleaning symbioses from the parasites' perspective.

On a larger scale, parasites are fundamental components of marine food webs and ecosystems. Marcogliese discusses the transmission patterns of parasites in marine food webs. His paper is followed by two reviews of the impact of parasitism on marine ecosystems: in intertidal habitats (Mouritsen &

Poulin) and in the Baltic Sea, where anthropogenic environmental changes are known to affect parasites (Zander & Reimer). The importance of parasites in ecosystems becomes clear when they are introduced to new areas; Torchin *et al.* examine the impact of introduced parasites on native ecosystems, and the role played by parasites in determining the invasion success of exotic, non-parasitic species. The ubiquitous nature of parasites in marine systems means that they will have commercial importance in fisheries and aquaculture. On the positive side, they can be used as biological tags for stock discrimination, as highlighted by MacKenzie. On the negative side, many parasites are serious pathogens of commercially important host species; the supplement ends with two case studies of parasites with major impacts on their hosts, salmon lice (Tully & Nolan) and sealworm (McClelland).

I thank the contributors for their efforts in putting together this collection of up-to-date reviews. I must also acknowledge the 23 referees (the 5 who are themselves contributors and the 18 others who are not) who put in some time to improve the reviews, and Les Chappell, the co-ordinating editor, who very efficiently oversaw the whole project. As a whole, the reviews in this supplement present an easy-to-read, comprehensive account of recent advances in the study of marine parasites. I hope that they will stimulate students and professional scientists interested in parasitology, fisheries, marine ecology, community and ecosystem biology, and evolutionary biology, to pursue the many open questions still left to be tackled.

ROBERT POULIN
January 2002

Revealing the faunal tapestry: co-evolution and historical biogeography of hosts and parasites in marine systems

E. P. HOBERG[1]* and G. J. KLASSEN[2]

[1] *Parasite Biology, Epidemiology and Systematics Laboratory, and the US National Parasite Collection, Agricultural Research Service, US Department of Agriculture, BARC East 1180, Beltsville, Maryland, USA 20705*
[2] *Department of Biology, University of New Brunswick, P.O. Box 5050, Saint John, New Brunswick, Canada, E2L 4L5*

SUMMARY

Parasites are integral components of marine ecosystems, a general observation accepted by parasitologists, but often considered of trifling significance to the broader community of zoologists. Parasites, however, represent elegant tools to explore the origins, distribution and maintenance of biodiversity. Among these diverse assemblages, host and geographic ranges described by various helminths are structured and historically constrained by genealogical and ecological associations that can be revealed and evaluated using phylogenetic methodologies within the context of frameworks and hypotheses for co-evolution and historical biogeography. Despite over 200 years of sporadic investigations of helminth systematics, knowledge of parasite faunal diversity in chondrichthyan and osteichthyan fishes, seabirds and marine mammals remains to be distilled into a coherent and comprehensive picture that can be assessed using phylogenetic approaches. Phylogenetic studies among complex host–parasite assemblages that encompass varying temporal and geographic scales are the critical context for elucidating biodiversity and faunal structure, and for identifying historical and contemporary determinants of ecological organization and biogeographic patterns across the marine biosphere. Insights from phylogenetic inference indicate (1) the great age of marine parasite faunas; (2) a significant role for colonization in diversification across a taxonomic continuum at deep and relatively recent temporal scales; and (3) a primary role for allopatric speciation. Integration of ecological and phylogenetic knowledge from the study of parasites is synergistic, contributing substantial insights into the history and maintenance of marine systems.

Key words: Marine parasites, co-evolution, historical biogeography, marine biodiversity.

This may be wrong and I would be glad to have anyone disprove the theory as what we want is knowledge, not the pride of proving something to be true.
Ernest Hemingway (1934) Out in the Stream: A Cuban Letter; *Esquire.*

INTRODUCTION

Parasite faunas characteristic of marine vertebrates have been assembled through an intricate interaction of history, ecology and geography, as the determinants of organismal evolution and distribution. Elucidation of pattern and process in the origin and maintenance of biodiversity in marine systems follows from studies that integrate phylogenetic approaches and an historical context for biogeography and ecology (e.g. Brooks & McLennan, 1991, 1993 a,b; Hoberg, 1996, 1997; Page & Charleston, 1998; Brooks & Hoberg, 2000).

Substantial knowledge about species diversity and both host and geographic distribution for a phylogenetically and ecologically diverse array of parasites among fishes, seabirds and marine mammals has been assembled over the past 200 years. Despite a rich base of fundamental knowledge, the parasite faunas of marine vertebrates have received uneven attention with respect to their co-evolution (encompassing co-speciation and co-adaptation) and historical biogeography within a current methodological framework (Table 1). Thus, although we continue to acquire new information about the distribution of species and higher taxa, we also continue to be challenged to define and understand broad patterns in geographic distribution, host-association and evolutionary history. It is necessary to build a database of comparable studies across an array of taxonomic and geographic scales.

Early studies on faunal distribution and evolution relied on inspection and intuition to define the evolutionary and biogeographic histories for complex host–parasite assemblages; e.g. among marine mammals (Deliamure, 1955) and marine teleosts (Manter, 1966; Lebedev, 1969; and reviewed in detail by Rohde, 1993). These and other monographic studies such as the comparative work on the deep-sea faunas by Campbell (1983) were the precursors or empirical foundations for identifying large-scale patterns in distribution (e.g. depth and latitudinal gradients as outlined by Rohde, 1992, 1993), biogeography or host associations. Often research focused on attempts to use parasites to reveal host evolutionary relationships, or centres of origin and were based on concepts for 'parasitological

* Corresponding author: Tel: 301-504-8588. Fax: 301-504-8979. E-mail: ehoberg@anri.barc.usda.gov

Table 1. Studies emphasizing an explicit phylogenetic foundation for examination of hypotheses for co-evolution and historical biogeography among marine host-parasite assemblages; including studies of conceptual importance

Date	Author(s)	Area(s)	Host(s)	Parasite(s)	Contribution*
1979	Brooks	General	General	General	Principles, methods, correlation with micro-evolutionary approaches such as Island Biogeography.
1980	Brooks	General	Rockfish	Digenea	Role of phylogeny in interpreting the co-evolutionary relationships between hosts and 'communities' of parasites.
1980	Holmes & Price	General	Rockfish	Digenea	Rebuttal to Brooks
1981	Brooks	General	General	General	Foundations for Brooks Parsimony Analysis
1981b	Brooks et al.	South America	Chondrichthyes, Potamotrygonidae	Helminths	Empirical tests of origin of freshwater stingrays based on phylogeny, distribution and co-evolutionary relationship of parasites; vicariance vs dispersal; history for marine and freshwater taxa.
1983	Cressey et al.	General	Teleostei, Scombridae	Copepoda	Co-evolutionary relationships; host specificity, biogeography.
1985	Collette & Russo	Pacific/Atlantic	Teleostei, Scombridae	Copepoda	Phylogenies of Spanish mackerels and their copepod parasites; empirical test of co-evolutionary and biogeographic relationships across Panamanian Isthmus.
1985	Brooks	General	General	General	Concept for historical ecology as a research programme.
1986	Hoberg	Holarctic, Beringia	Aves, Alcidae	Eucestoda, Cyclophyllidea, Alcataenia spp.	Co-speciation and historical biogeography of North Pacific basin in Pliocene/Pleistocene; role of colonization in parasite diversification; vicariance and climate; host-specificity decoupled from co-speciation.
1987	Deets	General	Chondrichthyes	Copepoda, Siphonostomatoida	Parasite taxonomy and phylogeny with discussion on congruence between host and parasite evolution.
1987a	Bandoni & Brooks	General	Holocephala	Gyrocotylidea	Phylogeny of Gyrocotylidea. Co-evolution history attributed to combination of co-speciation and colonization. Indicates origin of gyrocotylids (and host association) predated separation of continents.
1987b	Bandoni & Brooks	Southern Continents	Teleostei	Amphilinidea	Phylogeny of amphilinids. Co-evolution attributed to high degree of co-speciation. Pattern of geographic distribution consistent with vicariance.
1988	Benz & Deets	General	Chondrichthyes; Teleostei, Mobulidae	Copepoda, Cercopidae	Phylogenetic and biogeographic relationships for copepods on epipelagic fishes.
1988	Brooks & Deardorf	Cosmopolitan	Chondrichthyes, Dasyatididae	Eucestoda, Tetraphyllidea; Nematoda, Echinocephalus	Phylogenetic and biogeographic analyses indicate an ancient 'Tethys Sea-Circum-Pacific origin; supports hypothesis of Pacific origin of Potamotrygonidae.
1988	Brooks & Bandoni	General	Holocephala; Teleostei	Gyrocotylidea/Amphilinidea	Distinguish relictual (i.e., ancient and persistent) co-evolutionary associations from recent colonization matching host phylogeny.
1989	Boeger & Kritsky	General	Chondrichthyes	Monogenea	Co-evolutionary history for Hexabothriidae.
1990	Measures et al.	Australasian	Chondrichthyes, Rajiformes	Monogenea, Monoctyle spp.	Host specificity.
1991	McLennan & Brooks	General	Teleostei, Gasterosteidae	Helminths	Phylogeny and behavioural characters. Parasites and sexual selection. Macro-evolutionary tests of micro-evolutionary predictions such as the Hamilton-Zuk hypothesis.
1991	Paggi et al.	Arctic, Atlantic Basin	Pinnipedia, Phocidae	Nematoda, Ascaridoidea, Pseudoterranova	Pliocene/Pleistocene history of isolation and diversification, Atlantic sector of Arctic.

Year	Author	Host	Parasite	Location	Notes
1992	Brooks	Chondrichthyes, Potamotrygonidae	Helminths	South America	Revisiting of the origin of fw stingrays. Emphasis on the hypothesis that (contrary to previous assumptions of Atlantic in origin), fauna originated from Pacific when Amazon flowed West.
1992	Klassen	Teleostei; Ostraciidae	Monogenea	General	First simultaneous reconstruction of host and parasite phylogenies. Marine parasite fauna useful in reconstructing ancient biogeographic relationship between Pacific and Caribbean. Modification of BPA to allow for inclusion of population level data in interpreting co-evolutionary hypotheses.
1992	Hoberg & Adams	Pinnipedia, Phocidae, Otariidae	Eucestoda, Tetrabothriidea, *Anophryocephalus*	Holarctic, Beringia	Co-speciation and historical biogeography of Holarctic fauna; colonization and linkage to diversification during Pliocene/Pleistocene.
1992	Hoberg	Pinnipedia, Phocidae, Otariidae; Aves, Alcidae	Eucestoda, Tetrabothriidea, Cyclophyllidea	Holarctic, Beringia	Biogeographic and historical congruence in diversification of phylogenetically disparate avian and pinniped cestodes; Arctic Refugium Hypothesis; host-switching as driver for diversification in marine systems; models for allopatric speciation.
1993	Nascetti *et al.*	Pinnipedia, Phocidae	Nematoda: Ascaridoidea, *Contracaecum*	Arctic, Atlantic Basin	Pliocene/Pleistocene history of isolation and diversification.
1993	Marcogliese & Cone	Teleostei, Anguillidae	Metazoan parasites	North Atlantic	Use of parasite community structure to differentiate between competing hypotheses for the origin and divergence of American and European eels.
1994 *a*	Klassen	Teleostei; Ostraciidae	Monogenea	General	Examination of monophyly of host-parasite association; biogeography of Monogenea in tropical oceans.
1994 *b*	Klassen	Teleostei; Ostraciidae	Monogenea	Atlantic	Hypothesis of multiple invasions of parasites, and by implication their hosts, into Caribbean from Pacific.
1994 *a*	Page	General	General	General	Analytical methods for cospeciation & historical biogeography.
1994 *b* / 1995	Hoberg	Pinnipedia, Phocidae	Eucestoda, Tetrabothriidea, *Anophryocephalus*	Holarctic, Beringia	Models for allopatric speciation, peripheral isolates; history of Arctic Basin, Beringia.
1995	Arduino *et al.*	Pinnipedia, Monachinae	Nematoda, Ascaridoidea, *Contracaecum*	Antarctica	Geographic colonization by phocids and subsequent parasite diversification.
1995 *a*	Pérez-Ponce de León & Brooks	Chelonia	Digenea, Pronocephalidae	Cosmopolitan	Historical biogeography, co-evolution in marine turtles; complex and deep history for dispersal and vicariance.
1995 *b*	Pérez-Ponce de León & Brooks	Chelonia	Digenea, *Pyelosomum* spp.	Cosmopolitan	Host-switching from turtles to marine iguanas.
1996	Hoberg	Marine birds	Eucestoda, Tetrabothriidea, *Tetrabothrius*	General	Deep history, putative co-evolution with seabirds subsequent to colonization.
1996	León-Règagnon *et al.*	Teleostei	Digenea, *Opisthadena* spp.	Circum-Pacific	Historical biogeography; archaic circum-Pacific distribution.
1996	Thomas *et al.*	Marine fishes	Helminths	General	Parasites as phylogenetic indicators for hosts.
1997	Hoberg	General	General	General	Concepts for historical biogeography and application of host-parasite systems.
1997	Hoberg *et al.*	Aves, Alcidae	Eucestoda, Cyclophyllidea *Alcataenia* spp.	General	Comparison of BPA and Component methods for examination of co-evolution.
1997	Boeger & Kritsky	Chondrichthyes; Osteichthyes	Monogenea	General	Deep evolutionary history, extending to Paleozoic, involving complex colonization and co-evolution.
1997	Bullini *et al.*	Pinnipedia, Phocidae	Nematoda, Ascaridoidea	Antarctica; Arctic	Bipolar relationships for ascaridoids in phocids; history of isolation and diversification.
1997	Mattiucci *et al.*	Cetacea	Nematoda, Ascaridoidea, *Anisakis*	Boreal; Subantarctic	Allopatric speciation; diversification in Pliocene/Pleistocene.
1997	Hayward	Teleostei, Sillaginidae	Monogenea	Indo-West Pacific	Identification of regional centers of diversity for parasites and hosts.
1997	Lovejoy	Chondrichthyes, Potamotrygonidae	Helminths	South America	Re-evaluation of hypotheses for relationships and origins of fresh-water stingrays in Amazonia.

Table 1. (Contd.)

Date	Author(s)	Area(s)	Host(s)	Parasite(s)	Contribution*
1997	Pérez-Ponce de León et al.	Atlantic/Pacific	Teleostei, *Albula* spp.	Monogenea, Pterinotrematidae	Historical biogeography, co-evolution.
1997	Paterson & Gray	Southern Ocean	Aves, Sphenisciformes, Procellariiformes	Phthiraptera	Co-speciation analysis; methods & protocols for Component/Reconciliation-based studies.
1998	Bray et al.	Tethys Sea	Teleostei, reef fishes	Digenea, *Lepidapedoides* spp.	Host specificity; broad tropical distribution.
1998	Choudhury & Dick	Atlantic Basin	Osteichthyes, Acipenseridae	Digenea, Deropristiidae	Patterns of co-speciation and vicariance in Cretaceous; relationships of marine and freshwater taxa.
1998	Hoberg et al.	Eastern Pacific	Chondrichthyes, Myliobatiformes	Nematoda, *Echinocephalus*	Pacific origins for freshwater stingrays of Amazon basin.
1998a	Fernández et al.	General	Cetacea; Pinnipedia	Digenea, Campulidae, Nasitrematidae	Origin of pinniped parasites by colonization from odontocete cetaceans.
1998b	Fernández et al.	General	Cetacea; Pinnipedia	Digenea, Campulidae	Co-evolution and history of campulids and *Orthosplanchnus* among marine mammals.
1998	Mendoza-Garfias & Pérez-Ponce de León	Atlantic/Pacific	Teleostei, *Cynoscion* spp.	Monogenea, *Cynoscionicola*	Historical biogeography, vicariance; Panamian Isthmus, Pliocene isolation.
1998	León-Regagnon	Pacific	Teleostei	Digenea, Hemiuridae	Historical biogeography and evolution in Pacific and Indo-Pacific reef fishes; host switching & dispersal.
1998	Leon-Regagnon et al.	Tethys Sea, Pacific	Teleostei, Clupeidae	Hemiuridae Bunocotylinae	History of radiation and dispersal in clupeid fishes centered in the Tethys Sea.
1999a	Hoberg et al.	General	Chondrichthyes; Aves; Mammalia	Eucestoda	Deep history extending to Paleozoic for origins of major taxa of marine cestodes.
1999b	Hoberg et al.	General	Aves; Mammalia	Eucestoda, Tetrabothriidea	Origin of tetrabothriids by colonization of basal seabirds in Mesozoic.
1999	Olson et al.	General	Chondrichthyes	Eucestoda, Tetraphyllidea, Lecanicephalidea	Co-evolutionary history, demonstration of host-specific phylogenetic patterns.
1999	Paterson & Poulin	General	Teleostei	Copepoda, *Chondracanthus* spp.	Co-speciation analyses; deep history of diversification by co-speciation and colonization in a geographically widespread host and parasite assemblage. Testing co-evolutionary models.
1999	Zamparo et al.	Eastern Pacific	Chondrichthyes, Myliobatiformes	Eucestoda, Tetraphyllidea	Pacific origins for Potamotrygonidae in Amazonian freshwater habitats.
2000	Hoberg & Adams	General	Pinnipedia; Cetacea	Helminths	Co-evolution, colonization, temporal and geographic scale, and faunal history.
2000	Brooks et al.	Pacific Ocean	Teleostei, Kyphosidae	Digenea, Lepocreadiidae	Historical biogeography, origins of Pacific taxa; Host switching and geographical dispersal.
2000	Fernández et al.	General	Cetacea, Mysticete	Digenea, Campulidae, *Lecithodesmus*	Colonization of Mysticete; host-switching processes in diversification.
2000	Nadler et al.	General	Pinnipedia; Aves	Nematoda, Ascaridoidea, *Contracaecum*	Host-switching among seabirds and pinnipeds.
2000	Paterson et al.	New Zealand, Southern Ocean	Aves, Sphenisciformes, Procellariiformes	Phthiraptera	Co-speciation analyses; comparison BPA & Reconciliation/TreeMap; conceptual issues of co-speciation, intra-host speciation & host switching; deep history of co-speciation.
2000	Rohde & Hayward	Circum-tropical	Teleostei, Scombridae	Copepoda; Monogenea	Structure of tropical faunas; biogeographic barriers to dispersal; Tethys Sea relationships.
2001	Caira & Jensen	General	Chondrichthyes	Eucestoda, Onchobothriidae	Host specificity & coevolution; incongruence for host and parasite phylogenies.
2001	Cribb et al.	General	Fishes, Molluscs	Digenea	Deep evolutionary history, ancestral hosts.
2001	Paterson & Banks	General	General	General	Concepts for co-speciation analyses; comparisons of analytical methods.
2001	Brooks et al.	General	General	General	Current mechanics and applications of Brooks Parsimony Analysis.

* This category is not meant to be inclusive, merely to highlight specific points relevant to our discussion.

rules' and host specificity (see Rohde, 1993) that have to some extent been superseded (Klassen, 1991; Brooks & McLennan, 1993*a*; Hoberg, Brooks & Siegel-Causey, 1997; Paterson & Banks, 2001). Early empirical observations, however, have often become the focus for recent phylogenetically-based approaches (e.g. Hoberg, 1992; Klassen, 1992; Brooks & McLennan, 1993*a*; Rohde & Hayward, 2000). Still explicit here is the concept, articulated by Manter (1966), that parasites serve as keystones for understanding the history of biotas because of their critical value as phylogenetic, ecological and biogeographic indicators of their host groups (e.g. Brooks, 1985; Brooks & McLennan, 1993*a,b*; Hoberg, 1997; Brooks & Hoberg, 2000).

Parasite faunas of marine vertebrates have been assembled across varying temporal and geographic scales. Further, associations are historically constrained by genealogical and ecological associations (e.g. Brooks & McLennan, 1993*a*). Origins, temporal continuity and structure of marine parasite assemblages can be examined within the framework of hypotheses for co-evolution or colonization that are derived from the comparative study of phylogenies for hosts and parasites generated from analyses based on morphological or molecular data (e.g. Brooks, 1979, 1981; Klassen, 1992; Brooks & McLennan, 1993*a*; Page, 1994*a, b*; Brooks & Hoberg, 2000; Hoberg & Adams, 2000; Paterson & Banks, 2001).

Hoberg & Adams (2000) recently outlined some of the primary criteria for defining associations that have developed through co-evolution versus colonization. Co-evolution, or association by descent, is corroborated through examination and interpretation of host–parasite associations that demonstrate: (1) consistency or congruence in host–parasite phylogenies or area relationships; (2) a high degree of co-speciation or co-adaptation; (3) recognition of phylogenetic or numerical relicts; (4) often widespread geographic distributions, that in marine systems may be global or antitropical in extent. General congruence in biogeographic patterns among complex host–parasite assemblages indicates coincidental physical and biotic processes as determinants of distribution (e.g. Hoberg, 1986, 1992, 1997). In these instances, geographic scale may be linked to the relative age for the initial association of parasite and host taxa, vagility of the assemblage, and duration of their history for co-evolution. Additionally, Hafner & Nadler (1988) and Hafner *et al.* (1994) introduced the concept of temporal comparisons for molecular evolution between hosts and parasites, revealing an important facet to be considered in studies of co-phylogeny.

Faunas derived from a history of colonization contrast with co-evolutionary systems in the following ways: (1) incongruent and inconsistent phylogenies for parasites and hosts; (2) similarities

in host trophic ecology; (3) faunas that are geographically or regionally delimited; (4) parasite faunas in which diversification is temporally circumscribed in the context of the origin and duration of the host group; (5) faunas of low diversity that are depauperate as opposed to relictual; and (6) associations of variable temporal duration and varying degrees of co-speciation/co-adaptation linked to the time frame for colonization of the host clade(s). Page (1994*a*), Paterson *et al.* (2000) and Paterson & Banks (2001) would further suggest that incongruence can arise from events of (1) intra-host speciation (but here incongruence may be a function of scale as demonstrated in an analysis of monogeneans and teleost hosts outlined below), and (2) different patterns of sorting events including extinctions.

Criteria for co-evolutionary or colonizing faunas set a hypothesis-driven framework to evaluate faunal structure in marine systems. Although most current studies on marine helminth systems have applied parsimony mapping (including mapping on both host and parasite phylogenies), or Brooks Parsimony Analysis (BPA) (see Brooks, 1981, 1990; Brooks & McLennan, 1993*a*; Brooks, van Veller & McLennan, 2001; van Veller & Brooks, 2001), alternative analytical methods have been articulated. These include Component or Reconciliation-based approaches which to some degree are now yielding increasingly convergent results with BPA (see Paterson & Banks, 2001). It is not our intent to enter the methodological debate within the context of this paper, but consistently we apply BPA as a primary tool for discovery of underlying patterns an in addressing a range of issues in co-evolutionary and historical biology in marine systems.

The following review explores a range of complex determinants of genealogical and ecological diversity and faunal structure within a phylogenetic framework for host–parasite systems, focusing on helminths, in marine environments. Phylogenetic reconstruction is a powerful and synergistic tool to elucidate the history of marine biotas, and more generally the history of parasites, host–parasite associations and the biosphere (e.g. Brooks & Hoberg, 2000). Using a series of examples from recent phylogeny-based studies (Table 1) we will examine some overlying generalities and contrasting patterns for faunal structure, geographic and temporal scale, and the role of co-evolution and colonization, and articulate concepts for substantial driving mechanisms that influence diversity in marine ecosystems.

A DEEP HISTORY FOR MARINE PARASITE FAUNAS

A growing consensus based on phylogenetic studies among higher-level helminth taxa including Digenea, Monogenea, Gyrocotylidea, Amphilinidea and the Eucestoda across a diversity of host groups encompassing Chondrichthyes, Osteichthyes and the tetrapods clearly indicates a deep age for the origins of parasitic groups among vertebrates (Brooks, 1989; Rohde, 1994; Kearn, 1994; Boeger & Kritsky, 1997; Hoberg, Gardner & Campbell, 1999 a; Hoberg, Jones & Bray, 1999 b; Littlewood et al. 1999; Cribb, Bray & Littlewood, 2001). Tapeworms appear to have initially diversified among actinopterygian and neopterygian fishes 350–400 million years before present, and chondrichthyans were apparently colonized secondarily (Hoberg et al. 1999 a). Patterns of association for eucestodes appear to parallel those for both Monogenea and Digenea, suggesting that basal diversification for parasitic flatworms coincided with the origins and divergence of lineages for the Chondrichthyes and Osteichthyes prior to the Mesozoic (Brooks, 1989; Boeger & Kritsky, 1997; Cribb et al. 2001). This is compatible with a long period of diversification of such eucestode groups as the 'tetraphyllideans', Lecanicephalidea, Diphyllidea and Litobothriidae (Hoberg, Mariaux & Brooks, 2001; Olson et al. 1999, 2001) among chondrichthyans in marine and secondarily freshwater environments and more generally is indicative of the archaic nature of the faunas in sharks and rays (Euzet, 1959; Brooks, Thorson & Mayes, 1981 b; Bandoni & Brooks, 1987 a, b; Brooks & Deardorf, 1988; Brooks & McLennan, 1993 a; Nasin, Caira & Euzet, 1997). Concepts linked to recognition of a protracted history for tapeworms and various marine host taxa have been articulated by Hoberg et al. (1999 a, b), and emphasize the relictual nature of many groups (see Brooks & Bandoni, 1988). Diversity may have been influenced by radiation subsequent to colonization, or by secondary radiations in contemporary host taxa. A deep history of colonization is apparent, a further indication of the linkage between phylogeny and ecology as factors determining the historical and contemporary structure of parasite faunas in marine environments.

Global extinction and parasite diversity in deep time

Recognition of deep histories for major parasite taxa has substantial implications with respect to the role of global-level extinction events through Earth history as determinants of faunal structure and geographic distribution (Hoberg et al. 1999 a, b). Pertinent here is the idea that patterns of differential extinction for free-living taxa, across an array of potential intermediate or definitive hosts, have influenced genealogical or ecological diversity for parasites with complex indirect life cycles. Bush & Kennedy (1994) have suggested that extinction at the level of metapopulations would be unlikely but did not discuss this issue in the context of the 7–9 global events now documented for the Phanerozoic (Briggs, 1995).

In the marine environment, extinction horizons may be characterized by ecological perturbations of varying extent and duration leading to rapid elimination or turnover for many taxa (Briggs, 1995; Jin et al. 2000). Parasite lineages have persisted in time across a mosaic of ecological stability and disruption, and global-scale extinction events must be viewed as a series of episodic ecological transitions for host–parasite assemblages (e.g. Hoberg et al. 1999 a). Given the scope of past extinction events, e.g. loss of an estimated 90–96 % of marine species at the Permian-Triassic boundary 250 MA (Bowring et al. 1998; Jin et al. 2000), it is probable that the resilience and adaptive plasticity of parasites to respond to rapid environmental perturbation may have been insufficient to lead to temporal and geographic continuity for all lineages and populations (see Bush & Kennedy, 1994). Of particular interest at the P-T boundary is the decimation of late Permian reef communities with complete collapse at the ecosystem level (Briggs, 1995). Among the 7 documented episodes of extinction during the Phanerozoic, there was substantial variation in the diversity of benthic or pelagic taxa involved, regional effects, and the degree to which such environmental crises resulted in major ecological re-organizations (reviewed in Briggs, 1995).

We might ask the following: (1) has differential extinction or turnover of host taxa been an episodic driver of diversification for parasite taxa; and (2) how can we account for taxonomic (or lineage) persistence, and ecological continuity in evolutionary time? It is apparent that parasite lineages where species have complex life cycles dependent on predictable trophic relationships have been persistent; basically lineages have tracked across extinction events. There is a distinction between dependence on a 'specific' host, or host taxon, versus dependence on a particular ecological/trophic association, such that it may be transmission dynamics rather than host-association which is conservative in evolutionary time (Hoberg & Adams, 2000).

Lineage persistence and ecological continuity is linked to the interactive effects of differential extinction for intermediate hosts and definitive hosts, or for parasites through the dynamics of host-density effects or stage-specific mortality. Colonization may contribute to continuity through host-switches before, during or after the event horizon; such may involve a switch to an ecologically equivalent group with subsequent radiation (see Hoberg et al. 1999 b). Environmental disruption is predicted to be a driver for relaxation of ecological isolating mechanisms (ecological release) that enhance the potential for host-switching. Alternatively habitat shifts by potential hosts may lead to loss of an assemblage of dependent parasites. Episodic refugial effects and bottlenecks may further lead to punctuated cycles of diversification across a diversity of parasite–host assemblages, particularly when considered within the context of models for rapid speciation (Hoberg, 1995; Hoberg *et al.* 1999*a*).

A co-evolutionary component is also involved in lineage persistence. Parasite taxa may persist as (1) relics of once dominant groups through ancestor-descendant relationships (Bandoni & Brooks, 1987*a*, *b*; Brooks & Bandoni, 1988); or (2) as representatives of now extinct host taxa following colonization and co-evolutionary radiation in a novel host group (Hoberg *et al.* 1999*a*). Recognition of the potential impact of global extinction crises on genealogical and ecological diversity and structure of marine parasite faunas may eventually contribute to explanatory power for understanding patterns of helminth distribution at varying geographic and temporal scales. With refinement of application of molecular clock hypotheses, it may be possible to correlate divergence time for family or ordinal level taxa with particular periods of ecological disruption in Earth history.

GEOGRAPHIC AND TEMPORAL SCALE IN MARINE SYSTEMS

One of the questions arising from the issue of the connection between phylogeny and ecology is that of scale and emergent properties. By default, phylogeneticists tend to assume (either implicitly or explicitly) that when a parasite is found on a host it occurs over the entire range of that host. Ecologists have often pointed out the obvious shortcomings of this assumption in criticizing conclusions from phylogenetically-based biogeographic studies. However, with increasingly more detailed distributional data becoming available, it is not always necessary for this assumption to be made. Klassen (1992) has shown that by identifying geographic subpopulations of hosts within their range and identifying parasite distributions within those subpopulations, a data-set can be generated that is analyzable with secondary Brooks Parsimony Analysis (BPA) (see Brooks & McLennan, 1993*a*; Hoberg *et al.* 1997; Brooks *et al.* 2001), allowing for more refined interpretations of co-speciation and, in particular, colonization events. Klassen demonstrated, for instance, that through this approach, what might have been interpreted as sympatric speciation can actually be seen as allopatric speciation on geographically-isolated subpopulations of the same host species.

A related aspect of scale that has been understudied is seen when enlarging the scope of the analysis; the following example is based on a BPA analysis. That is when a particular group of parasites from a particular group of hosts is studied, one makes (again explicitly or not) the assumption that the basal-most node of the two lineages arose together through co-speciation [a side note is important here: although in BPA it is not critical nor even necessary to make this assumption, it is typically done; on the other hand Component Analysis or Reconciliation Analysis (e.g. Page, 1994*a*; Hoberg *et al.* 1997) cannot proceed without this assumption]. Many studies have hinted, *a posteriori*, that this may not be the case. Klassen's work on boxfish parasites indicated clearly two important conclusions about this assumption: (1) later identification of violation of this assumption in no way reduces the validity of the conclusions about the co-evolutionary relationship between the two lineages originally studied. That is to say, this assumption can be made comfortably if the question is specifically about co-evolution between the two lineages (host and parasite) as originally specified. (2) More often than not, one will find that when looking beyond the original two lineages, the assumption invariably becomes weakened. Klassen showed that by expanding the analysis between boxfish and their gill parasites to all teleosts that this group of parasites infect the pattern became both more complicated and more interesting. It became more complicated in that more and more colonization events were uncovered. It became more interesting in that, on a global scale, these colonization events tended to identify ecological association among hosts in specific geographic areas after major and collective vicariant events. This apparently punctuational pattern of co-evolution deserves further investigation.

An empirical analysis of scale

The most recent and, to some, most promising development toward defining the phylogenetic component of such associations comes with the development of methods for 'controlling for phylogeny' such as phylogenetically independent contrasts (e.g. Harvey & Pagel, 1991). Although this approach has been applied successfully both for free-living taxa and for parasites (e.g. Garland, Harvey & Ives, 1992; Sasal, Morand & Guegan, 1997), controlling for phylogeny is based on the desire to remove the effect of phylogeny (historical constraint) so that the truly interesting ecological questions may be addressed without the confounding effects implicit in differential evolutionary histories. But when the question is one of the interaction between phylogeny and ecology then controlling for, that is removing, the evolutionary variable is not, in our view, the correct approach. Phylogeny and ecology must then, as Brooks has long argued, be examined together. A way must be found to incorporate the one in the other. We present here an example of how this integration may be achieved. The essence of our example involves recognizing the influence that changing the scale of the analysis has on the interpretation of the pattern of association, irrespective of methodology.

Klassen (1992) examined the effect of incre-

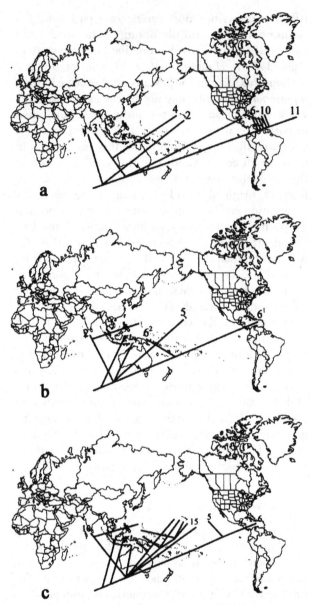

Fig. 1. Reconstruction of area relationship of Ostraciinae boxfishes based on the distribution of their parasites (only host groups including Ostraciidae are labelled). 1a. species-level analysis. 1 – *Ostracion rhinorhynchus*, 2 – *Lactoria* spp., 3 – *Ostracion cubicus* 1, 4 – *Ostracion cubicus*, 2 – *O. meleagris*, *O. cyanurus*, 5 to 11 – Atlantic Ostraciinae in the genera *Acanthostracion* and *Lactophrys*. 1b. family-level analysis, within Tetraodontiformes. 2 – Ostraciidae and triacanthidae, 5 – Ostraciidae and Balistidae, 6¹ – Ostraciidae and Balistidae, 6² – Ostraciidae. 1c. 5 – Ostraciidae, Labridae, Mullidae, Balistidae, Chaetodontidae, Ophidiidae, Apogonidae, Acanthuridae, 6 – Ostraciidae, Labridae, Balistidae, 10 – Triacanthidae, Ostraciidae, 15 – Ostraciidae, Pentapodidae, Balistidae, Chromidae, Pomacentridae, Holocentridae. Modified from Klassen (1992).

mentally expanding the scale of analysis from only boxfish and their parasites, through all Tetraodontiform fishes and their expanded set of parasites to considering the complete suite of Percomorpha hosts of *Haliotrema* parasites. For details of the data and

their analysis the reader is referred to Klassen (1992). Here we will briefly discuss only those aspects that contribute specifically to our changed perception of the co-evolutionary association as scale changes (see Fig. 1). Three specific points can be made about the importance of considering scale. (1) parasite taxa initially identified as species specific (terminal 1 of Fig. 1a) may be shown through subsequent analysis to belong to a parasite clade that parasitizes a larger clade of hosts (terminals 2 and 10 of Figs 1b and 1c, respectively). Thus, the presence of this species-specific parasite on its ostraciine host is the result of speciation through host-switching from an unrelated clade. (2) Scale can be shown to affect the biogeographic component of co-evolutionary interactions. For instance, the species level analysis (Fig. 1a) clearly shows the sister-area relationship between Caribbean and eastern Atlantic (and unresolved relationships within the Caribbean, which we will get back to in the second example) but cannot resolve area relationships between Atlantic and Indo-Pacific. Alternatively, the higher-level analyses (Figs 1b, 1c) cannot resolve within-Atlantic sister-area relationships (due to 'rounding-error', see Klassen, 1992) but provide evidence for a Pacific-Atlantic sister-area relationship, not available at the species level. The association, thus, indicates a vicariant event of great age. (3) Further, the association indicates that a whole ecology, not just two lineages, was involved in this event. The Pacific–Atlantic area relationship is supported by parasites from balistids and mullids in addition to ostraciids. The Indo–Pacific area relationship is repeatedly supported by several sister pairs of parasites from a variety of hosts. These repeated biogeographic patterns of co-evolutionary events not only point at various vicariance scenarios not seen in the species-level analysis but also identify groups of hosts likely to be of similar geological age and ecologically associated.

Klassen (1992) also indicated a new approach that would permit exploration of the historical component of host–parasite associations when both host and parasite species are found in, apparently unresolvable, sympatry, specifically the Caribbean clade of boxfish and their parasites. The basic approach is presented here as a 5-step procedure. (1) BPA I analysis. This requires three pieces of information, a host phylogeny, a parasite phylogeny and a host parasite list. The parasite phylogeny and list are then converted to a character matrix for the hosts (see Brooks & McLennan, 1991, 1993a; Brooks et al. 2001). This character matrix can then be treated in two ways. The data can be 'mapped' onto the host phylogeny. The CI becomes a rough measure of the degree of cospeciation with homoplasy requiring further a posteriori explanation (Fig. 2a). Alternately, the data can be used to generate a hypothesis of host relationship on the assumption of strict co-speci-

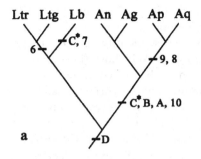

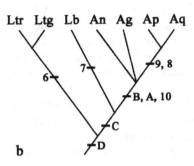

Fig. 2. Results of BPA I analysis for 9 species of *Haliotrema* on 7 species of Atlantic Ostraciinae. Labels at terminal nodes refer to host species. Numbers and letters at internodes refer to parasite taxa mapped as host characters, numbers are the 9 extant taxa, letters represent internodes on the parasite tree. Asterisks refer to events deviating from strict cospeciation, these require further explanation (* indicates potential host transfer, ** indicates sympatric speciation). 2a. Parasite data mapped onto existing host phylogeny. 2b. Parasite data allowed to present their own hypothesis of host relationships. Modified from Klassen (1992).

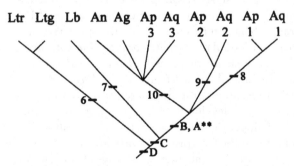

Fig. 3. Results of BPA II analysis for 9 species of *Haliotrema* on 7 species of Atlantic Ostraciinae. Labels as in Figure 2. *Acanthostracion quadricornis* and *A. polygonius* are represented by three terminal nodes each, reflecting the BPA treatment of hosts with multiple parasites (these are treated as separate 'populations' identified by each parasite). Modified from Klassen (1992).

ation. This result can then be compared with the 'true' host phylogeny with consistency indicating co-speciation and deviation requiring further *a posteriori* explanation (Fig. 2b). Typically supporters and detractors alike have focused only on the first of

these options for BPA I. We consistently advocate presenting both as their comparison can help in the first step of resolving the degree to which deviations from the assumption of cospeciation may indicate certain problems.

For instance, Fig. 2 indicates that at least two items require further explanation. In Fig. 2a 'character' C appears to arise twice. If we look at Fig. 2b we see that this case of homoplasy can be resolved if we hypothesize that parasite 7 on host Lb is a case of host transfer (involving speciation). Thus the presence of parasite 7 on its host is due to an ecological association and not co-speciation and the apparent paradox of the parallelism for C disappears. More problematic for BPA I is the presence of multiple parasite 'characters' at two of the hosts internodes. This is seen in both reconstructions (Figs 2a, 2b) and was once interpreted as potential evidence for sympatric speciation.

BPA II was introduced by Brooks (1990) specifically to deal with the coding artifact that seemed to lead to many of these instances of 'sympatric' speciation. Since this artifact is always associated with multiple parasites on a single host, BPA II splits each host taxon with two or more parasites into as many distinct 'populations' as parasites. Fig. 3 is the result of the BPA II reanalysis. Note that this step changes nothing about the presence of parasite 7. It does, however, reduce the apparent sympatry to two characters (A and B). This step requires the further hypothesis that the speciation of parasites 8, 9 and 10 occurred not in correspondence with speciation of their hosts but with the isolation of distinct host populations. The apparent sympatry of A and B remains unexplained. So far, this is a typical BPA analysis consistent with currently outlined protocols (Brooks *et al.* 2001).

One of the questions Klassen (1992) asked was whether the hypothetical host populations of BPA II had biogeographic reality. He added a further step to BPA that we will refer to as BPA II-D ('D' for distribution). The subsequent three steps illustrate how the distribution information together with two ecological assumptions may help provide meaningful and predictive hypotheses about each of the associations identified in BPA I as requiring explanation.

Fig. 4 is the BPA II reconstruction with parasite distribution data superimposed. Blanks indicate the absence of a particular parasite from a host for a particular local, 'x' and 'o' indicate presence. The 'o's, however, identify records that are questionable due to low abundances. That is these identify instances when 2 or fewer specimens of a particular parasite were found. Based on the ecological assumption that such rare occurrences indicate 'accidental' infections (an assumption with precedence both in the free-living and parasitic literature, see Esch, Bush & Aho, 1990) any population of host with only an 'o' is removed from the analysis. This results

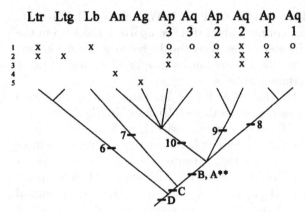

Fig. 4.

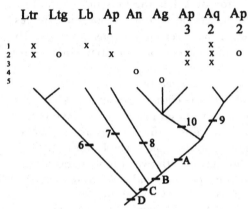

Fig. 5.

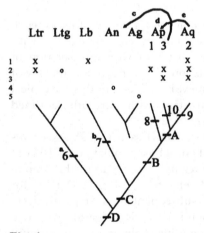

Fig. 6.

Figs 4–6. Results of BPA II-D analysis for 9 species of *Haliotrema* on 7 species of Atlantic Ostraciinae. Labels as in Fig. 2. Additionally, geographic distributions of each parasite for each terminal node are indicated. Areas are: 1 – Coastal North America, 2 – Caribbean, 3 – Coastal South America, 4 – eastern Atlantic (north), 5 – eastern Atlantic (south). X and O refer to relative abundances of parasite on hosts in a particular local. Fig. 4. Same reconstruction as for Fig. 3. 'x' indicates confirmed presence, 'o' indicates unconfirmed or 'accidental' presence (see text for explanation). Fig. 5. Reconstruction after nodes with unconfirmed presences have been removed. 'x' indicates 'core' host, 'o' indicates satellite host (see text for explanation). Fig 6. Reconstruction after population nodes with satellite presences have been removed. 'x' indicates 'core' host,

in the reconstruction of Fig. 5. Note now that all putative instances of 'sympatry' have been resolved.

A further step involves identifying, for each parasite species, the most likely host (host population) of origin. We have borrowed from Hanski (1982) by distinguishing between core and satellite populations for a species of parasite when found on more than one 'population' of host. Thus 'x' are core populations by virtue of being more widespread (conversely 'o's identify the isolated, stochastic distribution of satellite populations). Further borrowing from the source-sink concept of Island Biogeography (Rosenzweig, 1995) we hypothesize that only core (or source) populations are important in identifying the historical component of the association of these parasites with their host 'populations'.

The final step in this modified BPA II-D results in a simplified reconstruction of the relationship of host populations based on the combination of parasite phylogenetic and distributional data (Fig. 6). Accordingly, there are a total of five associations between the hosts and their parasites that deviate from the hypothesis of strict co-speciation. They are identified in Fig. 6 with labels a to e. 'a' – even though, throughout the unmodified BPA analysis the presence of parasite 6 on its hosts has been considered unproblematic, BPA II-D implies that *Lactophrys triqueter* may be the primary host. 'b' – nothing fundamental has changed about the interpretation of parasite 7 on its host. However, taking distribution data into account indicates that the presence of this parasite may be a peripheral isolate. 'c' – similar to 'a', the presence of parasite 10 on the eastern Atlantic hosts is interpreted as ecological transfer not historical association. 'd' – parasite 8 is interpreted as having speciated on a Caribbean population of *Acanthostracion polygonius*. 'e' – the presence of parasite 9 on *A. polygonius* is interpreted as dispersal from *A. quadricornis*.

We present this modified approach to BPA as a first step in developing a means of incorporating both phylogeny and ecology in the same analysis. We suggest that the interpretations arising out of such an analysis provide, at worst, a way of developing testable hypotheses that take account of both evolutionary past and ecological present.

General vicariant patterns

Another aspect of scale can be seen in the notion espoused by vicariance biogeographers, that multiple lineage comparisons are necessary for a 'general vicariant pattern' to be identified. So far, the empirical data for such a general pattern based on

'o' indicates satellite host. Labels 'a' to 'e' indicate events that require *a posteriori* explanation beyond strict co-speciation (see text for explanation).

parasite lineages among Osteichthyes is lacking. There are, however, indications that these data will soon be forthcoming. First, the study by Klassen (1992) on boxfishes has revealed tentative pattern repetitions when expanding the scale to all hosts of *Haliotrema*; consider the example presented in the previous section. Secondly, two independent studies on Chaetodontidae and Priacanthidae will provide comparative data with those from the Ostraciidae for a quantitative comparison (G. J. Klassen, unpublished data; S. Morand, personal communication).

In contrast, general level or congruent patterns have been demonstrated for phylogenetically disparate groups of tapeworms that infect Phocidae and Otariidae (Pinnipedia) and Alcidae (Charadriiformes) in the North Pacific Basin and across Holarctic seas (Hoberg, 1986, 1992, 1995; Hoberg & Adams, 1992). Similar patterns have also been recognized for ascaridoid nematodes (species of *Contracaecum* and *Pseudoterranova*) in phocids and otariids (e.g. Paggi *et al.* 1991; Nascetti *et al.* 1993; Bullini *et al.* 1997, among others). The underlying processes are linked to radiation of hosts and parasites in Subarctic and Arctic refugia during the late Pliocene and Quaternary where refugial effects, habitat fragmentation and isolation were significant determinants of faunal diversification (Hoberg, 1992, 1995; Hoberg & Adams, 2000). Although molecular clock hypotheses have been applied to studies of ascaridoid evolution and biogeography, the temporal setting for diversification among tapeworms, seabirds and pinnipeds has been estimated based on the physical history of the Holarctic region.

The concept for geminate species (Jordan, 1908) is also beginning to receive renewed attention with respect to studies of diversity and relationships of taxa across the Panamanian Isthmus. Although putative species pairs have been recognized and a vicariant history relative to closure of the Panamanian Seaway has been postulated, the phylogenetic context for sister-species has not been firmly established (Marques, Brooks & Monks, 1995; Goshroy & Caira, 2001). Clear patterns have been established for host taxa including marine stingrays, the degree to which the history and distribution of associated species of *Acanthobothrium* and other helminths is congruent remains to be examined (Marques, Centritto & Stewart, 1997).

Archaic taxa and widespread patterns, global and antitropical distributions

Geographically-widespread parasite faunas encompassing global or antitropical distributions may be indicative of early associations with specific host groups. Although White (1989) has provided for much discussion about antitropical distributions from the perspective of various piscine taxa, and

Briggs (1995) has provided a broader review, parasitological data have yet been applied to this question. One potential source of data would be the parasites of Aracaninae (the sistergroup to Ostraciinae); these fishes have a well understood antitropical distribution, a tropical sister-group and their gill parasites fall into a group that is reasonably well understood (Klassen, 1992). Estimates of divergence time from molecular sequence data for hosts and parasites could further contribute to addressing the relative age and associations for components of widespread faunas.

Among seabirds and marine mammals some parasite groups, including the Tetrabothriidea (species of *Tetrabothrius*) and some Diphyllobothriidae, are geographically widespread, and although some genera are antitropical, species with bipolar distributions have not been identified (Deliamure, 1955; Hoberg 1996). Some evidence suggests that for marine birds, the *Tetrabothrius* faunas characteristic of the Southern Ocean and Northern Hemisphere are distinct and segregated (Hoberg & Ryan, 1987) and that host-specific core faunas may be associated with each of the major orders of seabirds (Hoberg, 1996); monophyly for characteristic species groups remains to be established. Among pinnipeds, the Campulid digeneans are geographically widespread and phylogenetic hypotheses are consistent with a protracted history of colonization and cospeciation (Fernández *et al.* 1998 *a*, *b*; Hoberg & Adams, 2000).

As a generality, among faunas in marine birds and mammals, and those in chelonians, osteichthyans and chondrichthyans, geographic scale may be linked to age, duration and vagility of the assemblage (e.g. Brooks & McLennan, 1993*a*; Pérez-Ponce de Leòn & Brooks, 1995*a*.) Thus, archaic taxa are more often widespread, whereas recent associations may be regional in scale (see Beveridge, 1986; Hoberg & Adams, 2000). A burgeoning body of empirical data from largely descriptive biogeographic inventories provides the foundation for further evaluations of this concept. Rohde (1993) summarized substantial databases for the distribution of digeneans and monogeneans in marine teleosts, providing a comparative context for piscine parasites along latitudinal gradients, and between oceanic regions such as the Pacific and Atlantic basins. A phylogenetic context for such data is critical for understanding the history of faunal assemblage, the interaction of dispersal and vicariance, and the evolutionary relationships of taxa within and among identifiable faunal provinces.

Integration of biodiversity data

A model for integration of detailed survey and inventory with phylogenetic/historical biogeographic approaches has been exemplified by early and continuing research on relationships of the parasite fauna of the freshwater rays, Potamotrygonidae, of the Neotropics (e.g. Brooks, Mayes & Thorson, 1981*a*; Brooks *et al.* 1981*b*; Brooks & Amato, 1992; Brooks, 1995; Zamparo, Brooks & Barriga, 1999). These studies articulated an hypothesis for Pacific origins of the freshwater stingrays and their parasites. Most recently this theoretical framework has become the focus for ongoing research to reveal the fundamental processes for the history of a major component of the Amazonian biota (e.g. Lovejoy, 1997; Marques, 2000) that can contribute to a more detailed understanding of biogeography and speciation processes during the Tertiary (e.g. Webb, 1995; Räsänen *et al.* 1995). Additionally, regional studies of parasite biodiversity in chondrichthyans from the Gulf of California (e.g. Caira & Burge, 2001; Goshroy & Caira, 2001) are contributing insights to elucidating the broader distribution and history of cestodes in sharks and rays (Caira & Jensen, 2001).

Combined survey and phylogenetic reconstruction are further exemplified by studies of Australian reef fishes (e.g. Cribb, Bray & Barker, 1992; Barker *et al.* 1994; Bray, Cribb & Littlewood, 1998; Bray & Cribb, 2000). Also of note are the detailed biodiversity inventories for coastal waters of Mexico (e.g. Pérez-Ponce de Leòn *et al.* 1999), and their foundation for phylogenetic and biogeographic analyses.

These studies are critical in establishing accurate concepts for host and geographic distribution, and particularly ideas about host-specificity within and among assemblages (e.g. Gibson & Bray, 1994), but need to be considered in an explanatory framework derived from comparative phylogenetics. Although a number of faunal provinces and biotas have received focused attention, there has yet to be a synoptic and integrated approach linking survey, inventory and phylogenetic reconstruction. Such will continue to remain a challenge for any comprehensive work on chondrichthyans, given the exceptional diversity that remains to be discovered and described among the tetraphyllideans and other eucestodes (e.g. Caira & Jensen, 2001). Current methodological development for historical analyses appears to now coincide with increasing knowledge of biodiversity, factors that will promote resolution of history at a broad scale.

Further studies must extend beyond descriptive biogeography which focuses on documentation of distribution, ecological diversity and host association and include integrated approaches to phylogenetic and historical reconstruction (Brooks & Hoberg, 2000). In this manner such questions as how species are related within and between zones and regions or how higher taxonomic groups are distributed in time and space may be addressed. Evaluation of historical structure then becomes the context for identification of common mechanisms involved in distributional history for biotas including the relative roles of co-speciation or host-switching and vicariance or dispersal. Various facets of history are being increasingly addressed in current assessments of biodiversity and biogeography.

CO-EVOLUTION, COLONIZATION AND DIVERSIFICATION

Empirical tests of co-evolutionary scenarios for marine systems

Although not synoptic for any one host–parasite assemblage or taxon, there are sufficient empirical studies in the literature (e.g. Table 1) to derive a preliminary interpretation of the degree of contribution for co-speciation and dispersal to co-evolutionary scenarios. The majority of these studies have been conducted primarily by inspection and mapping and may benefit from reanalysis according to current comparative protocols, particularly with the potential insights based on inclusion of molecular-based data (see Brooks *et al.* 2001; Paterson & Banks, 2001). Most extensive of these are Brooks' work on stingrays, Collette & Russo (1985) on mackerel, Klassen's on boxfishes, and Hoberg's on the Beringian/North Pacific fauna. Where this has been accomplished, e.g. the studies of *Alcataenia*, the original conclusions have been strongly upheld (see Hoberg *et al.* 1997; Paterson & Banks, 2001). Additionally, Caira & Jensen (2001) reiterated the necessity in co-evolutionary studies to focus on monophyletic taxa and systems with a high level of specificity, accompanied by a robust understanding of host and parasite diversity (accurate taxonomy, identity and comprehensive sampling), and accurate estimates of both host and parasite phylogenies (see also Page, Paterson & Clayton, 1996). We would suggest, however, that the search for pattern and interpretation of process is an exploratory activity rather than an attempt to identify strictly co-evolving systems. Indeed it is discovery of the departures from strict co-speciation (and support for Fahrenholz's Rule) that reveal significant insights into the complex ecological history of faunal associations as indicated for example in the detailed study for *Haliotrema* and boxfishes (Klassen, 1992, 1994*a*).

The dominant recurring theme evident in diversification of helminth faunas among marine vertebrates including fishes, mammals, chelonians and birds has been colonization. For example, radiation of Trypanorhyncha and the tetraphyllidean assemblage in sharks and rays appears attributable to initial colonization, although a deep history of secondary co-speciation may be indicated by high levels of host-specificity for many species and higher taxa (e.g. Euzet, 1959; Hoberg *et al.* 1999*a*; Beveridge, Campbell & Palm, 1999; Caira & Jensen, 2001). In general, chondrichthyan faunas have yet to be examined in great detail based on phylogenetic methods other than through the development of hypotheses for the origins of the freshwater rays, Potamotrygonidae (Brooks *et al.* 1981*b*; Brooks, 1992, 1995), or otherwise in groups of limited scope (Nasin *et al.* 1997; Caira & Jensen, 2001). Host-switching by digeneans and monogeneans has been identified among different groups of teleosts (e.g. Klassen, 1992; Gibson & Bray, 1994; Barker *et al.* 1994; Bray & Cribb, 2000; Brooks, Pérez-Ponce de León & León-Règagnon, 2000). Considerable details, however, remain to be revealed with respect to the co-evolutionary histories of helminth faunas among osteichthyan and chondrichthyan fishes.

The pronocephalid digeneans characteristic in marine chelonians have also been demonstrated to have a complex history involving extensive colonization, and multiple marine-freshwater transitions (Pérez-Ponce de León & Brooks, 1995*a*). Colonization not only involved habitat shifts for turtles, but shifts by parasites from turtles to such phylogenetically disparate taxa as marine iguanas (Pérez-Ponce de León & Brooks, 1995*b*). The patterns indicated a deep and complex history including vicariance and dispersal.

Among marine homeotherms including cetaceans, pinnipeds and seabirds, few taxa are indicators of historical co-evolutionary linkages, or association by descent, between marine and terrestrial faunas (Deliamure, 1995). Among diphyllobothriids, there is broad evidence for diversification by what has been termed as 'hostal radiation' where ecologically-driven host-switching occurs among phylogenetically-unrelated pinniped or cetacean taxa (Iurakhno, 1991). Phylogenetic studies of the eucestodes have supported an hypothesis for the origin of Tetrabothriidea by host-switching, first to basal marine birds and secondarily to cetaceans and pinnipeds (Hoberg & Adams, 1992, 2000; Hoberg, 1996); co-speciation may have been critical in later diversification of *Tetrabothrius* among avian hosts, but phylogenetic studies have yet to be completed (Hoberg, 1996). Colonization has also been recognized as a significant driver of diversification among the Tetrabothriidea in marine mammals (Hoberg & Adams, 1992) and particularly for *Anophryocephalus* spp. among Phocidae (Hoberg, 1992, 1995). Fernández *et al.* (1998*a, b*) and Hoberg & Adams (2000) demonstrated a complex history involving colonization and co-speciation among odontocetes and pinnipeds for some campulid digeneans. Nadler *et al.* (2000) demonstrated that *Contracaecum* spp. associated with pinnipeds are not monophyletic, and that host-switching among seabirds and pinnipeds has occurred among the ascaridoids.

Additionally, Hoberg (1986, 1992) and Hoberg *et al.* (1997) documented the pervasive nature of colonization in the evolution of *Alcataenia* tapeworms among seabirds of the family Alcidae.

Significantly, the development of marked host-specificity was evident among species that had originated subsequent to relatively recent colonization of host taxa. These studies supported the concept that strict (or 'phylogenetic') specificity should be decoupled from the process of co-speciation, and that the former was not necessarily an unequivocal indicator of the temporal duration of an association (Brooks, 1979, 1985; Hoberg, 1986).

Interestingly, arthropod ectoparasites on both fishes and seabirds may represent a contrast to the histories of colonization being postulated for a variety of helminths and their hosts. The limited number of studies of copepods among teleosts have indicated substantial patterns of co-evolution and co-speciation (summarized in Paterson & Poulin, 1999). Such patterns have been demonstrated among parasite taxa that also exhibit relatively low levels of host-specificity (Poulin, 1992). Further detailed analyses of a wider diversity of copepod taxa and their hosts are necessary to establish this as a generality, but it would provide an interesting comparison to the monogeneans on the same spectrum of piscine hosts (see Rhode & Hayward, 2000).

Phithiraptera among seabirds also appear to have deep co-evolutionary histories with their avian hosts (Paterson, Gray & Wallis, 1993; Paterson & Gray, 1997; Paterson *et al.* 2000). Such may reflect the constraints on the potential for transmission among conspecifics, or for host-switching between phylogenetically unrelated seabirds in relative sympatry at large colony sites (e.g. Paterson *et al.* 2000); the degree of coloniality and the physical attributes of nest sites, and limited interactions during foraging in pelagic situations may serve as substantial controls on distribution. Among the assemblage of lice on both Procellariiformes and Sphenisciformes, co-speciation was postulated as a dominant driver for diversification with contributions from intra-host speciation; patterns of host association were further influenced by sorting events (Paterson, Palma & Gray, 1999; Paterson *et al.* 2000).

Hoberg & Adams (1992, 2000) discussed issues related to host-switching, particularly among marine homeotherms. It is important to note that, among those systems that have been thus far examined based on phylogenetic methods, recognition of widespread co-speciation has not been documented (see also Jackson, 1999). In marine and other systems, host-switching for parasites with complex life cycles is a stochastic process that may be linked to the predictably of guild associations or foodweb structure over extended evolutionary time frames. It is not clear that constraints to host-switching will be the same for parasites with direct versus indirect cycles, or whether ecto- and endo-parasites may be influenced differentially by variation in life history for their respective piscine, avian or mammalian hosts.

Vicariance versus colonization

One of the questions that has often been asked is whether marine systems show similar patterns to those in freshwater and terrestrial environments, or whether patterns in such large and seemingly uniform habitats can even be unraveled. Brooks & McLennan (1993 *a*, and references therein) and Hoberg (1986, 1992, 1995, 1997) have addressed this problem theoretically and concluded that there is no particular reason why this should not be the case. Empirically this has now been repeatedly demonstrated; e.g. Brooks *et al.* (1981 *b*) and Brooks & Deardorf (1988), Hoberg *et al.* (1998) for parasites of rays, Bandoni & Brooks (1987 *a*) for Holocephala, Bandoni & Brooks (1987 *b*) as well as Klassen (1992, 1994 *a*) for several teleost lineages, and Hoberg (1986, 1992, 1995) and Hoberg & Adams (2000) for seabirds and pinnipeds.

Collette & Russo (1985) seem to indicate, however, that caution must be taken with primarily open ocean pelagic species, the implication being that reconstructing clear patterns may be simplified for taxa in coastal settings. What is missing still is a synoptic work assessing what overall pattern, if any, can be retrieved from these studies collectively about, for instance, the biogeographic relationships between Indo-Pacific, Pacific and Atlantic Oceans. This is particularly important as there appears to be a lack of consensus within the ichthyological community (Briggs, 1995; Palumbi, 1997). Although temporally deep diversification times have been postulated for some parasite–host assemblages (e.g. Rhode & Hayward, 2000 for copepods and monogeneans on Scombridae), molecular divergence studies for such free-living taxa as echinoids and butterfly fishes (*Chaetodon* spp.) suggest active processes for speciation extending through the Pliocene and Pleistocene (Palumbi, 1997).

Rhode & Hayward (2000) examined hypotheses for the efficiency of oceanic barriers to dispersal based on detailed analyses of monogeneans and copepods among scombrid fishes. Centres of diversity for the contemporary fauna were recognized in the Indo-West Pacific and secondarily in the West Atlantic. Closure of the Tethys Sea and associated habitat fragmentation was postulated as a significant driver of isolation and speciation for both monogeneans and copepods on *Scomber* and *Scomberomorus* indicative of a relictual distribution for these assemblages. The East Pacific Barrier was recognized as a major control on the current distribution for these assemblages between the East and West Pacific; such suggests a role as a selective barrier for dispersal for a variety of phylogenetically disparate taxa at differing temporal scales.

A preliminary approach to examination of large patterns in the Pacific was taken by Klassen (1992) who indicated that gill parasites of coral reef teleosts favoured an Indo-Australian origin for these assemblages; comments on these relationships have been outlined by Marques *et al.* (1997) for species of *Acanthobothrium* in marine stingrays. Further tests of general patterns will be possible in the future as work is currently being conducted independently on the gill parasites of Chaetodontidae and Priacanthidae (G. J. Klassen, unpublished data; S. Morand, personal communication). A combination of these data with that from other coral reef fishes (Klassen, 1992, 1994 *a*, *b*) should permit the derivation of general conclusions about underlying vicariant patterns; an important adjunct to such studies will be inclusion of molecular data in refining ideas about the timing of divergence for populations and species and the physical/environmental determinants of speciation (Palumbi, 1997).

Allopatric speciation as a model

In those systems that have been examined, and particularly among faunas in marine birds and mammals, speciation has been largely allopatric. In these systems, speciation of cestodes and ascaridoid nematodes appears to be driven by the geographic ranges and a history for isolation of definitive hosts (Hoberg, 1995; Bullini *et al.* 1997). Thus, isolation and speciation among diverse assemblages of marine parasites may often proceed independently from that of populations of intermediate hosts. Although different mechanisms for allopatric speciation have been identified (e.g. microallopatry, peripheral isolates) in the speciation of cestodes in pinnipeds and seabirds, all appear to be driven by the particular history of the vertebrate hosts (Hoberg & Adams, 2000). Further for some parasites with direct cycles, the studies of *Haliotrema* outlined above show that many so called scenarios for sympatric speciation may actually represent examples of allopatric speciation for parasites on allopatric host populations. The degree to which allopatry and geographic isolation represent a general model for marine parasites and their hosts remains to be examined in greater detail. In contrast, Rohde (1993) has suggested a role for some form of sympatric speciation to account for the diversity of congeneric species that are encountered in some host individuals. The latter, as indicated above, could be a reflection of our limited understanding of scale in marine systems. Modifications to BPA such as BPA-D as outlined above provide a method for identifying scenarios that might be termed intra-host speciation and a tool for exploring the historical basis for such phenomenon (see also Paterson & Banks, 2001).

Intra-host speciation

Processes for intra-host speciation represent another form or facet of co-evolution (e.g. Paterson & Banks,

2001). They may be invoked based on the observation of the co-occurrence of multiple congeners in single host species, but there are few examples where such systems have been examined phylogenetically. A phylogenetic context is necessary to first demonstrate sister-species relationships and secondarily to discriminate between hypotheses for co-speciation versus forms of colonization. Examples of this phenomenon may be particularly common among genera and species of the Onchobothriidae and Phyllobthriidae in chondrichthyans (Caira, Jensen & Healy 2001) and appear to be commonly reported for species of *Acanthobothrium* and *Pedibothrium* (Caira, 1992; Marques *et al.* 1995, 1997; Caira & Burge, 2001; Caira & Zahner, 2001) and among *Rhinebothroides* spp. (Brooks & Amato, 1992). Paterson & Poulin (1999) identified intra-host speciation as an important process for diversification of copepods of the genus *Chondracanthus* on a variety of marine teleosts.

Assuming that allopatric speciation is a primary determinant for parasite diversification, it may be useful to consider if such intra-host patterns are indicators of punctuated or cyclical/periodic pulses or bouts of geographic isolation for hosts that drive divergence and speciation among parasite lineages (Hoberg, 1995). Duration of isolation may be insufficient to result in divergence for hosts, but may lead to speciation for parasites. Is this a phenomenon linked to the age or geographic extent of an assemblage, in that the influence may be most pronounced among geographically widespread taxa? The issue of geographic and temporal scale is important in this context as it is clear that considerable discrete variation, or species-level partitions that can be demonstrated through comparative molecular analyses are often masked by a similarity or uniformity in morphological characters. Paterson & Poulin (1999) considered that the relatively extensive level of intra-host speciation evident for species of copepods in the genus *Chondracanthus* could reflect allopatric speciation across a broad geographic range occupied by hosts.

Hoberg (1995) suggested that such intra-host patterns were important indicators of cryptic isolation events for components of a host–parasite assemblage. Parasites become cryptic indicators of a complex history of episodic isolation for hosts, and this may either be reflected in the speciose and host-specific nature of some parasite taxa in respective hosts; or may also reflect the facets of biogeographic history that can no longer be recognized for the host group. Of interest would be examination of patterns for episodic isolation linked to marine transgression/regression cycles in the Amazonian basin as drivers of diversification for parasites; speciation may be linked to marine transgression and isolation of discrete drainages over variable time frames since the Miocene (see Webb, 1995; Marques, 2000).

CONCLUSIONS AND FUTURE DEVELOPMENTS

Phylogenetic studies of parasites and hosts represent a critical context for revealing and understanding patterns in biodiversity, faunal structure and historical and contemporary biogeography (Brooks & McLennan, 1993 *a,b*; Brooks & Hoberg, 2000; Paterson & Poulin, 1999). Phylogeny-based approaches are powerful because hierarchical order constrains the range of explanations for faunal structure and history in a comparative context linking host and parasite taxa. Synergy is evident in integration of phylogenetic, biogeographic, and ecological history in the articulation of synoptic hypotheses for faunal development over often disparate spatial and temporal scales (e.g. Hoberg, 1997). In this regard, parasites constitute exquisite phylogenetic and historical ecological indicators that reveal substantial insights into the history of the marine biosphere. Phylogenetic hypotheses for hosts and parasites are the tapestry for revealing the interaction of co-evolution, colonization and extinction on patterns of faunal structure and ecological continuity across deep temporal and geographic scales in the marine environment.

The great potential for this research programme has been amply demonstrated (Table 1) by an array of studies across a phylogenetically diverse landscape of hosts and parasites (see also Brooks & McLennan, 1993 *a*; Brooks & Hoberg, 2000). Despite nearly 25 years of explicit co-evolutionary studies based on phylogenetic approaches, we still continue to lack critical information for most host and parasite taxa and in many respects the literature is diverse but fragmented. For example, there remains a single detailed historical study of helminths among seabirds (Hoberg, 1986, 1992), and our understanding of species diversity and phylogeny among the speciose tetraphyllidean taxa of chondrichthyan hosts remains to be expanded (Caira & Jensen, 2001). We continue to have relatively few robust species-level phylogenies for parasites within the context of a detailed understanding of relationships for higher inclusive taxa. Likewise, our knowledge of host phylogeny often is inadequate as the basis for modern comparative studies in co-evolution although our basic understanding for relationships among such groups as teleosts (e.g. Stiassny, Parouti & Johnson, 1996), chondrichthyans (reviewed in Caira & Jensen, 2001) and marine mammals (Berta & Sumich, 1999) has dramatically improved in the past decade. Continued expansion of a phylogenetic framework is necessary as a foundation for a detailed and rich comparative research programme; such a situation clearly represents a challenge and an opportunity.

Additionally, Brooks & Hoberg (2000) have emphasized the need to bridge the gap between phylogenetics and ecology, although little effort has so far been put into developing research programmes that are explicitly directed toward that goal (Brooks & McLennan, 1991). As yet there remains minimal overlap between parasite groups or assemblages for which we have extensive knowledge of community ecology and those which have been evaluated in a phylogenetic context (Poulin, 1998). Brooks (1980) attempted this in the early 1980s to great criticism. McLennan & Brooks (1991) accomplished this successfully in the context of behavioural ecology. Marcogliesie & Cone (e.g. 1993) have been directing their research increasingly in that direction. Klassen (unpublished data) is developing a programme dedicated to both building the empirical data-base needed by both fields and exploring methodological options for integrating micro- and macro-evolutionary approaches.

Burgeoning interest in biodiversity assessment and particularly the Global Taxonomy Initiative clearly indicates a place for parasitological survey and inventory linked to phylogenetic approaches as a cornerstone for future research (Brooks & Hoberg, 2000). Such integrated and comprehensive surveys in marine environments are exemplified in ongoing investigations of the Australian reef faunas, biodiversity survey and inventory in chondrichthyans from the Gulf of California, faunal assessments more widely along Mexican coastal waters, and historical studies of marine-freshwater transitions in Amazonia. We have reviewed an array of interesting examples from a diversity of host–parasite assemblages in marine habitats, but we are challenged to develop broad and synoptic coverage that is necessary to reveal truly general concepts for history across global seas. The time is appropriate for integrative approaches linking systematics, evolutionary biology and ecology in frameworks that can contribute to a more refined understanding of the history and structure of global marine systems and the biosphere.

ACKNOWLEDGEMENTS

The authors thank D. R. Brooks and D. A. McLennan of the University of Toronto for contributions leading to articulation of ideas and concepts outlined in the current manuscript. Discussions with B. Rosenthal and K. Galbreath at the Parasite Biology, Epidemiology and Systematics Laboratory provided useful insights about speciation processes and historical biogeography. We further thank A. Paterson and an anonymous reviewer for critical comments that created a focus for discussion and led to improvements of the manuscript.

REFERENCES

ARDUINO, P., NASCETTI, G., CIANCHI, R., PLÖTZ, J., MATTIUCCI, S., D'AMELIO, S., PAGGI, L., ORRECHIA, P. & BULLINI, L. (1995). Isozyme variation and taxonomic rank of *Contracaecum radiatum* (v. Linstow, 1907) from the Antarctic Ocean (Nematoda: Ascaridoidea). *Systematic Parasitology* 30, 1–9.

BANDONI, S. M. & BROOKS, D. R. (1987a). Revision and phylogenetic analysis of the Gyrocotylidea Poche, 1926 (Platyhelminthes: Cercomeria: Cercomeromorpha). *Canadian Journal of Zoology* 65, 2369–2389.

BANDONI, S. M. & BROOKS, D. R. (1987b). Revision and phylogenetic analysis of the Amphilinidea Poche, 1922 (Platyhelminthes: Cercomeria: Cercomeromorpha). *Canadian Journal of Zoology* 65, 1110–1128.

BARKER, S. C., CRIBB, T. H., BRAY, R. A. & ADLARD, R. D. (1994). Host–parasite associations on a coral reef: pomacentrid fishes and digenean trematodes. *International Journal for Parasitology* 24, 643–647.

BENZ, G. W. & DEETS, G. B. (1988). Fifty-one years later: an update on *Entepherus*, with a phylogenetic analysis of Cercopidae Dana, 1849 (Copepoda: Siphosotomatoidea). *Canadian Journal of Zoology* 66, 856–865.

BERTA, A. & SUMICH, J. L. (1999). *Marine Mammals Evolutionary Biology*. San Francisco, Academic Press.

BEVERIDGE, I. (1986). Biogeography of parasites in Australia. *Parasitology Today, Australian Supplement*, July S3–S7.

BEVERIDGE, I., CAMPBELL, R. A. & PALM, H. W. (1999). Preliminary cladistic analysis of genera of the cestode order Trypanorhyncha Diesing, 1863. *Systematic Parasitology* 42, 29–49.

BOEGER, W. A. & KRITSKY, D. C. (1989). Phylogeny, coevolution and revision of the Hexabothriidae Price, 1942 (Monogenea). *International Journal for Parasitology* 19, 425–440.

BOEGER, W. A. & KRITSKY, D. C. (1997). Coevolution of the Monogenoidea (Platyhelminthes) based on a revised hypothesis of parasite phylogeny. *International Journal for Parasitology* 27, 1495–1511.

BOWRING, S. A., ERWIN, D. H., JIN, Y. G., MARTIN, M. W., DAVIDEK, K. & WANG, W. (1998). U/Pb zircon geochronology and tempo of the End-Permian mass extinction. *Science* 280, 1039–1045.

BRAY, R. A. & CRIBB, T. H. (2000). Species of *Trifoliovarium* Yamaguti, 1940 (Digenea: Lecithasteridae) from Australian waters, with a description of *T. draconis* n. sp. and a cladistic study of the subfamily Trifoliovariinae Yamaguti, 1958. *Systematic Parasitology* 47, 183–192.

BRAY, R. A., CRIBB, T. H. & LITTLEWOOD, T. J. (1998). A phylogenetic study of *Lepidapedoides* Yamaguti, 1970 (Digenea) with a key and description of two new species from Western Australia. *Systematic Parasitology* 39, 183–197.

BRIGGS, J. C. (1995). *Global Biogeography*. Amsterdam, Elsevier.

BROOKS, D. R. (1979). Testing the context and extent of host–parasite coevolution. *Systematic Zoology* 28, 299–307.

BROOKS, D. R. (1980). Allopatric speciation and non-interactive parasite community structure. *Systematic Zoology* 29, 192–203.

BROOKS, D. R. (1981). Hennig's parasitological method: a proposed solution. *Systematic Biology* 30, 229–249.

BROOKS, D. R. (1985). Historical ecology: a new approach to studying the evolution of ecological associations. *Annals of the Missouri Botanical Garden* 72, 660–680.

BROOKS, D. R. (1989). A summary of the database pertaining to the phylogeny of the major groups of parasitic Platyhelminthes, with a revised classification. *Canadian Journal of Zoology* **67**, 714–720.

BROOKS, D. R. (1990). Parsimony analysis in historical biogeography and coevolution: methodological and theoretical update. *Systematic Zoology* **39**, 14–30.

BROOKS, D. R. (1992). Origins, diversification and historical structure of the helminth fauna inhabiting neotropical freshwater stingrays (Potamotrygonidae). *Journal of Parasitology* **78**, 588–595.

BROOKS, D. R. (1995). Neotropical freshwater stingrays and their parasites: a tale of an ocean and a river long ago. *Journal of Aquariculture & Aquatic Science* **7**, 52–61.

BROOKS, D. R. & AMATO, J. F. R. (1992). Cestode parasites in *Potamotrygon motoro* (Natterer) (Chondrichthyes: Potamotrygonidae) from southwestern Brazil, including *Rhinobothroides mclennanae* n. sp. (Tetraphyllidea: Phyllobthriidae), and a revised host-parasite checklist for helminths inhabiting neotropical freshwater stingrays. *Journal of Parasitology* **78**, 393–398.

BROOKS, D. R. & BANDONI, S. (1988). Coevolution and relicts. *Systematic Zoology* **37**, 19–33.

BROOKS, D. R. & DEARDORF, T. (1988). *Rhinobothrium devaneyi* n.sp. (Eucestoda: Tetraphyllidea) and *Echinocephalus overstreeti* Deardorf and Ko, 1983 (Nematoda: Gnathostomatidae) in a thorny back ray, *Urogymnus asperrimus*, from Enewetak atoll, with phylogenetic analysis of both species groups. *Journal of Parasitology* **74**, 459–465.

BROOKS, D. R. & HOBERG, E. P. (2000). Triage for the biosphere: the need and rationale for taxonomic inventories and phylogenetic studies of parasites. *Comparative Parasitology* **67**, 1–25.

BROOKS, D. R., MAYES, M. & THORSON, T. B. (1981*a*). Cestode parasites in *Myliobatis goodei* Garman (Myliobatiformes: Myliobatidae) from Rio de la Plata, Uruguay with a summary of cestodes collected from South American elasmobranchs during 1975–1979. *Proceedings of the Biological Society of Washington* **93**, 1239–1252.

BROOKS, D. R. & MCLENNAN, D. A. (1991). *Phylogeny, Ecology and Behavior, A Research Program in Comparative Biology.* Chicago, University of Chicago Press.

BROOKS, D. R. & MCLENNAN, D. A. (1993*a*). *Parascript: Parasites and the Language of Evolution.* Washington, DC, Smithsonian Institution Press.

BROOKS, D. R. & MCLENNAN, D. A. (1993*b*). Historical ecology: examining phylogenetic components of community evolution. In *Species Diversity in Ecological Communities: Historical and Geographical Perspectives* (ed. Ricklefs, R. E. & Schluter, D.), pp. 267–280. Chicago, USA, University of Chicago Press.

BROOKS, D. R., PÉREZ-PONCE DE LEÓN, G. & LEÓN-RÈGAGNON, V. (2000). Phylogenetic analysis of the Enenterinae (Digenea, Lepocreadiidae) and discussion of the evolution of the digenean fauna of kyphosid fishes. *Zoologica Scripta* **29**, 237–246.

BROOKS, D. R., THORSON, T. B. & MAYES, M. A. (1981*b*). Freshwater stingrays (Potamotrygonidae) and their helminth parasites: testing hypotheses of evolution

and coevolution. In *Advances in Cladistics* (ed. Funk, V. A. & Brooks, D. R.), pp. 147–175. New York, New York Botanical Garden.

BROOKS, D. R., VAN VELLER, M. G. P. & MCLENNAN, D. A. (2001). How to do BPA, really. *Journal of Biogeography* **28**, 345–358.

BULLINI, L., ARDUINO, P., CIANCHI, R., NASCETTI, G., D'AMELIO, S., MATTIUCCI, S., PAGGI, L., ORRECHIA, P., PLÖTZ, J., BERLAND, B., SMITH, J. W. & BRATTEY, J. W. (1997). Genetic and ecological research on anisakid endoparasites of fish and marine mammals in the Antarctic and Arctic-Boreal regions. In *Antarctic Communities: Species Structure and Survival* (ed. Battaglia, B., Valencia J., & Walton, D. W. H.), pp. 39–44. Cambridge, UK, Cambridge University Press.

BUSH, A. O. & KENNEDY, C. R. (1994). Host fragmentation and helminth parasites: hedging your bets against extinction. *International Journal for Parasitology* **24**, 1333–1343.

CAIRA, J. N. (1992). Verification of multiple species of *Pedibothrium* in the Atlantic nurse shark with comments on the Australasian members of the genus. *Journal of Parasitology* **78**, 289–308.

CAIRA, J. N. & BURGE, A. N. (2001). Three new species of *Acanthobothrium* (Cestoda: Tetraphyllidea) from the ocellated electric ray, *Diplobatis ommata*, in the Gulf of California, Mexico. *Comparative Parasitology* **68**, 52–65.

CAIRA, J. N. & JENSEN, K. (2001). An investigation of the co-evolutionary relationships between onchobothriid tapeworms and their elasmobranch hosts. *International Journal for Parasitology* **31**, 960–975.

CAIRA, J. C., JENSEN, K. & HEALY, C. J. (2001). Interrelationships among tetraphyllidean and lecanicephalidean cestodes. In *Interrelationships of the Platyhelminthes* (ed. Littlewood, D. T. J. & Bray, R.), pp. 135–158. London, UK, Taylor and Francis Publishers.

CAIRA, J. C. & ZAHNER, S. D. (2001). Two species of *Acanthobothrium* Beneden, 1849 (Tetraphyllidea: Onchobothriidae) from horn sharks in the Gulf of California, Mexico. *Systematic Parasitology* **50**, 219–229.

CAMPBELL, R. A. (1983). Parasitism in the deep sea. In *Deep Sea Biology* (ed. Rowe, G. T.), pp. 473–552. New York, USA, Wiley Interscience, John Wiley & Sons.

CHOUDHURY, A. & DICK, T. A. (1998). Systematics of the Deropristiidae Cable & Hunnien, 1942 (Trematoda) and biogeographical associations with sturgeons (Osteichthyes: Acipenseridae). *Systematic Parasitology* **41**, 21–39.

COLLETTE, B. B. & RUSSO, J. L. (1985). Interrelationships of the Spanish mackerels (Pisces: Scombridae: *Scomberomorus*) and their copepod parasites. *Cladistics* **1**, 141–158.

CRESSEY, R. F., COLLETTE, B. B. & RUSSO, J. L. (1983). Copepods and scombrid fishes: a study in host–parasite relationships. *Fisheries Bulletin* **81**, 227–265.

CRIBB, T. H., BRAY, R. A. & BARKER, S. C. (1992). Zoogonidae (Digenea) from southern Great Barrier Reef fishes with a description of *Steganoderma*

(*Lecithostaphylus*) *gibsoni* n. sp. *Systematic Parasitology*
23, 7–12.

CRIBB, T. H., BRAY, R. A. & LITTLEWOOD, D. T. J. (2001).
The nature and evolution of the association among
digeneans, molluscs and fishes. *International Journal
for Parasitology* **31**, 997–1011.

DEETS, G. B. (1987). Phylogenetic analysis and revision of
Kroeyerina Wilson, 1932 (Siphonostomatoida:
Kroeyeriidae), copepods parasitic on chondrichthyans,
with descriptions of four new species and the erection
of a new genus, *Prokroeyeria. Canadian Journal of
Zoology* **65**, 2121–2148.

DELIAMURE, S. L. (1955) *Helminthofauna of Marine
Mammals* (*Phylogeny and Ecology*). Izdatel'stvo Akad
Nauk SSSR, Moscow. [English Translation, Israel
Program for Scientific Translations, Jerusalem].

ESCH, G. W., BUSH, A. O. & AHO, J. M. (eds.) (1990).
Parasite Communities : Patterns and Processes. London,
Chapman and Hall.

EUZET, L. (1959). Recherches sur les cestodes
tétraphyllides des sélaciens des côtes de France.
Doctoral Dissertation, Montpellier, France.

FERNÁNDEZ, M., AZNAR, F. J., LATORRE, A. & RAGA, J. A.
(1998*a*). Molecular phylogeny of the families
Campulidae and Nasitrematidae (Trematoda) based
on mtDNA sequence comparison. *International
Journal for Parasitology* **28**, 767–775.

FERNÁNDEZ, M., AZNAR, F. J., RAGA, J. A. & LATORRE, A.
(2000). The origin of *Lecithodesmus* (Digenea:
Campulidae) based on ND3 gene comparison. *Journal
of Parasitology* **86**, 850–852.

FERNÁNDEZ, M., LITTLEWOOD, D. T. J., LATORRE, A., RAGA,
J. A. & ROLLINSON, D. (1998*b*). Phylogenetic
relationships of the family Campulidae (Trematoda)
based on 18s rRNA sequences. *Parasitology* **117**,
383–391.

GARLAND, T. JR., HARVEY, P. H. & IVES, A. R. (1992).
Procedures for the analysis of comparative data using
phylogenetically independent contrasts. *American
Naturalist* **41**, 18–32.

GIBSON, D. I. & BRAY, R. A. (1994). The evolutionary
expansion and host–parasite relationships of the
Digenea. *International Journal for Parasitology* **24**,
1213–1226.

GOSHROY, S. & CAIRA, J. N. (2001). Four new species of
Acanthobothrium (Cestoda: Tetraphyllidea) from the
whiptail stingray *Dasyatis brevis* in the Gulf of
California, Mexico. *Journal of Parasitology* **87**,
354–372.

HAFNER, M. S. & NADLER, S. A. (1988). Phylogenetic trees
support the coevolution of parasites and their hosts.
Nature **332**, 258–259.

HAFNER, M. S., SUDMAN, P. D., VILLABLANCA, F. X.,
SPRADLING, T. A., DEMASTES, J. W. & NADLER, S. A.
(1994). Disparate rates of molecular evolution in
cospeciating hosts and parasites. *Science* **265**,
1087–1090.

HANSKI, I. (1982). Dynamics of regional distribution: the
core and satellite species hypothesis. *Oikos* **38**,
210–221.

HARVEY, P. H. & PAGEL, M. (1991). *The Comparative
Method in Evolutionary Biology*. Oxford, UK, Oxford
University Press.

HAYWARD, C. J. (1997). Distribution of external parasites
indicates boundaries to dispersal of sillaginid fishes in
the Indo-West Pacific. *Marine Freshwater Research* **48**,
391–400.

HOBERG, E. P. (1986). Evolution and historical
biogeography of a parasite-host assemblage :
Alcataenia spp. (Cyclophyllidea: Dilepididae) in
Alcidae (Charadriiformes). *Canadian Journal of
Zoology* **64**, 2576–2589.

HOBERG, E. P. (1992). Congruent and synchronic patterns
in biogeography and speciation among seabirds,
pinnipeds and cestodes. *Journal of Parasitology* **78**,
601–615.

HOBERG, E. P. (1995). Historical biogeography and modes
of speciation across high-latitude seas of the
Holarctic: concepts for host–parasite coevolution
among the Phocini (Phocidae) and Tetrabothriidae.
Canadian Journal of Zoology **73**, 45–57.

HOBERG, E. P. (1996). Faunal diversity among avian
parasite assemblages: the interaction of history,
ecology, and biogeography in marine systems. *Bulletin
of the Scandinavian Society of Parasitology* **6**, 65–89.

HOBERG, E. P. (1997). Phylogeny and historical
reconstruction: host–parasite systems as keystones in
biogeography and ecology. In *Biodiversity II :
Understanding and Protecting Our Biological Resources*
(ed. Reaka-Kudla, M., Wilson, D. E. & Wilson,
E. O.), pp. 243–261. Washington, DC, USA, Joseph
Henry Press.

HOBERG, E. P. & ADAMS, A. (1992). Phylogeny, historical
biogeography, and ecology of *Anophryocephalus* spp.
(Eucestoda: Tetrabothriidae) among pinnipeds of the
Holarctic during the late Tertiary and Pleistocene.
Canadian Journal of Zoology **70**, 703–719.

HOBERG, E. P. & ADAMS, A. (2000). Phylogeny, history and
biodiversity: understanding faunal structure and
biogeography in the marine realm. *Bulletin of the
Scandinavian Society of Parasitology* **10**, 19–37.

HOBERG, E. P., BROOKS, D. R., MOLINA-UREÑA, H. & ERBE, E.
(1998). *Echinocephalus janzeni* n. sp. (Chondrichthyes:
Myliobatiformes) from the Pacific coast of Costa Rica
and Mexico, with historical biogeographic analysis of
the genus. *Journal of Parasitology* **84**, 571–581.

HOBERG, E. P., BROOKS, D. R. & SIEGEL-CAUSEY, D. (1997).
Host–parasite cospeciation: history, principles and
prospects. In *Host–parasite Evolution : General
Principles and Avian Models* (ed. Clayton, D. H. &
Moore, J.), pp. 212–235. Oxford, UK, Oxford
University Press.

HOBERG, E. P., GARDNER, S. L. & CAMPBELL, R. A. (1999*a*).
Systematics of the Eucestoda: advances toward a new
phylogenetic paradigm, and observations on the early
diversification of tapeworms and vertebrates.
Systematic Parasitology **42**, 1–12.

HOBERG, E. P., JONES, A. & BRAY, R. (1999*b*). Phylogenetic
analysis among families of the Cyclophyllidea
(Eucestoda) based on comparative morphology, with
new hypotheses for co-evolution in vertebrates.
Systematic Parasitology **42**, 51–73.

HOBERG, E. P., MARIAUX, J. & BROOKS, D. R. (2001).
Phylogeny among orders of the Eucestoda
(Cercomeromorphae): integrating morphology,
molecules and total evidence. In *Interrelationships of*

the *Platyhelminthes* (ed. Littlewood, D. T. J. & Bray, R.), pp. 112–126. London, UK, Taylor and Francis Publishers.

HOBERG, E. P. & RYAN, P. G. (1987). Ecology of helminth parasitism in *Puffinus gravis* (Procellariiformes) on the breeding grounds at Gough Island. *Canadian Journal of Zoology* **67**, 220–225.

HOLMES, J. C. & PRICE, P. W. (1980). Parasite communities: the roles of phylogeny and ecology. *Systematic Zoology* **29**, 203–215.

IURAKHNO, M. V. (1991). (On the evolution of helminths of marine mammals). In *Evoliutsiya Parazitov*. pp. 124–127. Materialy Pergovo Vsesoyuznogo Sympoziuma. Insitut Ekologii Volzhskogo Basseina AN SSR.

JACKSON, J. A. (1999). Analysis of parasite host-switching: limitations on the use of phylogenies. *Parasitology* **119**, S111–S123.

JIN, Y. G., WANG, Y., WANG, W., SHANG, Q. H., CAO, C. Q. & ERWIN, D. H. (2000). Pattern of marine mass extinction near the Permian-Triassic boundary in South China. *Science* **289**, 432–436.

JORDAN, D. S. (1908). The law of geminate species. *American Naturalist* **42**, 73–80.

KEARN, G. C. (1994). Evolutionary expansion of the Monogenea. *International Journal for Parasitology* **24**, 1227–1271.

KLASSEN, G. J. (1991). Coevolution: a history of the macroevolutionary approach to studying host–parasite associations. *Journal of Parasitology* **78**, 573–587.

KLASSEN, G. J. (1992). Phylogeny and biogeography of ostracin boxfishes (Tetraodontiformes: Ostraciidae) and their gill parasites *Haliotrema* sp. (Monogenea: Ancyrocephalidae): a study in host–parasite coevolution. Ph.D. Dissertation, University of Toronto, Canada. 366 pp.

KLASSEN, G. J. (1994*a*). Phylogeny of *Haliotrema* species (Monogenea: Ancyrocephalidae) from boxfishes (Tetraodontiformes: Ostraciidae): are *Haliotrema* species from boxfishes monophyletic? *Journal of Parasitology* **80**, 596–610.

KLASSEN, G. J. (1994*b*). On the monophyly of *Haliotrema* species (Monogenea: Ancyrocephalidae) from boxfishes (Tetraodontiformes: Ostraciidae): relationships within the *bodiani* group. *Journal of Parasitology* **80**, 611–619.

LEBEDEV, B. I. (1969). Basic regularities in the distribution of monogeneans and trematodes of marine fishes in the world ocean. *Zoologicheskii Zhurnal* **48**, 41–50.

LEÓN-RÈGAGNON, V. (1998). *Machidatrema* n. gen. (Digenea: Hemiuridae: Bunocotylinae) and phylogenetic analysis of its species. *Journal of Parasitology* **84**, 140–146.

LEÓN-RÈGAGNON, V., PÉREZ-PONCE DE LEÓN, G. & BROOKS, D. R. (1996). Phylogenetic analysis of *Opisthadena* (Digenea: Hemiuridae). *Journal of Parasitology* **82**, 1005–1010.

LEÓN-RÈGAGNON, V., PÉREZ-PONCE DE LEÓN, G. & BROOKS, D. R. (1998). Phylogenetic analysis of the Bunocotylinae Dollfus, 1950 (Digenea: Hemiuridae). *Journal of Parasitology* **84**, 147–152.

LITTLEWOOD, D. T. J., RHODE, K., BRAY, R. A. & HERNIOU, E. A. (1999). Phylogeny of the Platyhelminthes and the

evolution of parasitism. *Biological Journal of the Linnaean Society* **68**, 257–287.

LOVEJOY, N. (1997). Stingrays, parasites, and Neotropical biogeography: a closer look at Brooks *et al.*'s hypotheses concerning the origins of Neotropical freshwater stingrays (Potamotrygonidae). *Systematic Biology* **46**, 218–230.

MANTER, H. W. (1966). Parasites of fishes as biological indicators of ancient and recent conditions. In *Host Parasite Relationships* (ed. McCauley, J. E.), pp. 59–71. Corvallis, USA, Oregon State University Press.

MARCOGLIESE, D. & CONE, D. (1993). What metazoan parasites tell us about the evolution of American and European eels. *Evolution* **47**, 1632–1635.

MARQUES, F. (2000). Evolution of Neotropical freshwater stingrays and their parasites: taking into account space and time. Ph.D. Dissertation, University of Toronto, Canada.

MARQUES, F., BROOKS, D. R. & MONKS, S. (1995). Five new species of *Acanthobothrium* Van Beneden, 1849 (Eucestoda: Tetraphyllidea: Onchobothriidae) in stingrays from the Gulf of Nicoya, Costa Rica. *Journal of Parasitology* **81**, 942–951.

MARQUES, F., CENTRITTO, R. & STEWART, S. A. (1997). Two new species of *Acanthobothrium* in *Narcine entemedor* (Rajiformes: Narcinidae) from the northwest coast of Guancaste Peninsula, Costa Rica. *Journal of Parasitology* **83**, 927–931.

MATTIUCCI, S., NASCETTI, G., CIANCHI, R., PAGGI, L., ARDUINO, P., MARGOLIS, L., BRATTEY, J., WEBB, S., D'AMELIO, S., ORECCHIA, P. & BULLINI, L. (1997). Genetic and ecological data on the *Anisakis simplex* complex, with evidence for a new species (Nematoda, Ascaridoidea, Anisakidae). *Journal of Parasitology* **83**, 401–416.

McLENNAN, D. A. & BROOKS, D. R. (1991). Parasites and sexual selection: a macroevolutionary perspective. *Quarterly Review of Biology* **66**, 255–286.

MEASURES, L. N., BEVERLEY-BURTON, M. & WILLIAMS, A. (1990). Three new species of *Monocotyle* (Monogenea: Monocotylidae) from the stingray, *Himantura uranak* (Rajiformes: Dasyatidae) from the Great Barrier Reef: phylogenetic reconstruction, systematics, and emended diagnoses. *International Journal for Parasitology* **20**, 755–767.

MENDOZA-GARFIAS, M. B. & PÉREZ-PONCE DE LEÓN, G. (1998). Relaciones filogenéticas entre las especies del género *Cynoscionicola* (Monogenea: Microcotylidae). *Revista de Biología Tropical* **46**, 335–368.

NADLER, S. A., D'AMELIO, S., FAGERHOLM, H.-P., BERLAND, B. & PAGGI, L. (2000). Phylogenetic relationships among species of *Contracaecum* Railliet & Henry, 1912 and *Phocascaris* Høst, 1932 (Nematoda: Ascaridoidea) based on nuclear rDNA sequence data. *Parasitology* **121**, 455–463.

NASCETTI, G., CIANCHI, R., MATTIUCI, S., D'AMELIO, S., ORRECHIA, P., PAGGI, L., BRATTEY, J., BERLAND, B., SMITH, J. W. & BULLINI, L. (1993). Three sibling species within *Contracaecum osculatum* (Nematoda: Ascaridida: Ascaridoidea) from the Atlantic Arctic-Boreal region: reproductive isolation and host-preferences. *International Journal for Parasitology* **23**, 105–120.

NASIN, C., CAIRA, J. N. & EUZET, L. (1997). A revision of *Calliobothrium* (Tetraphyllidea: Onchobothriidae) with descriptions of three new species and a cladistic analysis of the genus. *Journal of Parasitology* **83**, 714–733.

OLSON, P. D., LITTLEWOOD, D. T. J., BRAY, R. A. & MARIAUX, J. (2001). Interrelationships and evolution of the tapeworms (Platyhelminthes: Cestoda). *Molecular Phylogenetics & Evolution* **19**, 443–467.

OLSON, P. D., RHUNKE, T. R., SANNEY, J. & HUDSON, T. (1999). Evidence for host-specific clades of tetraphyllidean tapeworms (Platyhelminthes: Eucestoda) revealed by analysis of 18S ssrDNA. *International Journal for Parasitology* **29**, 1465–1476.

PAGE, R. D. M. (1994a). Maps between trees and cladistic analysis of historical associations among genes, organisms and areas. *Systematic Biology* **43**, 58–77.

PAGE, R. D. M. (1994b). Parallel phylogenies, reconstructing the history of host–parasite assemblages. *Cladistics* **10**, 155–173.

PAGE, R. D. M. & CHARLESTON, M. A. (1998). Trees within trees: phylogeny and historical associations. *Trends in Ecology & Evolution* **13**, 356–359.

PAGE, R. D. M., PATERSON, A. M. & CLAYTON, D. H. (1996). Lice and cospeciation: a response to Barker. *International Journal for Parasitology* **26**, 213–218.

PAGGI, L., NASCETTI, G., CIANCHI, R., ORECCHIA, P., MATTIUCCI, S., D'AMELIO, S., BERLAND, B., BRATTEY, J., SMITH, J. W. & BULLINI, L. (1991). Genetic evidence for three species within *Pseudoterranova decipiens* (Nematoda: Ascaridida, Ascaridoidea) in the North Atlantic and Norwegian and Barents Seas. *International Journal for Parasitology* **21**, 195–212.

PALUMBI, S. R. (1997). Molecular biogeography of the Pacific. *Coral Reefs* **16**, (Suppl.), S47–S52.

PATERSON, A. M. & BANKS, J. (2001). Analytical approaches to measuring cospeciation of host and parasites: through the glass darkly. *International Journal for Parasitology* **31**, 1012–1022.

PATERSON, A. M. & GRAY, R. D. (1997). Host-parasite co-speciation, host switching, and missing the boat. In *Host-parasite Evolution: General Principles and Avian Models* (ed. Clayton, D. H. & Moore, J.), pp. 236–250. Oxford, UK, Oxford University Press.

PATERSON, A. M., GRAY, R. D. & WALLIS, G. P. (1993). Parasites, petrels and penguins: does louse phylogeny reflect seabird phylogeny? *International Journal for Parasitology* **23**, 515–526.

PATERSON, A. M., PALMA, R. L. & GRAY, R. D. (1999). How frequently do avian lice miss the boat? Implications for coevolutionary studies. *Systematic Biology* **48**, 214–223.

PATERSON, A. M. & POULIN, R. (1999). Have chondracanthid copepods co-speciated with their teleost hosts? *Systematic Parasitology* **44**, 79–85.

PATERSON, A. M., WALLIS, G. P., WALLIS, L. J. & GRAY, R. D. (2000). Seabird and louse coevolution: complex histories revealed by 12S rDNA sequences and reconciliation analyses. *Systematic Biology* **49**, 383–399.

PÉREZ-PONCE DE LEÓN, G. & BROOKS, D. R. (1995a). Phylogenetic relationships of the genera of the Pronocephalidae Looss, 1902 (Digenea: Paramphistomiformes). *Journal of Parasitology* **81**, 267–277.

PÉREZ-PONCE DE LEÓN, G. & BROOKS, D. R. (1995b). Phylogenetic relationships among the species of *Pyelosomum* Looss, 1899 (Digenea: Pronocephalidae). *Journal of Parasitology* **81**, 278–280.

PÉREZ-PONCE DE LEÓN, G., GARCIA-PRIETO, L., MENDOZA-GARFIAS, B., LEÓN-RÈGAGNON, V., PULIODO-FLORES, G., ARANDA-CRUZ, C. & GARCIA-VARGAS, F. (1999). *Listados Faunisticos de México IX. Bioversidad de Helmintos Parásitos de Peces Marinos y Estuarinos de la Bahía de Chamela, Jalisco.* Universidad Nacional Autónoma de México.

PÉREZ-PONCE DE LEÓN, G., LEÓN-RÈGAGNON, V. & MENDOZA-GARFIAS, B. (1997). Análisis filogenético de la familia Pterinotrematidae (Platyhelminthes: Cercomeromorphae: Monogenea). *Anales Instituto Biologie Universidad Autónoma México, Series Zoology* **68**, 193–205.

POULIN, R. (1992). Determinants of host-specificity in parasites of freshwater fishes. *International Journal for Parasitology* **22**, 753–758.

POULIN, R. (1998). *Evolutionary Ecology of Parasites.* London, Chapman & Hall.

RÄSÄNEN, M, E., LINNA, A. M., SANTOS, J. C. R. & NEGRI, F. R. (1995). Late Miocene tidal deposits in the Amazonian foreland basin. *Science* **269**, 386–390.

ROHDE, K. (1992). Latitudinal gradients in species diversity: the search for the primary cause. *Oikos* **65**, 514–527.

ROHDE, K. (1993). *Ecology of Marine Parasites.* 2nd Edition. Wallingford, UK, CAB International.

ROHDE, K. (1994). The origins of parasitism in the Platyhelminthes. *International Journal for Parasitology* **24**, 1099–1115.

ROHDE, K. & HAYWARD, C. J. (2000). Oceanic barriers as indicated by scombrid fishes and their parasites. *International Journal for Parasitology* **30**, 579–583.

ROSENZWEIG, M. L. (1995). *Species Diversity in Space and Time.* Cambridge, UK, Cambridge University Press.

SASAL, P., MORAND, S. & GUEGAN, J.-F. (1997). Determinants of parasite species richness in Mediterranean marine fishes. *Marine Ecology Progress Series* **149**, 61–71.

STIASSNY, M. L. J., PARENTI, L. R. & JOHNSON, C. D. (eds.) (1996). *Interrelationships of Fishes.* San Francisco, USA, Academic Press.

THOMAS, F., VERNEAU, O., DE MEEÛS, T. & RENAUD, F. (1996). Parasites as to host evolutionary prints: insights into host evolution from parasitological data. *International Journal for Parasitology* **26**, 677–686.

VAN VELLER, M. G. P. & BROOKS, D. R. (2001). When simplicity is not parsimonious: a priori and a posteriori methods in historical biogeography. *Journal of Biogeography* **28**, 1–11.

WEBB, S. D. (1995). Biological implications of the Middle Miocene Amazon seaway. *Science* **269**, 361–362.

WHITE, B. N. (1989). Antitropicality and vicariance: a reply to Briggs. *Systematic Zoology* **38**, 77–79.

ZAMPARO, D., BROOKS, D. R. & BARRIGA, M. (1999). *Pararhinebothroides hobergi* n. gen. n. sp. (Eucestoda: Tetraphyllidea) in *Urobatis tumbesensis* (Chondrichthyes: Myliobatiformes) from coastal Ecuador. *Journal of Parasitology* **85**, 534–539.

The trematodes of groupers (Serranidae: Epinephelinae): knowledge, nature and evolution

T. H. CRIBB[1]*, R. A. BRAY[2], T. WRIGHT[1] *and* S. PICHELIN[1]

[1] *Centre for Marine Studies and Department of Microbiology and Parasitology, The University of Queensland, Brisbane 4072, Australia*
[2] *Department of Zoology, The Natural History Museum, Cromwell Road, London SW7 5BD, UK*

SUMMARY

Groupers (Epinephelinae) are prominent marine fishes distributed in the warmer waters of the world. Review of the literature suggests that trematodes are known from only 62 of the 159 species and only 9 of 15 genera; nearly 90 % of host–parasite combinations have been reported only once or twice. All 20 families and all but 7 of 76 genera of trematodes found in epinephelines also occur in non-epinephelines. Only 12 genera of trematodes are reported from both the Atlantic–Eastern Pacific and the Indo–West Pacific. Few (perhaps no) species are credibly cosmopolitan but some have wide distributions across the Indo–West Pacific. The hierarchical 'relatedness' of epinephelines as suggested by how they share trematode taxa (families, genera, species) shows little congruence with what is known of their phylogeny. The major determinant of relatedness appears to be geographical proximity. Together these attributes suggest that host-parasite co-evolution has contributed little to the evolution of trematode communities of epinephelines. Instead, they appear to have arisen through localized episodes of host-switching, presumably both into and out of the epinephelines. The Epinephelinae may well be typical of most groups of marine fishes both in the extent to which their trematode parasites are known and in that, apparently, co-evolution has contributed little to the evolution of their communities of trematodes.

Key words: Serranidae, Epinephelinae, trematodes, co-evolution, host-specificity, biogeography.

INTRODUCTION

Many thousands of papers describe the parasites of teleost fishes. Surprisingly, there have been few attempts to synthesize ideas about all the parasites from particular host groups (as opposed to producing checklists e.g. Williams & Bunkley-Williams (1996)). Barker *et al.* (1994) analysed the distribution of trematodes in 39 species of sympatric Pomacentridae on the Great Barrier Reef, and Rohde & Hayward (2000) analysed the ectoparasites of 26 species of Scombridae, but we are aware of few other such studies. Perhaps the difficulty relates to the fact that the literature is now so large and diffuse. Whatever the reason, we suspect that there has been a serious under-analysis of broad patterns of distributions (geographical, host, ecological) of parasites of whole taxonomic assemblages of fishes. Here we analyse the Epinephelinae, the groupers or gropers, coral cods, coral trout and others. The epinephelines are a major group of commercially significant fishes concentrated in the warmer oceans of the world. The literature abounds with reports of their parasites (they are the tenth most frequently reported host group in the *Catalogue of Trematodes of Fishes* maintained at the University of Queensland) but they have never been considered as a whole. Here we seek to answer questions from the prosaic to the

profound. How well is the parasite fauna known? What is the nature of the parasite fauna in terms of its taxonomic components, host-specificity and geographical distribution? What evolutionary processes explain the patterns of parasite distribution?

MATERIALS AND METHODS

We used Yamaguti (1971), the records of the Host-Parasite Catalogue of the Parasitic Worms Division of The Natural History Museum, London, *Helminthological Abstracts* (1984–2000) and *Biological Abstracts* (1985–2000) as sources for original reports. We searched for all the presently accepted genera of Epinephelinae together with *Garrupa* and *Promicrops* (synonyms of *Epinephelus*) and *Serranus*. *Serranus* is the type genus of the Serraninae but there are many early records of epinephelines recorded under this genus. We checked the validity of all the names that we found using the comprehensive synonymies provided by Heemstra & Randall (1993). This enabled us to update many records from *Serranus*, *Garrupa* and *Promicrops* to currently recognized epinepheline names. All records were incorporated into our data-base *Catalogue of Trematodes of Fishes* which presently summarises 22828 records of trematodes from fishes from over 2000 published papers.

We accepted the identifications of parasites given in the literature uncritically in the initial compilation of the main set of records. We subsequently updated names as they have been recombined or corrected in

* Corresponding author: Department of Microbiology and Parasitology, The University of Queensland, Brisbane 4072, Australia. Tel.: 61 7 3365 2581. Fax: 61 7 3365 1588. E-mail: T.Cribb@mailbox.uq.edu.au.

the literature by later authors, assuming that there was a reasonable basis for such changes. The exception to this was some species of *Helicometra*. Sekerak & Arai (1974) proposed comprehensive synonymies within this genus. In particular, they proposed that many species should be synonyms of *Helicometra pulchella* (this name itself subsequently shown by Bray (1987) to be best considered a junior synonym of *H. fasciata*). These synonymies must be considered controversial because molecular studies (Reversat, Renaud & Maillard, 1989; Reversat, Maillard & Silan, 1991; Reversat & Silan, 1991) have suggested that the diversity of species within this genus is easily underestimated. As a consequence, we took the conservative approach of recognizing all the species of *Helicometra* mentioned in the literature on the Epinephelinae.

We found 546 individual published records of trematode species in fishes identifiable as epinepheline serranids. This figure excludes records from 'Serranus sp.' which might well have represented either species of *Epinephelus* (Epinephelinae) or true *Serranus* (Serraninae). The count reduces to 434 records when incompletely identified hosts and trematodes (frequently the genus is reported but the species is unknown) are also excluded. All the following analyses (Tables 2–5) relate to this reduced data set of 434 records. It is worth emphasizing the loss of information incurred by the incomplete identification of host or parasite; about 20% of the 546 records cannot be evaluated critically. We did not attempt to distinguish between the value of records in the literature which might have reported a single specimen of a trematode from a single host individual and those that might have been of large numbers of individuals from multiple host individuals. The basic host–parasite list and the references from which they were derived are given in the Appendix. The full data-base is available from the authors on request.

Parasite-based relatedness of hosts was inferred by the approach used recently by Feliu *et al.* (1997) for Iberian rodents and their parasites and by Beveridge & Chilton (2001) for Australian macropodid marsupials and their cloacinine nematodes. We created a matrix of trematode taxa and their epinepheline hosts. In different analyses we incorporated not only parasite species, but parasite genera and families because this level of inclusiveness has the potential to be informative where parasite species that occur in a single host species are uninformative. Each taxon was coded as absent (0) or present (1). We created an 'outgroup' fish with no parasites ('0' for each parasite); Beveridge & Chilton (2001) used as an outgroup a primitive macropodoid which lacks cloacinine nematodes. We analysed the data set using heuristic searches in PAUP* (Swofford, 1998).

A problem with analysis of this kind is that many of the taxa have been studied inadequately so that

their lack of parasites (characters) is artificial. This is certainly the case here – most species of epinephelines have no trematodes reported from them at all. Such poorly studied species can only lead to spurious conclusions if treated uncritically. In the absence of explicit published information about sample sizes (which is almost completely lacking) this problem is difficult to solve. We attempted to ameliorate the problem by progressively reducing the data set. Several different matrices were created (see Table 5). The first matrix contained all 'informative characters' (= trematode taxa) for all fish that shared at least 1 character with another epinepheline. Subsequent matrices were progressively restricted to include hosts sharing at least 2 informative characters and so on. There are two advantages to such restriction. First, poorly studied species are excluded. Second, the ratio of characters to taxa increases (see Table 5) so that the relationships that are inferred are likely to be progressively more robust. There are two problems with restriction of the data set. First, as the criterion for inclusion becomes more exclusive, the number of host species retained in the analysis plunges so that the analysis becomes informative about a smaller and smaller proportion of the fauna. Second, the restriction may exclude taxa that genuinely lack a large range of parasites (instead of being simply little studied). The data set was analysed without records deemed 'suspect' (discussed below). 'Suspect' records were of *Helicometrina nimia*, *Hamacreadium mutabile*, *Stephanostomum dentatum*, *Prosorhynchus ozakii*, *P. pacificus*, *Lepidapedoides levenseni* and *Opecoelus mexicanus* from the Indo–West Pacific (IWP) and of *Prosorhynchus epinepheli* from the Atlantic–Eastern Pacific (AEP).

THE EPINEPHELINAE

The Serranidae is a major family of mainly marine fishes. It contains about 449 species (Nelson, 1994) making it one of the largest families of vertebrates. The Serranidae are presently considered to comprise 5 subfamilies: Serraninae, Anthiinae, Epinephelinae, Grammistinae and Niphoniinae (Nelson, 1994). Our study is restricted to the Epinephelinae which comprises 15 genera and 159 species according to Heemstra & Randall (1993). There continues to be considerable activity on the taxonomy of the Epinephelinae although no new species have been described since the revision mentioned above. Phylogenetic relationships between the genera are largely unresolved although Craig *et al.* (2001) used analysis of 16S rDNA of 42 of the 159 species to suggest that the dominant genus, *Epinephelus*, is paraphyletic. Individual species range from maximum sizes as small as about 26 cm (e.g. *Cephalopholis boenak*) to some of the largest of teleost fishes; *Epinephelus lanceolatus* and *E. itajara* reach at least 230 cm and well over 200 kg. Epinephelines range from shallow

Table 1. The 30 families of fishes most frequently reported as hosts of trematodes. The Epinephelinae is the tenth most frequently reported host taxon

Host family	Rank	No. host/parasite records	% all records	No. host species	Records per host	No. trematode families
Carangidae	1	1425	6·2	140	10·2	28
Scombridae	2	1100	4·8	49	22·4	21
Pleuronectidae	3	1066	4·6	93	11·5	21
Cyprinidae	4	696	3·0	2010	0·3	26
Serranidae (all)	5	687	3·0	449	1·5	21
Sparidae	6	654	2·8	100	6·5	28
Haemulidae	7	616	2·7	150	4·1	19
Sciaenidae	8	592	2·6	270	2·2	25
Lutjanidae	9	590	2·6	125	4·7	25
[Serranidae (Epinephelinae)		547	2·4	159	3·4	20]
Labridae	10	426	1·9	500	0·9	24
Scorpaenidae	11	421	1·8	388	1·1	17
Gadidae	12	384	1·7	30	12·8	16
Salmonidae	13	375	1·6	66	5·7	23
Clupeidae	14	360	1·6	181	2·0	20
Centrarchidae	15	355	1·5	29	12·2	14
Cottidae	16	327	1·4	300	1·1	16
Mugilidae	17	327	1·4	80	4·1	20
Acanthuridae	18	308	1·3	72	4·3	18
Pomacentridae	19	273	1·2	315	0·9	11
Tetraodontidae	20	256	1·1	121	2·1	21
Nototheniidae	21	248	1·1	50	5·0	11
Percidae	22	247	1·1	162	1·5	14
Mullidae	23	224	1·0	55	4·1	18
Gobiidae	24	217	0·9	1875	0·1	19
Balistidae	25	213	0·9	40	5·3	15
Chaetodontidae	26	212	0·9	89	2·4	14
Anguillidae	27	209	0·9	15	13·9	16
Bagridae	28	197	0·9	210	0·9	21
Congridae	29	190	0·8	150	1·3	12
Monacanthidae	30	186	0·8	95	1·95	17
Totals/means		13381	58·1	8209	1·6	19

water-dwelling species that can be found on reef flats (e.g. *Epinephelus quoyanus*) to relatively deep-water species (e.g. many species of *Epinephelus* have been recorded from depths of well over 100 metres). Epinephelines are predators, usually using an ambush approach to food capture and where it has been studied, fishes, crustaceans and cephalopods dominate the diet. A few species are capable of plankton feeding (*Paranthias* spp. and *Epinephelus undulosus* according to Heemstra & Randall (1993)).

The species of Epinephelinae are distributed widely but are concentrated mainly in warmer waters of the world. The distributions given by Heemstra & Randall (1993) suggest two main areas of distribution. These are the Indo–West Pacific (IWP), from the east coast of Africa eastward to the Tuamotus and Hawaii in the Pacific, and the Atlantic–Eastern Pacific (AEP). Three genera, *Dermatolepis*, *Epinephelus* and *Cephalopholis*, occur in both areas; the remaining genera are each found in only one area. A few AEP species of epinephelines span the Isthmus of Panama and are found on both sides of the Atlantic, but none is known that crosses the Pacific from east to west. Only 1 species,

Epinephelus marginatus which extends just around the southern tip of South Africa, has a distribution that incorporates the Atlantic Ocean and the Indian Ocean. There appear, thus, to be natural boundaries in the distribution of epinephelines that run between the eastern and western halves of the Pacific and at the southern tip of Africa. 110 species of epinephelines are known from the IWP and 50 are known from the AEP (*E. marginatus* occurs in both regions).

Analysis of our *Catalogue of Trematodes of Fishes* shows that as well as being the fifth largest family of fishes, the Serranidae has the fifth largest count of records of trematodes. The sub-set of the Epinephelinae is between the ninth and tenth largest families. The ratio of number of parasite records to number of host species for families of fishes varies dramatically. For the 30 most reported families (Table 1) it ranges from 0·1 for the Gobiidae to 22·4 for the Scombridae. Reasons for the differences relate to a combination of how heavily parasitised these taxa are and how thoroughly they have been studied. Overall, the Serranidae is just below the mean ratio for the 30 most reported families and the Epinephelinae is a little more than double it. These

Table 2. Extent to which epinephelines have been reported to harbour trematodes

Genus	No. of fish species	No. from which trematodes are reported
Aethaloperca	1	0
Alphestes	3	2
Anyperodon	1	0
Cephalopholis	22	7
Cromileptes	1	1
Dermatolepis	3	0
Epinephelus	97	38
Gonioplectrus	1	0
Gracila	1	0
Mycteroperca	15	8
Paranthias	2	1
Plectropomus	7	2
Saloptia	1	0
Triso	1	1
Variola	2	2
Totals	159	62

Table 3. Frequency with which host–parasite combination of trematodes of epinephelines have been reported in the literature

Frequency	Records	% all records
1	211	72·8
2	45	15·5
3	22	7·6
4	4	1·4
5	4	1·4
6	2	0·6
8	1	0·3
11	1	0·3
Total	290	

figures suggest that the Epinephelinae are a little better studied than is typical for most families of fishes.

HOW WELL ARE THE TREMATODES OF EPINEPHELINES KNOWN?

We found records of parasites from only 62 (< 40 %) of the 159 species of Epinephelinae (Table 2). There are reports of trematodes from species of 9 genera – *Alphestes*, *Cephalopholis*, *Cromileptes*, *Epinephelus*, *Mycteroperca*, *Paranthias*, *Plectropomus*, *Triso* and *Variola* but none from species of *Aethaloperca*, *Anyperodon*, *Dermatolepis*, *Gonioplectrus*, *Gracila* or *Saloptia* which are monotypic except for *Dermatolepis* which has 3 species. Only a few epinepheline species appear to have been studied in detail. *Epinephelus striatus*, the Nassau grouper, from the western Atlantic (Caribbean to southern Brazil) has 18 species of trematodes; *E. akaara*, from the waters of southern Japan, the coast of China and Taiwan has 17 species including 7 species of Opecoelidae; *E. fasciatus*, which is widespread in the

IWP, has 13 species; and *E. adscensionis* from the mid-Atlantic and the coasts of North and South America is reported to harbour 10 species of trematodes. In contrast, in addition to the 97 species from which no trematodes have been reported, a further 13 species are reported as harbouring only a single species. We found 147 identified species of trematodes and 290 individual combinations of trematode species and host. Of these, 211 combinations (73 %) have been reported only once and a further 45 combinations have been recorded only twice (see Table 3). Only two combinations have been reported frequently; *Bivesicula claviformis* (Bivesiculidae) has been reported in *Epinephelus fasciatus* from the IWP 11 times and *Helicometra torta* (Opecoelidae) in *E. striatus* from the Caribbean 8 times.

Overall we must conclude that the fauna is incompletely known. Trematodes are recorded from fewer than half the species of potential hosts, most of these hosts have been examined inadequately, and most of the parasites have been reported only rarely. Full life-cycles are known for almost none of the parasites. Despite these inadequacies, certain types of parasites are reported sufficiently frequently for us to be able to interpret them as characteristic and others are sufficiently infrequently reported that we can be confident that they are exceptional.

NATURE AND HOST-SPECIFICITY

The trematode fauna of epinephelines is rich. The 62 species of epinephelines from which trematodes are reported are infected with 147 species of trematodes. These trematodes are distributed in 76 genera and 20 families (Table 4). Despite this overall richness, a handful of families and genera account for most of the records. Thus, 11 genera (*Allonematobothrium*, *Allopodocotyle*, *Cainocreadium*, *Gonapodasmius*, *Helicometra*, *Helicometrina*, *Lecithochirium*, *Lepidapedoides*, *Neolepidapedoides*, *Prosorhynchus* and *Stephanostomum*) and just 6 families (Acanthocolpidae, Bucephalidae, Didymozoidae, Hemiuridae, Lepocreadiidae and Opecoelidae) comprise 46 % and 81 % of the records, respectively.

The host-specificity of the parasites can be considered conveniently on three levels – family, genus and species. Every digenean family represented in the Epinephelinae also has a wide host distribution outside the subfamily; no trematode family is linked solely or even mainly with the Epinephelinae (or even with the Serranidae). At the level of genus, we find that of the 76 genera, remarkably only 7 are restricted to the Epinephelinae. Further, these genera (*Indoglomeritrema*, *Pacificreadium*, *Pearsonellum*, *Postporus*, *Chabaudtrema*, *Mitotrema*, *Paraprosorhynchus*) are all monotypic. At the level of species, just 66 of 147 trematode species (45 %) are restricted to epinepheline ser-

Table 4. Classification of genera of trematodes reported from epinepheline serranids and the number of records of each genus from Atlantic-Eastern Pacific (AEP) and Indo-West Pacific (IWP)

Family	Genus	AEP	IWP
Acanthocolpidae	*Stephanostomum*	24	5
	Tormopsolus		1
Apocreadiidae	*Postporus*	13	
Bivesiculidae	*Bivesicula*		23
Bucephalidae	*Bucephalus*	2	
	Folliculovarium		1
	Neoprosorhynchus		2
	Paraprosorhynchus	1	
	Prosorhynchus	39	27
	Pseudoprosorhynchus		1
	Rhipidocotyle		3
	Telorhynchus		1
Bunocotylidae	*Aphanurus*		1
Cryptogonimidae	*Mitotrema*		4
	Paracryptogonimus	1	
	Pseudometadena		1
Derogenidae	*Derogenes*		1
Didymozoidae	*Allonematobothrium*	2	8
	Atalostrophion	1	
	Didymozoon	2	
	Gonapodasmius	1	16
	Indoglomeritrema		1
	Lobatozoum	1	
	Maccallozoum	1	
Fellodistomidae	*Monascus*		1
	Proctoeces		1
Gorgoderidae	*Phyllodistomum*	1	3
Hemiuridae	*Brachyphallus*		1
	Dissosaccus	1	
	Ectenurus	2	
	Elytrophallus	2	
	Erilepturus		4
	Lecithochirium	19	
	Lecithocladium	3	
	Plerurus		5
	Tubulovesicula		3
Hirudinellidae	*Hirudinella*		1
Lecithasteridae	*Hysterolecithoides*		1
	Lecithophyllum	1	
	Thulinia		1
Lepocreadiidae	*Bianium*		1
	Hypocreadium	1	
	Lepidapedoides	21	14
	Lepidapedon	1	
	Multitestis		1
	Myzoxenus	1	
	Neolepidapedoides	7	
	Prodistomum	1	
Monorchiidae	*Lasiotocus*		2
	Proctotrematoides		2
	Retractomonorchis		3
Opecoelidae	*Allopodocotyle*		9
	Apertile		2
	Apopodocotyle	1	
	Cainocreadium	9	16
	Coitocaecum		1
	Crowcrocaecum		1
	Dactylostomum		1
	Hamacreadium	10	5
	Helicometra	17	20
	Helicometrina	9	4

Table 4. contd.

Family	Genus	AEP	IWP
	Opecoeloides	1	
	Opecoelus	2	6
	Pacificreadium		7
	Peracreadium	1	
	Plagioporus		2
	Podocotyle	1	
	Pseudopecoelina		1
	Pseudopecoelus		2
	Pseudoplagioporus		1
	Vesicocoelium		1
Pronocephalidae	*Barisomum*	1	
Sanguinicolidae	*Pearsonellum*		9
Sclerodistomidae	*Prosogonotrema*		2
	Sclerodistomum	1	
Zoogonidae	*Deretrema*	1	

ranids. The sharing is occasionally with other serranid subfamilies (e.g. *Lepidapedoides angustus* is shared by 6 epinephelines and the grammistine *Diploprion bifasciatum* on the Great Barrier Reef), but usually with lutjanids, lethrinids, carangids, mullids and many other families. Just 16 species of trematodes are both restricted to and shared among epinephelines.

The figures for species may mislead in a number of ways. First, we predict that the incompleteness of present records (most species have been reported only once) means that there are more parasites shared with other epinephelines and non-epinephelines. Second, there are a number of species of trematodes that, although clearly principally parasites of serranids, are known to occur in other host groups occasionally (and *vice versa*). For example, *Bivesicula claviformis* has been reported from epinephelines 21 times but has also been found once each in a scombrid, a holocentrid, a carangid and a haemulid. It is impossible to determine whether the records from non-epinephelines were made in error or if they are genuine. Finally, species that are said to be shared may represent complexes of species with higher levels of host-specificity than are recognized presently.

Levels of host-specificity vary dramatically between trematode families. The Hemiuridae has the lowest specificity. All 9 genera and 14 species have been reported from non-serranids. Acanthocolpids of epinepheline serranids occur in 2 genera – a single species of *Tormopsolus* and 7 of *Stephanostomum*. Both genera are large, containing over 100 species in the case of *Stephanostomum*; five of the 8 species have been reported from non-serranids. The Opecoelidae are represented by 20 genera. Of these, all but the monotypic *Pacificreadium* are found in non-serranids. Of the remaining 19 genera, only *Allopodocotyle* (4 species), *Cainocreadium* (4 species), *Helicometra* (6 species) and *Helicometrina* (4 species)

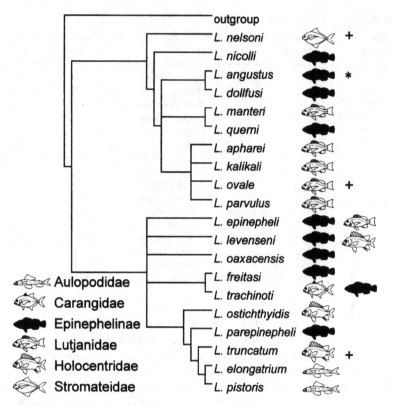

Fig. 1. Host group distribution mapped on a phylogeny of the lepocreadiid genus *Lepidapedoides* from Bray *et al.* (1998). + = other families of fishes are minor host groups for this species; * = grammistine serranids also.

have significant numbers of species within epinephelines. About half the species of *Helicometrina* occur in epinephelines and a far smaller proportion for all the other genera. Fifteen of 39 opecoelid species are restricted to epinephelines. Seven genera of didymozoids are reported as having representatives within the Epinephelinae. Apart from the monotypic *Indoglomeritrema*, all are also represented in non-epinephelines but 15 of the 17 species are restricted to epinephelines. The genus most closely allied with the epinephelines is *Allonematobothrium*; five of the 6 species are restricted entirely to epinephelines. The Bucephalidae are represented by 8 genera. Six of these genera are represented only by single species in epinephelines; *Rhipidocotyle* has 2 species. Eleven of the 21 species are restricted to epinephelines. *Prosorhynchus* is the only genus apparently strongly radiated within the Epinephelinae where it is represented by 13 species but the genus as a whole includes about 60 species. Eight genera of lepocreadiids are reported from epinepheline serranids. Ten of 19 species are restricted to epinephelines. Six of the genera are represented by only single species from genera that are relatively large and widely distributed in other families of fishes. Two genera of lepocreadiids are strongly concentrated in the Epinephelinae. The 5 species of *Neolepidapedoides* are restricted to serranids. One (*N. hypoplectri*) occurs in a serranine and the others are restricted to epinephelines; all occur in the Western Atlantic. *Lepidapedoides* comprises 20

species (Bray, Cribb & Littlewood, 1998), many of which occur at least partly in epinepheline serranids. Of the 20 species, 10 are reported from epinephelines, 7 are restricted to epinephelines and another is restricted to serranids (Bray, Cribb & Barker, 1996; Bray *et al.* 1998). If hosts are mapped on the preliminary phylogeny of *Lepidapedoides* presented by Bray *et al.* (1998) then the epinepheline parasites appear scattered throughout the tree (Fig. 1). Bray *et al.* (1998) considered the host-specificity of the genus and found that the species ranged from oioxenic (strictly specific to a single species) through to euryxenic species (distribution defined, apparently, only by ecology).

However the data are considered, overall the host-specificity of trematodes of serranids is remarkably low at all levels and in all but a handful of cases. This implies that there has been almost no classical host–parasite co-evolution in the sense of clades of parasites radiated with and restricted to the Epinephelinae. In only a handful of cases (*Lepidapedoides*, *Neolepidapedoides*, *Prosorhynchus* and *Allonematobothrium*) is there evidence for fidelity between a lineage of parasites and the epinephelines. If the host distribution of *Lepidapedoides* is a guide, however, even in these genera the case for strict co-evolution is weak.

Instead of having arisen through co-evolution, we conclude that the communities of trematodes of epinephelines have developed through repeated independent cases of host-switching. The host-switching has involved many families and genera of trematodes and has presumably been both into and out of the Epinephelinae. The host-switching has probably been accompanied by sorting (extinction), inertia (lack of parasite speciation) and 'missing the boat' events and, perhaps, duplication (intra-host speciation) (Paterson & Banks, 2001). The broad, opportunistic carnivorous diet of many serranids has presumably facilitated what we suspect is the dominant process of host-switching. In this connection it is interesting to observe on the trematodes that epinephelines lack. Most striking among these is the Cryptogonimidae. Only 3 species of cryptogonimids are known from serranids. These trematodes are transmitted by the ingestion of metacercariae encysted in the flesh of fishes. This is the same route of infection that is used by all Bucephalidae, most Didymozoidae and some Opecoelidae and Hemiuridae – all of which are common in the Epinephelinae. Further, cryptogonimids are some of the most abundant trematodes in the Lutjanidae. Lutjanids are perciform fishes usually common in much the same environments as epinephelines (especially coral reefs) and, indeed, share some other trematode species with epinephelines. A general absence of cryptogonimids in serranids suggests that the development of parasite communities of epinephelines must be determined by more than just what they eat.

GEOGRAPHICAL DISTRIBUTION

The most useful geographical analysis that can be made is of contrasts between the IWP and AEP, the 2 major areas of host distribution. Overall, the IWP is distinctly richer than the AEP. Seventeen families and 50 genera are reported from this region compared with 13 families and 37 genera from the AEP (Table 4). This contrast is perhaps predictable in that the epinepheline fauna in the IWP is larger and the fauna of other fishes and parasites is likely to be richer than that in the AEP, giving it a larger pool of parasites upon which to draw. Ten of the parasite families, including all the larger taxa, occur in both areas but strikingly, 64 of 76 (84%) genera have been reported from only 1 area. A few of the families that occur in only 1 area are noteworthy. The Bivesiculidae occurs in epinephelines only in the IWP although it does occur in other families of fishes in the Atlantic. Although this family is represented presently in this area by only 2 fully identified species, Shimazu & Machida (1995) suggested that there may be further unrecognized species in epinephelines and we have unpublished evidence that supports this view. Thus, it appears that this family has colonized and radiated in the Epinephelinae only in the Indo-western Pacific. There are reports of 6 species of Monorchiidae from epinephelines. All are from species of *Epinephelus* and

Cephalopholis in the Indian Ocean. These 6 species include 2 pairs of congeners in *Cephalopholis sonnerati*. They appear to represent repeated and highly localised colonisation and radiation events. Similarly the Pronocephalidae appear to have adopted parasitism of fishes (their origins are apparently with marine reptiles) only in the AEP. We interpret each of these cases as identifiable components of the wider pattern of repeated independent parasite adoptions by the epinephelines.

Trematode species occur overwhelmingly in either the AEP or the IWP; just 10 of 147 species of trematodes have distributions that are said to incorporate both the IWP and the AEP. In most of these 10 cases it appears likely that errors have been made in parasite identification. For example, *Helicometrina nimia* has been reported, both from serranids and other families of fishes, from the Bahamas, Baja California, Jamaica, the Pacific coast of Panama, Yucatan Peninsula, Mexico and, anomalously, twice from the Karachi Coast of Pakistan. The Pakistani records are isolated in the IWP and should certainly be reconsidered despite the conclusion of Hafeezullah (1971) that they were indistinguishable from *H. nimia*. Another anomalous distribution is that of the opecoelid *Hamacreadium mutabile*. The 14 records of this species from epinephelines are from the Caribbean and Gulf of Mexico, the Galapagos Islands and, anomalously, the Red Sea. However, *H. mutabile* has not only been reported from a wide range of other carnivorous fishes (especially Lutjanidae) from these same regions but also widely in the Indo-West Pacific. This anomalous distribution is demonstrated by our records for the species on the Great Barrier Reef. We (Bray & Cribb, 1989) reported *H. mutabile* as abundant in 8 species of *Lutjanus* but from none of 20 species of epinephelines from the same region. This disparity in host and geographical distribution is strongly suggestive of biological (and presumably taxonomic) distinction and requires further consideration. We also consider suspect records of *Stephanostomum dentatum*, *Prosorhynchus ozakii*, *P. pacificus*, *Lepidapedoides levenseni*, *Hamacreadium mutabile* and *Opecoelus mexicanus* from the IWP and of *Prosorhynchus epinepheli* from the AEP. In each case, the records are rendered suspect by suggesting anomalously large disjunct distributions and being supported by taxonomic analysis that may be wanting (e.g. perfunctory descriptions, inadequate comparisons or absence of figures).

Two species, *Prosorhynchus caudovatus* and *Lepidapedoides nicolli*, have been reported principally from the AEP but also once each from the IWP on the Indian Ocean coast of South Africa. In addition to the Indian Ocean report, *L. nicolli* is known from Africa (Atlantic Coast), the Galapagos, Ghana, Mexico (Pacific Coast), Florida, Venezuela and the east coast of South Africa and *P. caudovatus* is

Table 5. *Statistics of hosts and parasite characters for analysis of parasite-based relatedness of hosts*

	All taxa				Genera and species				Species only			
A	B	C	D	E	B	C	D	E	B	C	D	E
1	91	61	1·49	552	76	60	1·27	671	56	47	0·84	177
2	91	60	1·52	551	75	56	1·34	366	37	40	1·08	151
3	90	56	1·61	542	66	40	1·65	324	21	33	1·57	112
4	87	48	1·81	515	64	37	1·73	314	15	31	2·07	93
5	80	40	2·00	477	61	30	2·03	277	10	24	2·4	67
6	78	37	2·11	460	59	25	2·36	256	6	15	2·5	43
7	76	32	2·38	428	55	18	3·06	211				
8	74	30	2·47	412	46	13	3·54	168				
9	71	25	2·84	369	45	12	3·75	159				
10	67	20	3·35	321	43	11	3·91	148				
11	63	15	4·20	267	40	10	4·00	135				
12	54	12	4·50	225	40	10	4·00	135				
13	49	10	4·90	194								
14	49	10	4·90	194								

A – Min. no. shared characters (= trematode taxa); B – characters in analysis; C – host species in analysis; D – characters/taxa; E – total number informative characters in analysis.

known from the Mediterranean, Cape Verde and Ghana. We have no reason to suspect the identifications of these species (unless they are cryptic species). The distributions may imply that their two host species, *Epinephelus albomarginatus* and *E. andersoni*, have affinities with the AEP fauna rather than the IWP or that southern Africa is a region where the two faunas mix.

Although cosmopolitan distributions probably never occur, it appears that some species have broad distributions in the IWP. *Bivesicula claviformis* has been reported from the northern and southern Great Barrier Reef, Japan, Borneo, China, Fiji and the Red Sea. This distribution reflects that of what is apparently the most important host of this species, *Epinephelus fasciatus*, which is widespread in the IWP. Many other less well-known IWP trematode species have been reported from a number of sites which probably imply the same sort of wide IWP distribution. Thus, *Pacificreadium serrani* is known from the Great Barrier Reef and the Red Sea, *Cainocreadium epinepheli* is known from the Gulf of Arabia, Japan and the Great Barrier Reef, and *Mitotrema anthostomatum* is known from the Great Barrier Reef and Fiji. None of the trematodes of serranids that have putatively widespread distributions have been examined by molecular methods to test this. Lo *et al.* (2001) used a molecular approach to suggest that 3 species of trematodes of fishes were identical between French Polynesia and the Great Barrier Reef so that Indo-Pacific wide distributions are at least plausible. A note of caution should be applied, however, in assuming that broadly IWP distributions will prove the norm for trematodes of epinephelines. There is a growing realisation that widespread fish species in the IWP, long thought to represent single species, may themselves have

complex structures of species or sub-species (Gill, 1999).

Rigby *et al.* (1997) demonstrated that there may be a striking increase in parasite richness from French Polynesia west to the Great Barrier Reef (Australia) for the parasites (including the trematodes) of *Epinephelus merra*. They invoked principles of island biogeography to explain this. How general this phenomenon may be, whether it affects different taxa of parasites differently, and where the greatest and lowest species-richness occurs are all matters to be determined.

PARASITE-BASED RELATEDNESS OF HOSTS

We analysed all of the matrices summarised in Table 5 that included at least 15 epinepheline species. We found that analysis of species only restricted the data set unduly but that incorporation of parasite families was probably generally misleading because it was clear that most families had been adopted by epinephelines more than once. The relationships for fish sharing 5 or more genera and species (50% majority rule consensus tree of 6 equal length trees) are shown in Fig. 2. This analysis included just 30 of 159 species of Epinephelinae (Table 5). The analysis identifies 4 major clades, 2 are of 3 and 7 species entirely from the IWP and the other 2 are of 4 and 16 species entirely from the AEP except for *Epinephelus quernus* (an IWP species). These relationships are broadly typical of those suggested by most of the analyses – clades dominated by fishes from just one area.

Hierarchies produced by analyses of this sort do not attempt to represent a phylogeny of the host animals concerned. Brooks (1981) discussed the dangers in assuming that this kind of analysis might

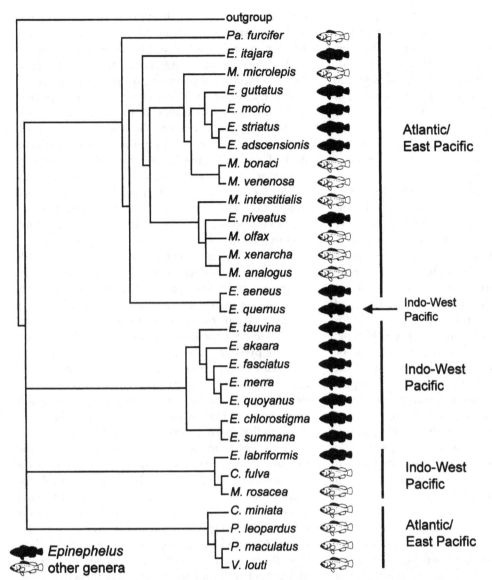

Fig. 2. Parasite-based relatedness of 30 species of Epinephelinae. 50% majority-rule consensus tree of 6 equal length trees.

do this. Instead, the 'tree' provides a visual description of the parasitological similarity of the hosts. Parasitological similarity is generated by the interaction of the ecology, physiology and evolutionary history of both parasites and hosts. Thus, the sharing of a parasite species by 2 fishes typically implies that they share an overlapping or similar diet and similar gastrointestinal physiology; the only clear exception in the trematodes of epinephelines is the single species of sanguinicolid whose cercaria is presumed to penetrate the definitive host directly. Sharing of genera or families of parasites implies a more generalised level of similarity and might imply that co-evolution has led to the sharing of the taxon. It is where the relationships inferred by an analysis of this kind *differs* from the phylogeny of the hosts that they are informative. A hierarchy congruent with the true host phylogeny would imply that strict host–parasite co-evolution had occurred.

The principal conclusion that we draw from this analysis is that geographical location is apparently

the first determinant of the composition of the community of trematodes in epinephelines. Thus, although the dominant genus, *Epinephelus*, occurs in both the IWP and the AEP, its species are parasitologically more closely related to other genera in the areas in which they are found than they are to species of *Epinephelus* from other areas. This conclusion is not refuted by the conclusion of Craig *et al.* (2001) that *Epinephelus* is paraphyletic because the distinct clades that they identified included species of *Epinephelus* from the AEP and the IWP. Unfortunately their study is not particularly helpful to the present analysis because it was dominated heavily by taxa from the AEP. If the AEP clades are examined more closely, we see the same pattern on a smaller scale. *Mycteroperca*, which is restricted to the AEP, fails to form an independent clade. Rather, it is distributed in both major clades that we can recognise within the AEP.

The only anomalous distribution in Fig. 2 is that of *E. quernus*, a Hawaiian species that occurs in a clade otherwise entirely from the AEP. The position of this species was generated by its sharing of two species and four genera with other epinephelines. The two species occur only in the IWP but the four genera all occur in both areas. Overall, 21 of the 40 instances in which these 6 taxa occurred elsewhere in the matrix were in epinephelines from the AEP. This apparent anomaly demonstrates the fallibility of this approach and emphasizes that the patterns that it reveals should be treated as an aid to the understanding of the underlying processes.

The finding of distinct AEP and IWP clades was enhanced by the exclusion of the 'suspect' records. When included the results were less clear. We have two observations about the exclusion of these data. First, we conclude that it is always more unsatisfactory to include than to exclude records that are dubious a priori. Identification of the records as suspect raises as a general problem the reliability of published taxonomic work and the need, not only for much of the fauna to be explored for the first time, but for the veracity of the existing records to be confirmed. Second, we argue that the nature of the analysis should overcome the elimination of suspect records. This is because comparable results were found when families and genera were incorporated as characters. Thus, if the trematodes of Epinephelus had radiated with it since the genus arose, we would expect to find characteristic genera associated with it throughout the world and that these would be absent from other genera of epinephelines. This is not the case; 84% of genera have been reported from only 1 area. Broadly, we find that epinephelines in either the AEP or the IWP tend to share similar families, genera and species and that, on balance they differ from those found in the other area. The separation of the fauna into groups characterized by geographical area leads to the same conclusion as that from the consideration of patterns of host-specificity – the communities of trematodes of serranids have evolved, predominantly, through localised and independent cases of parasite adoption so that co-evolutionary processes appear to have contributed little to the present-day communities of trematodes in epinephelines.

CONCLUSIONS

The Epinephelinae are abundant, frequently commercially important, conspicuous and usually occur at only moderate depths. It is striking, therefore, that there should be reports of trematodes for only 62 of the 159 species. There is no suggestion anywhere in the literature (or from our studies) that there are species or groups of epinephelines that lack trematodes nor geographical areas where they are lacking. To the contrary, wherever they are studied in any detail, epinephelines appear to harbour rich assemblages of digeneans. The first conclusion of this

analysis is, therefore, disappointingly, that parasitological studies have not yet advanced sufficiently for this group to allow a definitive analysis to be made. Our data-base suggests that the Epinephelinae are better known than most groups of marine fishes with a cosmopolitan distribution. We see practical value in this demonstration of the limitations of the available literature. Prior to the study we had no real sense of just how thoroughly these fishes had been studied, except for the observation that records were abundant. We can probably now conclude that not a single species of Epinephelinae has been studied comprehensively. Further, most of the records remain unconfirmed and a significant proportion are unconvincing.

The reader might question whether it is possible to conclude anything about the evolution of the communities of trematodes in epinephelines when the fauna is manifestly so poorly known. We have argued that no (or almost no) clades of trematodes are restricted to the Epinephelinae or the Serranidae, that the present-day communities of trematodes are the result of extensive host-switching in and out of epinephelines, and that in the context of host-switching, geographical proximity is an identifiable marker of the origins of such switches. Could new data modify these interpretations?

The conclusion that no (or almost no) clades of trematodes are restricted to the Epinephelinae or the Serranidae is based on the observation that all the families and all the genera of trematodes that have more than one species also occur in other families of fishes. This interpretation could be modified in two ways. The revision of existing taxonomy to show that there are indeed many epinepheline-specific higher taxa of trematodes would have this effect – but such a change in taxonomy seems unlikely. Alternatively, it might be shown that where trematode species are shared by more than one epinepheline that in fact they routinely comprise complexes of closely related species and that these species are not shared with non-epinephelines. This possibility may well be borne out for some taxa, but it appears unlikely to be a general phenomenon. This conclusion cannot be falsified simply by the discovery of new host–parasite associations unless these were extremely numerous and all the taxa were shown to be restricted to epinephelines.

The conclusion that present-day communities of trematodes are the result of extensive host-switching in and out of epinephelines follows from the patterns of host-specificity discussed above. This interpretation will only be rejected by the unlikely sets of circumstances mentioned above or by the equally unconvincing explanation that present-day distributions have been arrived at through co-evolution and impossibly complex patterns of localized extinction of species and genera from selected lineages of fishes.

The conclusion that, in the context of host-switching, geographical proximity is an identifiable marker of the origins of such switches arises from the observation that Atlantic and Pacific epinephelines tend to share parasite taxa with other epinephelines from their own region. This interpretation could be rejected if new collecting showed that distributions of trematode genera and species is much wider than present records suggest. However, there seems no reason to predict such a change when taxa such as *Bivesicula* have been reported 23 times from the IWP and never from the AEP and *Postporus* has been reported 13 times, only from the AEP.

Thus, despite the incompleteness of the data, our principal evolutionary conclusion is unlikely to be changed by further study. We conclude that, in the main, the trematode parasites of epinephelines have been adopted by repeated independent host-switching events (presumably both into and out of the Epinephelinae). The handful of possible exceptions (genera with multiple species in the Epinephelinae) are exceptional. Is this disappointing? There is a widespread and understandable tendency for evolutionary parasitologists to be attracted by 'systems' in which host-specificity is high and host–parasite co-evolution may have been an important, if not the dominant force in the evolution of parasite as-

semblages; the extensive studies of lice (e.g. Hafner & Page, 1995) and tetraphyllidean cestodes (e.g. Caira & Jensen, 2001) are prime examples. Host–parasite systems such as the epinephelines and their trematodes that are apparently more chaotic are seemingly less inherently attractive. It is noteworthy that in total there are only 17 families of lice (Barker, 1994) and 8 families of tetraphyllidean cestodes (Euzet, 1994) and that only exceptional hosts have more than two or three of these. Strikingly, the 30 most studied families of fishes (Table 1) have 11–28 families of trematodes recorded from them. Thus, most trematode–teleost systems have enormous potential for the type of complexity seen here for the Epinephelinae.

There is now a strong body of work that analyses parasite species-richness in terms of ecological attributes of hosts (e.g. Morand & Poulin, 1998; Morand *et al.* 2000). Understanding of the communities of trematodes of fishes such as epinephelines will clearly require these approaches in association with more taxonomic, phylogenetic, distributional and life-cycle studies and an integration of knowledge of host and parasite phylogeny and host and parasite ecology. We will know that we have succeeded once we can predict the parasites of as yet unexamined host species.

APPENDIX

Parasite-host list of fully identified trematodes reported from fully identified epinepheline serranids. Numbers in brackets refer to numbered references given at foot of list. Aca. – Acanthocolpidae; Apo. – Apocreadiidae; Biv. – Bivesiculidae; Buc. – Bucephalidae; Bun. – Bunocotylidae; Cry. – Cryptogonimidae; Der. – Derogenidae; Did. – Didymozoidae; Fel. – Fellodistomidae; Gor. – Gorgoderidae; Hem. – Hemiuridae; Hir. – Hirudinellidae; Lec. – Lecithasteridae; Lep. – Lepocreadiidae; Mon. – Monorchiidae; Ope. – Opecoelidae; Pro. – Pronocephalidae; San. – Sanguinicolidae; Scl. – Sclerodistomidae; Zoo. – Zoogonidae.

HOST	PARASITE FAMILY PARASITE
Alphestes afer	Aca: *Stephanostomum microstephanum* [30]
A. multiguttatus	Buc: *Bucephalus heterotentaculatus* [9]
	Lec: *Lecithophyllum intermedium* [9]
	Lep: *Lepidapedon elongatum* [21]
Cephalopholis boenak	Did: *Gonapodasmius branchialis* [114, 115]; *G. pacificus* [44]
C. cruentata	Ope: *Cainocreadium lintoni* [32]; *Hamacreadium mutabile* [144]
C. fulva	Aca: *Stephanostomum casum* [89]
	Lep: *Neolepidapedoides equilatum* [117]
	Ope: *Apopodocotyle oscitans* [32]; *Cainocreadium lintoni* [117]; *Hamacreadium mutabile* [32, 92]; *Helicometrina nimia* [20, 89, 120]; *Opecoelus mexicanus* [92]
C. miniata	Gor: *Phyllodistomum mamaevi* [98, 99]
	Lep: *Lepidapedoides angustus* [18]
	Ope: *Hamacreadium mutabile* [97, 99]; *Opecoelus mexicanus* [97, 99]; *Pacificreadium serrani* [86]
C. panamensis	Ope: *Helicometra torta* [102]
C. sonnerati	Mon: *Lasiotocus bengalensis* [7]; *L. puriensis* [7]; *Retractomonorchis gibsoni* [6]; *R. madhavae* [5]
C. urodeta	Ope: *Opecoelus mexicanus* [139]
Cromileptes altivelis	Cry: *Mitotrema anthostomatum* [25, 59]
Epinephelus adscensionis	Aca: *Stephanostomum dentatum* [74]
	Apo: *Postporus epinepheli* [88]
	Hem: *Lecithochirium musculus* [88]; *L. parvum* [88]
	Lep: *Myzoxenus lachnolaimi* [32]; *Neolepidapedoides epinepheli* [117]
	Ope: *Cainocreadium longisaccum* [117]; *Hamacreadium confusum* [89]; *Helicometra torta* [29, 96]; *Opecoeloides vitellosus* [88]

Appendix continued

E. aeneus   Buc: *Prosorhynchus caudovatus* [36, 121]; *P. epinepheli* [125]
                   Did: *Allonematobothrium ghanensis* [39]
                   Hem: *Lecithochirium musculus* [42]; *Lecithocladium aegyptensis* [36]
                   Lep: *Lepidapedoides nicolli* [40, 43]

E. akaara   Aca: *Tormopsolus orientalis* [137]
                   Biv: *Bivesicula claviformis* [135, 137]
                   Buc: *Prosorhynchus epinepheli* [137]
                   Bun: *Aphanurus stossichii* [131]
                   Der: *Derogenes epinepheli* [131]
                   Did: *Gonapodasmius pristipomatis* [136]
                   Fel: *Proctoeces maculatus* [134]
                   Hem: *Tubulovesicula magnacetabulum* [137]
                   Lec: *Hysterolecithoides epinepheli* [134]
                   Lep: *Bianium plicitum* [130]
                   Ope: *Cainocreadium epinepheli* [134, 137]; *Coitocaecum gymnophallum* [134];
                        Crowcrocaecum epinepheli [130]; *Dactylostomum epinepheli* [130]; *Helicometra
                        epinepheli* [130, 132, 134, 141]; *Opecoelus lobatus* [134, 138]; *Pseudopecoelus
                        epinepheli* [130]

E. albomarginatus Lep: *Lepidapedoides nicolli* [13]

E. analogus   Buc: *Prosorhynchus gonoderus* [118]; *P. ozakii* [118]; *P. pacificus* [133]
                   Lep: *Lepidapedoides epinepheli* [11]; *Lepidapedoides nicolli* [133]
                   Ope: *Helicometrina nimia* [118]

E. andersoni   Buc: *Prosorhynchus caudovatus* [12]

E. areolatus   Aca: *Stephanostomum dentatum* [97, 99]
                   Buc: *Prosorhynchus chorinemi* [99]; *P. epinepheli* [46]; *P. ozakii* [97]
                   Fel: *Monascus filiformis* [99]
                   Hir: *Hirudinella ventricosa* [99]
                   Lep: *Lepidapedoides levenseni* [99]
                   Ope: *Cainocreadium epinepheli* [99, 108]
                   Scl: *Prosogonotrema bilabiatum* [99]

E. awara   Did: *Gonapodasmius pacificus* [114]
                   Ope: *Helicometra aposinuata* [132]; *Plagioporus oligolecithosus* [132]

E. bleekeri   Buc: *Prosorhynchus mcintoshi* [126, 127, 128]

E. bruneus   Hem: *Erilepturus hamati* [51]
                   Ope: *Allopodocotyle epinepheli* [64]

E. chlorostigma   Aca: *Stephanostomum nagatyi* [109]
                   Buc: *Prosorhynchus epinepheli* [46, 110]
                   Ope: *Allopodocotyle epinepheli* [139]; *Cainocreadium epinepheli* [108, 139];
                        Hamacreadium mutabile [105]

E. cyanopodus   Buc: *Prosorhynchus epinepheli* [44]
                   Lep: *Lepidapedoides angustus* [18]; *L. dollfusi* [18]; *Multitestis pyriformis* [15]
                   Ope: *Plagioporus epinepheli* [112]; *Pseudopecoelina xishaense* [44]

E. diacanthus   Buc: *Prosorhynchus epinepheli* [46]
                   Mon: *Proctotrematoides diacanthi* [10, 145]
                   Ope: *Helicometrina nimia* [10, 145]

E. fasciatus   Biv: *Bivesicula claviformis* [23, 24, 38, 44, 85, 106, 113]
                   Buc: *Folliculovarium xishaense* [44]; *Rhipidocotyle angusticolle* [111]
                   Did: *Gonapodasmius pacificus* [114]
                   Hem: *Erilepturus hamati* [134]; *Lecithochirium parvum* [31]; *Tubulovesicula
                        magnacetabulum* [31]
                   Lep: *Lepidapedoides angustus* [18, 106]
                   Ope: *Allopodocotyle epinepheli* [106]; *Helicometra borneoensis* [38]; *H. epinepheli*
                        [134]; *H. fasciata* [14, 59, 97, 99, 106]; *H. nasae* [87]

E. fuscoguttatus   Cry: *Mitotrema anthostomatum* [25]

E. goreensis   Buc: *Prosorhynchus caudovatus* [39]
                   Ope: *Podocotyle temensis* [41]

E. guttatus   Aca: *Stephanostomum dentatum* [33]
                   Apo: *Postporus epinepheli* [88]
                   Buc: *Paraprosorhynchus jupe* [53]
                   Did: *Atalostrophion promicrops* [62]
                   Hem: *Lecithochirium floridense* [92]
                   Ope: *Cainocreadium longisaccum* [19, 29]; *Hamacreadium mutabile* [92]

E. itajara   Aca: *Stephanostomum promicropsi* [49, 74, 119]
                   Buc: *Prosorhynchus promicropsi* [73, 74, 88]
                   Hem: *Lecithochirium microstomum* [74]
                   Lep: *Hypocreadium myohelicatum* [103]

E. labriformis   Hem: *Elytrophallus mexicanus* [72]
                   Lep: *Lepidapedoides oaxacensis* [56]
                   Ope: *Hamacreadium mutabile* [57]; *Helicometra torta* [72]

E. lanceolatus	Buc: *Neoprosorhynchus purius* [22, 26]
E. latifasciatus	Did: *Gonapodasmius branchialis* [81]
E. longispinis	Biv: *Bivesicula claviformis* [64]
	Ope: *Vesicocoelium solenophagum* [122]
E. maculosus	Lep: *Lepidapedoides levenseni* [60]
E. malabaricus	Buc: *Prosorhynchus pacificus* [46, 58]; *Telorhynchus arripidis* [111]
	Cry: *Pseudometadena celebesensis* [58]
	Hem: *Erilepturus hamati* [58]
	Ope: *Allopodocotyle serrani* [58]
E. marginatus	Did: *Didymozoon serrani* [79]
	Hem: *Lecithocladium aegyptensis* [36, 37]
E. merra	Biv: *Bivesicula claviformis* [76, 106]
	Buc: *Prosorhynchus epinepheli* [27]
	Hem: *Tubulovesicula angusticauda* [52]
	Lep: *Lepidapedoides angustus* [18, 106]
	Ope: *Cainocreadium epinepheli* [14, 59, 106]; *Hamacreadium mutabile* [84];
	Helicometra epinepheli [120]; *H. fasciata* [14, 59, 106]; *Opecoelus sphaericus* [31];
	Pacificreadium serrani [28]; *Pseudoplagioporus interruptus* [28]
	San: *Pearsonellum corventum* [59, 94]
E. morio	Aca: *Stephanostomum dentatum* [3, 74, 80]
	Apo: *Postporus epinepheli* [3, 34, 74, 80, 88]
	Did: *Allonematobothrium yucatanense* [80]
	Hem: *Lecithochirium floridense* [3, 80]; *L. musculus* [88]
	Lep: *Lepidapedoides levenseni* [3, 61, 74, 80]; *L. trachinoti* [88]
	Ope: *Helicometra torta* [3, 61, 70, 74, 80]; *Helicometrina nimia* [80]
	Pro: *Barisomum erubescens* [80]
E. mystacinus	Aca: *Stephanostomum microstephanum* [100]
	Did: *Lobatozoum bilobatum* [50]
E. niveatus	Aca: *Stephanostomum microstephanum* [71, 74, 120]
	Buc: *Prosorhynchus ozakii* [71, 73, 74]
	Lep: *Lepidapedoides nicolli* [71, 74]
E. ongus	Lep: *Lepidapedoides angustus* [18]
	San: *Pearsonellum corventum* [59, 94]
E. polyphekadion	Biv: *Bivesicula palauensis* [116]
	Ope: *Apertile overstreeti* [123, 124]
E. quernus	Buc: *Prosorhynchus epinepheli* [143]
	Did: *Allonematobothrium epinepheli* [142, 143]; *Gonapodasmius branchialis* [143];
	G. haemuli [104]
	Lep: *Lepidapedoides querni* [143]
E. quoyanus	Biv: *Bivesicula claviformis* [24, 106]
	Lep: *Lepidapedoides angustus* [18, 106]
	Ope: *Allopodocotyle epinepheli* [14, 59, 106]; *Cainocreadium epinepheli* [14, 59,
	106]; *Helicometra fasciata* [106]
	San: *Pearsonellum corventum* [59, 94]
E. radiatus	Did: *Gonapodasmius reticulum* [81]
E. septemfasciatus	Hem: *Erilepturus hamati* [138]
	Ope: *Cainocreadium epinepheli* [64]
E. striatus	Aca: *Stephanostomum casum* [61]; *S. dentatum* [88, 117]; *S. pagrosomi* [144]
	Apo: *Postporus epinepheli* [88, 117]
	Did: *Gonapodasmius tomex* [60]; *Maccallozoum epinepheli* [63]
	Hem: *Ectenurus americanus* [88]; *E. virgulus* [96]; *Lecithochirium floridense* [71];
	L. microstomum [96]; *L. parvum* [74]
	Lep: *Lepidapedoides levenseni* [60, 61, 129]; *L. trachinoti* [88]; *Neolepidapedoides*
	epinepheli [29, 117]
	Ope: *Cainocreadium lintoni* [29, 32, 117]; *Hamacreadium mutabile* [32, 118];
	Helicometra torta [34, 61, 68, 70, 74, 96, 117, 118]
	Scl: *Sclerodistomum diodontis* [129]
E. summana	Ope: *Cainocreadium epinepheli* [108]; *Hamacreadium mutabile* [105]; *Helicometrina*
	qatarensis [109]; *Pseudoplagioporus manteri* [107]
E. tauvina	Aca: *Stephanostomum nagatyi* [109]
	Biv: *Bivesicula claviformis* [116]
	Buc: *Prosorhynchus pacificus* [66]
	Did: *Allonematobothrium epinepheli* [44, 67, 82, 114]; *A. xishaense* [44, 114];
	Gonapodasmius epinepheli [1, 2]; *G. pacificus* [44, 114]; *Indoglomeritrema*
	epinepheli [65];
	Gor: *Phyllodistomum unicum* [93]
	Ope: *Cainocreadium epinepheli* [108]; *Helicometrina qatarensis* [109]
E. undulosus	Buc: *Prosorhynchus epinepheli* [46]
	Mon: *Retractomonorchis nahhasi* [4]
Mycteroperca bonaci	Aca: *Stephanostomum dentatum* [88]

Appendix continued

	Apo: *Postporus epinepheli* [96]
	Buc: *Prosorhynchus ozakii* [88]; *P. pacificus* [34, 73, 88, 90, 96]; *P. promicropsi* [144]
	Hem: *Lecithochirium microstomum* [96]; *L. parvum* [96]
	Lep: *Lepidapedoides nicolli* [91]; *Neolepidapedoides mycteropercae* [88]
	Ope: *Helicometrina execta* [70]
	Zoo: *Deretrema fusillus* [88]
M. interstitialis	Aca: *Stephanostomum casum* [129]; *S. dentatum* [29]
	Buc: *Prosorhynchus pacificus* [8, 88, 129]
	Ope: *Peracreadium mycteropercae* [118]
M. microlepis	Buc: *Prosorhynchus pacificus* [8, 73, 74, 96]
	Hem: *Dissosaccus laevis* [69]; *Lecithochirium floridense* [71]; *L. parvum* [96]
	Lep: *Neolepidapedoides macrum* [96]
M. olfax	Aca: *Stephanostomum multispinosum* [72]
	Buc: *Prosorhynchus ozakii* [72]; *P. pacificus* [72]
	Lep: *Lepidapedoides nicolli* [72]
M. rosacea	Gor: *Phyllodistomum marinae* [11]
	Hem: *Lecithochirium microstomum* [55]
	Ope: *Hamacreadium mutabile* [11]; *Helicometrina nimia* [9]
M. tigris	Buc: *Prosorhynchus pacificus* [32]
	Lep: *Lepidapedoides parepinepheli* [118]
M. venenosa	Aca: *Stephanostomum dentatum* [74, 118]
	Apo: *Postporus epinepheli* [74]
	Buc: *Prosorhynchus pacificus* [34, 35, 73, 74, 88, 118]
	Cry: *Paracryptogonimus americanus* [34]
	Hem: *Lecithochirium floridense* [101]; *L. monticelli* [60]
	Lep: *Neolepidapedoides mycteropercae* [88]
	Ope: *Helicometrina execta* [74]; *H. mirzai* [32]
M. xenarcha	Buc: *Bucephalus heterotentaculatus* [9]; *Prosorhynchus ozakii* [72]; *P. pacificus* [72]
	Lep: *Lepidapedoides nicolli* [72]
	Ope: *Hamacreadium mutabile* [72]
Paranthias furcifer	Hem: *Elytrophallus mexicanus* [72]; *Lecithochirium microstomum* [72]
	Lep: *Prodistomum orientalis* [72]
	Ope: *Opecoelus mexicanus* [72]
Plectropomus leopardus	Buc: *Pseudoprosorhynchus hainanensis*[115]; *Rhipidocotyle clavivesiculatum* [54, 115]
	Hem: *Plerurus digitatus* [16, 59]
	Lec: *Thulinia microrchis* [17]
	Ope: *Pacificreadium serrani* [14, 47, 59]; *Pseudopecoelus elongatus* [115]
	San: *Pearsonellum corventum* [59, 94, 95]
P. maculatus	Buc: *Prosorhynchus freitasi* [27]; *P. thapari* [75, 99]
	Cry: *Mitotrema anthostomatum* [78]
	Ope: *Allopodocotyle plectropomi* [77]; *Cainocreadium serrani* [77]; *Pacificreadium serrani* [28]
Triso dermopterus	Did: *Gonapodasmius hainanensis* [45, 114, 115]
Variola albimarginata	Buc: *Prosorhynchus platycephali* [31]
V. louti	Biv: *Bivesicula palauensis* [116]
	Buc: *Prosorhynchus crucibulum* [83]; *P. platycephali* [31, 48]; *P. serrani* [27]
	Hem: *Plerurus digitatus* [16, 106]
	Ope: *Pacificreadium serrani* [105]

SOURCE OF RECORD. 1. Abdul-Salam & Sreelatha (1992); 2. Abdul-Salam, Sreelatha & Farah (1990); 3. Aguirre-Macedo & Bray (in press); 4. Ahmad (1984*a*); 5. Ahmad (1984*b*); 6. Ahmad (1991); 7. Ahmad & Gupta (1985); 8. Amato (1982); 9. Arai (1963); 10. Bilqees (1981); 11. Bravo-Hollis & Manter (1957); 12. Bray (1984); 13. Bray (1985); 14. Bray & Cribb (1989); 15. Bray & Cribb (1998); 16. Bray, Cribb & Barker (1993*a*); 17. Bray, Cribb & Barker (1993*b*); 18. Bray, Cribb & Barker (1996); 19. Bunkley-Williams, Dyer & Williams (1996); 20. Caballero (1990); 21. Caballero, Bravo-Hollis & Grocott (1955); 22. Chauhan (1953); 23. Cribb *et al.* (1998); 24. Cribb, Bray & Barker (1994); 25. Cribb *et al.* (1996); 26. Dayal (1948); 27. Durio & Manter (1968*a*); 28. Durio & Manter (1968*b*); 29. Dyer, Williams & Bunkley-Williams (1985); 30. Dyer, Williams & Bunkley-Williams (1986); 31. Dyer, Williams & Bunkley-Williams (1988); 32. Dyer, Williams & Bunkley-Williams (1992); 33. Dyer, Williams & Bunkley-Williams (1998); 34. Fischthal (1977); 35. Fischthal (1978); 36. Fischthal (1980); 37. Fischthal (1982); 38. Fischthal & Kuntz (1965); 39. Fischthal & Thomas (1968); 40. Fischthal & Thomas (1970*a*); 41. Fischthal & Thomas (1970*b*); 42. Fischthal & Thomas (1972); 43. Gaevskaya & Aljoshkina (1983); 44. Gu & Shen (1983*a*); 45. Gu & Shen (1983*b*); 46. Hafeezullah & Siddiqi (1970); 47. Hall, Cribb & Barker (1999); 48. Hasegawa, Williams & Bunkley-Williams (1991); 49. Hutton & Sogandares-Bernal (1960); 50. Hyman (1963); 51. Ichihara (1974); 52. King (1964); 53. Kohn (1967); 54. Ku & Shen (1975); 55. Lamothe-Argumedo (1966); 56. Lamothe-Argumedo (1969*a*); 57. Lamothe-Argumedo (1969*b*); 58. Leong & Wong (1988); 59. Lester & Sewell (1990); 60. Linton (1907); 61. Linton (1910); 62. MacCallum (1915); 63. MacCallum (1917); 64. Machida, Ichihara & Kamegai (1970); 65. Madhavi & Hanumantha Rao (1983); 66. Madhavi (1974); 67. Madhavi (1982); 68. Manter (1930); 9. Manter (1931); 70. Manter (1933); 71. Manter (1934); 72. Manter (1940*a*); 73. Manter (1940b); 74. Manter (1947);

75. Manter (1953); 76. Manter (1961); 77. Manter (1963*a*); 78. Manter (1963*b*); 79. Monticelli (1889); 80. Moravec *et al.* (1997); 81. Murugesh & Madhavi (1994); 82. Murugesh, Krishna Sai Ram & Madhavi (1992); 83. Nagaty (1937); 84. Nagaty (1941); 85. Nagaty (1948); 86. Nagaty & Abdel Aal (1962*a*); 87. Nagaty & Abdel Aal (1962*b*); 88. Nahhas & Cable (1964); 89. Nahhas & Carlson (1994); 90. Nahhas & Short (1965); 91. Nasir & Gomez (1977); 92. Nikolaeva & Parukhin (1968); 93. Odhner (1910); 94. Overstreet & Køie (1989); 95. Overstreet & Thulin (1989); 96. Overstreet (1969); 97. Parukhin (1970); 98. Parukhin (1971); 99. Parukhin (1976); 100. Pérez-Vigueras (1955); 101. Pérez Vigueras (1958); 102. Pérez-Ponce de León *et al.* (1999); 103. Pérez-Ponce de León, León-Règagnon & Monks (1998); 104. Pozdnyakov (1994); 105. Ramadan (1983); 106. Rigby *et al.* (1997); 107. Saoud & Ramadan (1984); 108. Saoud, Ramadan & Al Kawari (1986); 109. Saoud, Ramadan & Al Kawari (1988*a*): 110. Saoud, Ramadan & Al Kawari (1988*b*); 111. Shalaby & Hassanine (1996); 112. Shen (1985*a*); 112. Shen (1985*b*); 114. Shen (1990*a*); 115. Shen (1990*b*); 116. Shimazu & Machida (1995); 117. Siddiqi & Cable (1960); 118. Sogandares-Bernal (1959); 119. Sogandares-Bernal & Hutton (1959); 120. Sparks (1957); 121. Szuks (1981); 122. Tang & Xu (1979); 123. Toman (1992*a*); 124. Toman (1992*b*); 125. Vassiliadès (1982); 126. Velasquez (1959); 127. Velasquez (1966); 128. Velasquez (1975); 129. Vélez (1978); 130. Wang (1982*a*); 131. Wang (1982*b*); 132. Wang, Wang & Zhang (1992); 133. Winter (1960); 134. Yamaguti (1934*a*); 135. Yamaguti (1938*a*); 136. Yamaguti (1938*b*); 137. Yamaguti (1939); 138. Yamaguti (1940); 139. Yamaguti (1942); 140. Yamaguti (1953); 141. Yamaguti (1958); 142. Yamaguti (1965); 143. Yamaguti (1970); 144. Yamaguti (1971); 145. Zaidi & Khan (1977).

REFERENCES

ABDUL-SALAM, J. & SREELATHA, B. S. (1992). Observations on the tissue response of the grouper, *Epinephelus tauvina* to *Gonapodasmius epinepheli* (Trematoda: Didymozoidae). *Rivista di Parassitologia* **53**, 203–213.

ABDUL-SALAM, J., SREELATHA, B. & FARAH, M. (1990). *Gonapodasmius epinepheli* n. sp. (Didymozoidae) from the grouper *Epinephelus tauvina* from the Arabian Gulf. *Systematic Parasitology* **17**, 67–74.

AGUIRRE-MACEDO, L. M. & BRAY, R. A. (in press). Some trematodes from *Epinephelus morio* (Pisces: Serranidae) from the coast of the Yucatan Peninsula, Mexico. *Studies on the Natural History of the Caribbean Region*.

AHMAD, J. (1984*a*). Digenetic trematodes from marine fishes from the Arabian Sea, off the Bombay coast, India. On four new species of the genus *Retractomonorchis* Madhavi, 1977 (Digenea: Monorchiidae). *Rivista di Parassitologia* **45**, 19–28.

AHMAD, J. (1984*b*). Studies on five new digenetic trematodes from marine fishes from the Arabian sea, off the Bombay coast, India. *Pakistan Journal of Zoology* **16**, 45–59.

AHMAD, J. (1991). A new genus and three new species of digenetic trematodes from marine fishes of Arabian Sea. *Pakistan Journal of Zoology* **23**, 99–104.

AHMAD, J. & GUPTA, V. (1985). Studies on new monorchid and gorgoderid trematodes (Trematoda: Digenea) from marine fishes from the Bay of Bengal, off the Puri coast, Orissa. *Rivista di Parassitologia* **46**, 45–59.

AMATO, J. F. R. (1982). Digenetic trematodes of percoid fishes of Florianópolis, southern Brasil – Bucephalidae. *Rivista Brasileira de Biologia* **42**, 667–680.

ARAI, H. P. (1963). Trematodos Digeneos de peces marinos de Baja California, Mexico. *Anales del Instituto de Biología. Universidad Nacional Autonóma de México* **33**, 113–130.

BARKER, S. C. (1994). Phylogeny and classification, origins, and evolution of host associations of lice. *International Journal for Parasitology* **24**, 1285–1291.

BARKER, S. C., CRIBB, T. H., BRAY, R. A. & ADLARD, R. D. (1994). Host–parasite associations on a coral reef: pomacentrid fishes and digenean trematodes. *International Journal for Parasitology* **24**, 643–647.

BEVERIDGE, I. & CHILTON, N. B. (2001). Co-evolutionary relationships between the nematode subfamily Cloacininae and its macropodid marsupial hosts. *International Journal for Parasitology* **31**, 976–996.

BILQEES, F. M. (1981). *Digenetic Trematodes Fishes of Karachi Coast*. Karachi, Kifayat Academy.

BRAVO-HOLLIS, M. & MANTER, H. W. (1957). Trematodes of marine fishes of Mexican waters. X. Thirteen Digenea including nine new species and two new genera from Pacific coast. *Proceedings of the Helminthological Society of Washington* **24**, 35–48.

BRAY, R. A. (1984). Some helminth parasites of marine fishes and cephalopods of South Africa: Aspidogastrea and the digenean families Bucephalidae, Haplosplanchnidae, Mesometridae and Fellodistomidae. *Journal of Natural History* **18**, 271–292.

BRAY, R. A. (1985). Some helminth parasites of marine fishes of South Africa: families Gorgoderidae, Zoogonidae, Cephaloporidae, Acanthocolpidae and Lepocreadiidae (Digenea). *Journal of Natural History* **19**, 377–405.

BRAY, R. A. (1987). Some helminth parasites of marine fishes of South Africa: family Opecoelidae (Digenea). *Journal of Natural History* **21**, 1049–1075.

BRAY, R. A. & CRIBB, T. H. (1989). Digeneans of the family Opecoelidae Ozaki, 1925 from the southern Great Barrier Reef, including a new genus and three new species. *Journal of Natural History* **23**, 429–473.

BRAY, R. A. & CRIBB, T. H. (1998). Lepocreadiidae (Digenea) of Australian coastal fishes: new species of *Opechona* Looss, 1907, *Lepotrema* Ozaki, 1932 and *Bianium* Stunkard, 1930 and comments on other species reported for the first time or poorly known in Australian waters. *Systematic Parasitology* **41**, 123–148.

BRAY, R. A., CRIBB, T. H. & BARKER, S. C. (1993*a*). Hemiuridae (Digenea) from marine fishes of the Great Barrier Reef, Queensland, Australia. *Systematic Parasitology* **25**, 37–62.

BRAY, R. A., CRIBB, T. H. & BARKER, S. C. (1993*b*). The Hemiuroidea (Digenea) of pomacentrid fishes (Perciformes) from Heron Island, Queensland, Australia. *Systematic Parasitology* **24**, 159–184.

BRAY, R. A., CRIBB, T. H. & BARKER, S. C. (1996). Four species of *Lepidapedoides* Yamaguti, 1970 (Digenea: Lepocreadiidae) from fishes of the southern Great Barrier Reef, with a tabulation of host–parasite data on the group. *Systematic Parasitology* **34**, 179–195.

BRAY, R. A., CRIBB, T. H. & LITTLEWOOD, D. T. J. (1998). A phylogenetic study of *Lepidapedoides* Yamaguti, 1970 (Digenea) with a key and descriptions of two new species from Western Australia. *Systematic Parasitology* **39**, 183–197.

BROOKS, D. R. (1981). Hennig's parasitological method: a proposed solution. *Systematic Zoology* **30**, 229–249.

BUNKLEY-WILLIAMS, L., DYER, W. G. & WILLIAMS, E. H. (1996). Some aspidogastrid and digenean trematodes of Puerto Rican marine fishes. *Journal of Aquatic Animal Health* **8**, 87–92.

CABALLERO, R. G. (1990). Tremátodos de peces marinos del Golfo de México y del mar Caribe. II. Familias Haplosplanchnidae y Opecoelidae. *Anales del Instituto de Ciencias del Mar y Limnologia Universidad Nacional Autónoma de México* **17**, 191–203.

CABALLERO, Y. C., E., BRAVO-HOLLIS, M. & GROCOTT, R. G. (1955). Helmintos de la Republica de Panama. XIV. Trematodos Monogeneos y Digeneos de peces marinos del Oceano Pacifico del norte, con descripcion de nuevas formas. *Anales del Instituto de Biología. Universidad Nacional Autonóma de México* **26**, 117–147.

CAIRA, J. N. & JENSEN, K. (2001). An investigation of the co-evolutionary relationships between onchobothriid tapeworms and their elasmobranch hosts. *International Journal for Parasitology* **31**, 960–975.

CHAUHAN, B. S. (1953). Studies on the trematode fauna of India. Part III. Subclass Digenea (Gasterostomata). *Records of the Indian Museum* **51**, 231–287.

CRAIG, M. T., PONDELLA, D. J., FRANCK, J. P. C. & HAFNER, J. C. (2001). On the status of the serranid fish genus *Epinephelus*: evidence for paraphyly based upon 16S rDNA sequence. *Molecular Phylogenetics and Evolution* **19**, 121–130.

CRIBB, T. H., ANDERSON, G. R., ADLARD, R. D. & BRAY, R. A. (1998). A DNA-based demonstration of a three-host life-cycle for the Bivesiculidae (Platyhelminthes: Digenea). *International Journal for Parasitology* **28**, 1791–1795.

CRIBB, T. H., BRAY, R. A. & BARKER, S. C. (1994). Bivesiculidae and Haplosplanchnidae (Digenea) from fishes of the southern Great Barrier Reef, Australia. *Systematic Parasitology* **28**, 81–97.

CRIBB, T. H., BRAY, R. A., BARKER, S. C. & ADLARD, R. D. (1996). Taxonomy and biology of *Mitotrema anthostomatum* Manter, 1963 (Digenea: Cryptogonimidae) from fishes of the southern Great Barrier Reef, Australia. *Journal of the Helminthological Society of Washington* **63**, 110–115.

DAYAL, J. (1948). Trematode parasites of Indian fishes, Part I. New trematodes of the family Bucephalidae Poche, 1907. *Indian Journal of Helminthology* **1**, 47–62.

DURIO, W. O. & MANTER, H. W. (1968 a). Some digenetic trematodes of marine fishes of New Caledonia. Part I. Bucephalidae, Monorchiidae, and some smaller families. *Proceedings of the Helminthological Society of Washington* **35**, 143–153.

DURIO, W. O. & MANTER, H. W. (1968 b). Some digenetic trematodes of marine fishes of New Caledonia. Part II. Opecoelidae and Lepocreadiidae. *Journal of Parasitology* **54**, 747–756.

DYER, W. G., WILLIAMS, E. H. & BUNKLEY-WILLIAMS, L. (1985). Digenetic trematodes of marine fishes of the western and southwestern coasts of Puerto Rico. *Proceedings of the Helminthological Society of Washington* **52**, 85–94.

DYER, W. G., WILLIAMS, E. H. & BUNKLEY-WILLIAMS, L. (1986). Some trematodes of marine fishes of southwestern and northwestern Puerto Rico. *Transactions of the Illinois Academy of Science* **79**, 141–143.

DYER, W. G., WILLIAMS, E. H. & BUNKLEY-WILLIAMS, L. (1988). Digenetic trematodes of marine fishes of Okinawa, Japan. *Journal of Parasitology* **74**, 638–645.

DYER, W. G., WILLIAMS, E. H. & BUNKLEY-WILLIAMS, L. (1992). *Homalometron dowgialloi* sp. n. (Homalometridae) from *Haemulon flavolineatum* and additional records of digenetic trematodes of marine fishes in the West Indies. *Journal of the Helminthological Society of Washington* **59**, 182–189.

DYER, W. G., WILLIAMS, E. H. & BUNKLEY-WILLIAMS, L. (1998). Some digenetic trematodes of marine fishes from Puerto Rico. *Caribbean Journal of Science* **34**, 141–146.

EUZET, L. (1994). Order Tetraphyllidea Carus, 1863. In *Keys to the Cestode Parasites of Vertebrates* (ed. Khalil, G. M., Jones, A. & Bray, R. A.), pp. 149–194. Cambridge, CAB International.

FELIU, C., RENAUD, F., CATZEFLIS, F., HUGOT, J.-P., DURAND, P. & MORAND, S. (1997). A comparative analysis of parasite species richness of Iberian rodents. *Parasitology* **115**, 453–466.

FISCHTHAL, J. H. (1977). Some digenetic trematodes of marine fishes from the Barrier Reef and Reef Lagoon of Belize. *Zoologica Scripta* **6**, 81–88.

FISCHTHAL, J. H. (1978). Allometric growth in three species of digenetic trematodes of marine fishes from Belize. *Journal of Helminthology* **52**, 29–39.

FISCHTHAL, J. H. (1980). Some digenetic trematodes of marine fishes from Israel's Mediterranean coast and their zoogeography, especially those from Red Sea immigrant fishes. *Zoologica Scripta* **9**, 11–23.

FISCHTHAL, J. H. (1982). Additional records of digenetic trematodes of marine fishes from Israel's Mediterranean coast. *Proceedings of the Helminthological Society of Washington* **49**, 34–44.

FISCHTHAL, J. H. & KUNTZ, R. E. (1965). Digenetic trematodes of fishes from North Borneo (Malaysia). *Proceedings of the Helminthological Society of Washington* **32**, 63–71.

FISCHTHAL, J. H. & THOMAS, J. D. (1968). Digenetic trematodes of marine fishes from Ghana: families Acanthocolpidae, Bucephalidae, Didymozoidae. *Proceedings of the Helminthological Society of Washington* **35**, 237–247.

FISCHTHAL, J. H. & THOMAS, J. D. (1970 a). Digenetic trematodes of marine fishes from Ghana: family Lepocreadiidae. *Journal of Helminthology* **44**, 365–386.

FISCHTHAL, J. H. & THOMAS, J. D. (1970 b). Digenetic trematodes of marine fishes from Ghana: family Opecoelidae. *Proceedings of the Helminthological Society of Washington* **37**, 129–141.

FISCHTHAL, J. H. & THOMAS, J. D. (1972). Additional hemiurid and other trematodes of fishes from Ghana. *Bulletin de l'Institut Fondamental d'Afrique Noire, Series A, Science Naturelles* **34**, 9–25.

GAEVSKAYA, A. V. & ALJOSHKINA, L. D. (1983). [New finds of fish trematodes on the Atlantic coast of Africa]. *Parazitologiya* **17**, 12–17.

GILL, A. C. (1999). Subspecies, geographic forms and widespread Indo-Pacific coral-reef fish species: a call for change in taxonomic practice. In *Proceedings of the 5th Indo-Pacific Fish Conference, Nouméa 1997* (ed. Séret, B. & Sire, J.-Y.), pp. 79–87. Société Française d'Ichthyologie & Institut de Recherche pour le Développement, Paris.

GU, C. D. & SHEN, J. (1983*a*). Digenetic trematodes of fishes from the Xisha Islands, Guangdong Province, China I. *Studia Marina Sinica* **20**, 157–184.

GU, C. D. & SHEN, J. (1983*b*). Four new species of didymozoid trematodes from China. *Acta Zoologica Sinica* **8**, 17–23.

HAFEEZULLAH, M. (1971). A review on the validity of *Helicometrina* Linton, 1910 and *Stenopera* Manter, 1933 (Trematoda). *Acta Parasitologica Polonica* **19**, 133–139.

HAFEEZULLAH, M. & SIDDIQI, A. H. (1970). Digenetic trematodes of marine fishes of India. Part I. Bucephalidae and Cryptogonimidae. *Indian Journal of Helminthology* **22**, 1–22.

HAFNER, M. S. & PAGE, R. D. M. (1995). Molecular phylogenies and host–parasite cospeciation: gophers and lice as a model system. *Philosophical Transactions of the Royal Society (Series B)* **349**, 77–83.

HALL, K. A., CRIBB, T. H. & BARKER, S. C. (1999). V4 region of small subunit rDNA indicates polyphyly of the Fellodistomidae (Digenea) which is supported by morphology and life-cycle data. *Systematic Parasitology* **43**, 81–92.

HASEGAWA, H., WILLIAMS, E. H. & BUNKLEY-WILLIAMS, L. (1991). Nematode parasites from marine fishes of Okinawa, Japan. *Journal of the Helminthological Society of Washington* **58**, 186–197.

HEEMSTRA, P. C. & RANDALL, J. E. (1993). *FAO Species Catalogue. Groupers of the World (Family Serranidae, Subfamily Epinephelinae). An Annotated and Illustrated Catalogue of the Grouper, Rockcod, Hind, Coral Grouper and Lyretail Species*. Rome, FAO.

HUTTON, R. F. & SOGANDARES-BERNAL, F. (1960). A list of parasites from marine and coastal animals of Florida. *Transactions of the American Microscopical Society* **79**, 287–292.

HYMAN, L. H. (1963). Notes on a didymozoid trematode from the Bahama Islands. *Bulletin of Marine Science of the Gulf and Caribbean* **13**, 193–196.

ICHIHARA, A. (1974). Hemiurid trematodes from marine fishes near the Tsushima Islands in the Sea of Japan. *Proceedings. Third International Congress of Parasitology*, 1614–1615. Munich, Facta Publication.

KING, R. E. (1964). Three hemiurid trematodes from South Viet Nam. *Transactions of the American Microscopical Society* **83**, 435–439.

KOHN, A. (1967). Sôbre um nôvo gênero de Prosorhynchinae Nicoll, 1914 e novos dados sôbre *Prosorhynchus bulbosus* Kohn, 1961 e *Rhipidocotyle quadriculatum* Kohn, 1961 (Trematoda, Bucephaliformes). *Memórias do Instituto Oswaldo Cruz, Rio de Janeiro* **65**, 107–114.

KU, C. & SHEN, J. (1975). Studies on the genus *Rhipidocotyle* Diesing (Bucephalidae, Trematoda)

from some marine fishes of China. *Acta Zoologica Sinica* **21**, 205–211.

LAMOTHE-ARGUMEDO, R. (1966). Trematodos de peces (II). Presencia de los trematodos *Bianium plicitum* (Linton, 1928) Stunkard, 1931, y *Lecithochirium microstomum* Chandler, 1935, en peces del Pacífico Mexicano. *Anales del Instituto de Biología. Universidad Nacional Autonóma de México* **36**, 147–157.

LAMOTHE-ARGUMEDO, R. (1969*a*). Tremátodos de peces III. Cuatro especies nuevas de tremátodos parásitos de peces del Pacífico Mexicano. *Anales del Instituto de Biología. Universidad Nacional Autonóma de México. Serie Zoología* **40**, 21–42.

LAMOTHE-ARGUMEDO, R. (1969*b*). Tremátodos de peces IV. Registro de cuatro especies de tremátodos de peces marinos de la costa del Pacífico Mexicano. *Anales del Instituto de Biología. Universidad Nacional Autonóma de México. Serie Zoología* **40**, 179–194.

LEONG, T. S. & WONG, S. Y. (1988). A comparative study of the parasite fauna of wild and cultured grouper (*Epinephelus malabaricus* Bloch et Schneider) in Malaysia. *Aquaculture* **68**, 203–207.

LESTER, R. J. G. & SEWELL, K. B. (1990). Checklist of parasites from Heron Island, Great Barrier Reef. *Australian Journal of Zoology* **37**, 101–128.

LINTON, E. (1907). Notes on parasites of Bermuda fishes. *Proceedings of the United States National Museum* **33**, 85–126.

LINTON, E. (1910). Helminth fauna of the Dry Tortugas II. Trematodes. *Papers from the Tortugas Laboratory of the Carnegie Institute of Washington* **4**, 11–98.

LO, C. M., MORGAN, J. A. T., GALZIN, R. & CRIBB, T. H. (2001). Identical digeneans in coral reef fishes from French Polynesia and the Great Barrier Reef (Australia) demonstrated by morphology and molecules. *International Journal for Parasitology* **31**, 1573–1578.

MacCALLUM, G. A. (1915). Some new species of ectoparasitic trematodes. *Zoologica* **1**, 395–410.

MacCALLUM, G. A. (1917). Some new forms of parasitic worms. *Zoopathologica* **1**, 43–75.

MACHIDA, M., ICHIHARA, A. & KAMEGAI, S. (1970). Digenetic trematodes collected from the fishes in the sea north of the Tsushima Islands. *Memoirs of the National Science Museum, Tokyo* **3**, 101–112.

MADHAVI, R. (1974). Digenetic trematodes from marine fishes of Waltair Coast, Bay of Bengal. Family Bucephalidae. *Rivista di Parassitologia* **35**, 189–199.

MADHAVI, R. (1982). Didymozoid trematodes (including new genera and species) from marine fishes of the Waltair Coast, Bay of Bengal. *Systematic Parasitology* **4**, 99–124.

MADHAVI, R. & HANUMANTHA RAO, K. (1983). A new didymozoid trematode *Indoglomeritrema epinepheli* gen. n., sp. n. from the marine fish *Epinephelus tauvina* from Bay of Bengal. *Acta Parasitologica Polonica* **28**, 261–265.

MANTER, H. W. (1930). Studies on the trematodes of Tortugas fishes. *Carnegie Institution of Washington* **29**, 338–340.

MANTER, H. W. (1931). Further studies on trematodes of Tortugas fishes. *Carnegie Institution Year Book* **30**, 386–387.

MANTER, H. W. (1933). The genus *Helicometra* and related trematodes from Tortugas, Florida. *Papers from the Tortugas Laboratory of the Carnegie Institute of Washington* **435**, 167–182.

MANTER, H. W. (1934). Some digenetic trematodes from deep-water fish of Tortugas, Florida. *Carnegie Institution of Washington* **435**, 257–345.

MANTER, H. W. (1940a). Digenetic trematodes of fishes from the Galapagos Islands and the neighboring Pacific. *Allan Hancock Pacific Expeditions* **2**, 325–497.

MANTER, H. W. (1940b). Gasterostomes (Trematoda) of Tortugas, Florida. *Papers from the Tortugas Laboratory of the Carnegie Institute of Washington* **33**, 1–19.

MANTER, H. W. (1947). The digenetic trematodes of marine fishes of Tortugas, Florida. *American Midland Naturalist* **38**, 257–416.

MANTER, H. W. (1953). Two new species of Prosorhynchinae (Trematoda: Gasterostomata) from the Fiji Islands. *Thapar Commemoration Volume*, **1953**, 193–200.

MANTER, H. W. (1961). Studies on digenetic trematodes of fishes of Fiji. I. Families Haplosplanchnidae, Bivesiculidae, and Hemiuridae. *Proceedings of the Helminthological Society of Washington* **28**, 67–74.

MANTER, H. W. (1963a). Studies on digenetic trematodes of fishes of Fiji. II. Families Lepocreadiidae, Opistholebetidae, and Opecoelidae. *Journal of Parasitology* **49**, 99–113.

MANTER, H. W. (1963b). Studies on digenetic trematodes of fishes of Fiji. III. Families Acanthocolpidae, Fellodistomatidae, and Cryptogonimidae. *Journal of Parasitology* **49**, 443–450.

MONTICELLI, F. S. (1889). Notes on some Entozoa in the collection of the British Museum. *Proceedings of the Zoological Society of London*, 321–325.

MORAND, S., CRIBB, T. H., KULBICKI, M., RIGBY, M. C., CHAUVET, C., DUFOUR, V., FALIEX, E., GALZIN, R., LO, C. M., LO-YAT, A., PICHELIN, S. & SASAL, P. (2000). Endoparasite species richness of New Caledonian butterfly fishes: host density and diet matter. *Parasitology* **121**, 65–73.

MORAND, S. & POULIN, R. (1998). Density, body mass and parasite species richness of terrestrial mammals. *Evolutionary Ecology* **12**, 717–727.

MORAVEC, F., VIDAL-MARTÍNEZ, V. M., VARGAS-VÁZQUEZ, J., VIVAS-RODRÍGUEZ, C., GONZÁLEZ-SOLÍS, D., MENDOZA-FRANCO, E., SIMÁ-ALVAREZ, R. & GÜEMEZ-RICALDE, J. (1997). Helminth parasites of *Epinephelus morio* (Pisces: Serranidae) of the Yucatan Peninsula, southeastern Mexico. *Folia Parasitologica* **44**, 255–266.

MURUGESH, M., KRISHNA SAI RAM, B. & MADHAVI, R. (1992). Nematobothriine didymozoid trematodes from marine teleost fish of the coast of Visakhapatnam, Bay of Bengal. *Rivista di Parassitologia* **53**, 79–86.

MURUGESH, M. & MADHAVI, R. (1994). Gonapodasmid didymozoids (Digenea: Didymozoidae) from serranid fishes of the Visakhapatnam coast, Bay of Bengal. *Rivista di Parassitologia* **55**, 47–55.

NAGATY, H. F. (1937). *Trematodes of Fishes from the Red Sea Part I. Studies on the Family Bucephalidae Poche, 1907*. Cairo, Egyptian University.

NAGATY, H. F. (1941). Trematodes of fishes from the Red Sea. Part 2. The genus *Hamacreadium* Linton, 1910

(Fam. Allocreadiidae) with a description of two new species. *Journal of the Egyptian Medical Association* **24**, 300–310.

NAGATY, H. F. (1948). Trematodes of fishes from the Red Sea Part 4. On some new and known forms with a single testis. *Journal of Parasitology* **34**, 355–363.

NAGATY, H. F. & ABDEL AAL, T. M. (1962a). Trematodes of fishes from the Red Sea. Part 15. Four new species of *Hamacreadium* family Allocreadiidae. *Journal of Parasitology* **48**, 384–386.

NAGATY, H. F. & ABDEL AAL, T. M. (1962b). Trematodes of fishes from the Red Sea Part 17. On three allocreadiid sp. and one schistorchiid sp. *Journal of the Arab Veterinary and Medical Association* **22**, 307–314.

NAHHAS, F. M. & CABLE, R. M. (1964). Digenetic and aspidogastrid trematodes from marine fishes of Curacao and Jamaica. *Tulane Studies in Zoology* **11**, 169–228.

NAHHAS, F. M. & CARLSON, K. (1994). Digenetic trematodes of marine fishes of Jamaica, West Indies. *Publications of the Hofstra University Marine Laboratory, Ecological Survey of Jamaica* **2**, 1–60.

NAHHAS, F. M. & SHORT, R. B. (1965). Digenetic trematodes of marine fishes from Apalachee Bay, Gulf of Mexico. *Tulane Studies in Zoology* **12**, 39–50.

NASIR, P. & GOMEZ, Y. (1977). Digenetic trematodes from Venezuelan marine fishes. *Rivista di Parassitologia* **38**, 53–73.

NELSON, J. S. (1994). *Fishes of the World*. New York, John Wiley & Sons.

NIKOLAEVA, V. M. & PARUKHIN, A. M. (1968). To the study of fish helminths in the Gulf of Mexico. In *Explorations of Central American Seas* (ed. Jankovskaya, E. B.), pp. 126–149. Kiev, Naukova Dumka.

ODHNER, T. (1910). Nordostafrikanische trematoden, grösstenteils vom Weissen Nil. *Results of the Swedish Zoological Expedition to Egypt and White Nile, 1901. Jägerskiöld Expedition* **23**, 1–170.

OVERSTREET, R. M. (1969). Digenetic trematodes of marine teleost fishes from Biscayne Bay, Florida. *Tulane Studies in Zoology and Botany* **15**, 119–176.

OVERSTREET, R. M. & KØIE, M. (1989). *Pearsonellum corventum*, gen. et. sp. nov. (Digenea: Sanguinicolidae), in serranid fishes from the Capricornia section of the Great Barrier Reef. *Australian Journal of Zoology* **37**, 71–79.

OVERSTREET, R. M. & THULIN, J. (1989). Response by *Plectropomus leopardus* and other serranid fishes to *Pearsonellum corventum* (Digenea: Sanguinicolidae), including melanomacrophage centres in the heart. *Australian Journal of Zoology* **37**, 129–142.

PARUKHIN, A. M. (1970). [Study of the trematode fauna of fish in the Red Sea and Gulf of Aden]. *Biologiya Morya, Kiev* **20**, 187–213.

PARUKHIN, A. M. (1971). [Study of the trematode fauna of fishes of the Red Sea and Gulf of Aden]. *Biologiya Morya, Kiev* **25**, 136–146.

PARUKHIN, A. M. (1976). [*Parasitic Worms of Bottom Fishes of the Southern Seas*]. Kiev, Naukova Dumka.

PATERSON, A. M. & BANKS, J. (2001). Analytical approaches to measuring cospeciation of host and parasites: through a glass, darkly. *International Journal for Parasitology* **31**, 1012–1022.

PÉREZ VIGUERAS, I. (1955). Descripcion de seis especies nuevas dé trématodes de la familia Acanthocolpidae y division del género *Stephanostomum* en subgéneros. *Revista Ibérica de Parasitologia Tomo Extraordinario, Libro-Homenja al Prof. Lopez-Neyra*, 421–441.

PÉREZ VIGUERAS, I. (1958). Contribuciónal conocimiento de la fauna Helminthológia Cubana. *Memorias de la Sociedad Cubana de Historia Natural* 24, 17–38.

PÉREZ-PONCE DE LEÓN, G., GARCÍA-PRIETO, L., MENDOZA-GARFIAS, B., LEÓN-RÈGAGNON, V., PULIDO-FLORES, G., ARANDA-CRUZ, C. & GARCÍA-VARGAS, F. (1999). *Listados Faunísticos de México. IX. Biodiversidad de Helmintos Parásitos de Peces Marinos y Estuarinos de la Bahía de Chamela, Jalisco. Mexico, Universidad Nacional Autónoma de México, Instituto de Biología.*

PÉREZ-PONCE DE LEÓN, G., LEÓN-RÈGAGNON, V. & MONKS, S. (1998). *Theletrum lamothei* sp. nov. (Digenea), parasite of *Echidna nocturna* from Cuajiniquil, Guanacaste, and other digenes of marine fishes from Costa Rica. *Revista de Biologia Tropical. Universidad de Costa Rica* 46, 345–354.

POZDNYAKOV, S. E. (1994). [Revision of the genus *Gonapodasmius* (Trematoda: Didymozoidae)]. *Izvestiya Tikhookeanskogo Nauchno Issledovatel'skogo Instituta Rybnogo Khozyaistva i Okeanografii (TINRO) [Izvestiya of the Pacific Research Institute of Fisheries and Oceanography (TINRO)]* 117, 174–181.

RAMADAN, M. M. (1983). A review of the trematode genus *Hamacreadium* Linton, 1910 (Opecoelidae), with descriptions of two new species from the Red Sea fishes. *Japanese Journal of Parasitology* 32, 531–539.

REVERSAT, J., MAILLARD, C. & SILAN, P. (1991). Polymorphismes phénotypique et enzymatique: intérêt et limites dans la description d'espèces d'*Helicometra* (Trematoda: Opecoelidae), mésoparasites de téléostéens marins. *Systematic Parasitology* 19, 147–158.

REVERSAT, J., RENAUD, F. & MAILLARD, C. (1989). Biology of parasite populations: the differential specificity of the genus *Helicometra* Odhner, 1902 (Trematoda: Opecoelidae) in the Mediterranean Sea demonstrated by enzyme electrophoresis. *International Journal for Parasitology* 19, 885–890.

REVERSAT, J. & SILAN, P. (1991). Comparative population biology of digenes and their first intermediate host mollusc: the case of three *Helicometra* (Trematoda: Opecoelidae) endoparasites of marine prosobranchs (Gastropoda). *Annales de Parasitologie Humaine et Comparée* 66, 219–225.

RIGBY, M. C., HOLMES, J. C., CRIBB, T. H. & MORAND, S. (1997). Patterns of species diversity in the gastrointestinal helminths of a coral reef fish, *Epinephelus merra* (Serranidae), from French Polynesia and the South Pacific Ocean. *Canadian Journal of Zoology* 75, 1818–1827.

ROHDE, K. & HAYWARD, C. J. (2000). Oceanic barriers as indicated by scombrid fishes and their parasites. *International Journal for Parasitology* 30, 579–583.

SAOUD, M. F. A. & RAMADAN, M. M. (1984). On two trematodes of genus *Pseudoplagioporus* Yamaguti, 1938 from Red Sea fishes. *Veterinary Medical Journal* 32, 340–352.

SAOUD, M. F. A., RAMADAN, M. M. & AL KAWARI, K. S. R. (1986). Helminth parasites of fishes from the Arabian Gulf. 2. The digenetic trematode genera *Hamacreadium* Linton, 1910 and *Cainocreadium* Nicoll, 1909. *Qatar University Science Bulletin* 6, 231–245.

SAOUD, M. F. A., RAMADAN, M. M. & AL KAWARI, K. S. R. (1988a). Helminth parasites of fishes from the Arabian Gulf. 5. On *Helicometria qatarensis* n. sp. (Digenea: Opecoelidae) and *Stephanostomum nagatyi* n. sp. (Digenea: Acanthocolpidae); parasites of *Epinephelus* spp. from Qatari waters. *Qatar University Science Bulletin* 8, 173–185.

SAOUD, M. F. A., RAMADAN, M. M. & AL KAWARI, K. S. R. (1988b). Helminth parasites of fishes from the Arabian Gulf VI. On three species of digenetic trematodes: *Prosorhynchus epinepheli* Yamaguti, 1939; *Paraproctotrema qatarensis* n. sp. and *Prosorchis breviformis* Srivastava, 1936. *Rivista di Parassitologia* 49, 79–85.

SEKERAK, A. D. & ARAI, H. P. (1974). A revision of *Helicometra* Odhner, 1902 and related genera (Trematoda: Opecoelidae), including a description of *Neohelicometra sebastis* n. sp. *Canadian Journal of Zoology* 52, 707–738.

SHALABY, I. M. I. & HASSANINE, R. M. E. (1996). On the rhynchus and body surface of three digenetic trematodes; family: Bucephalidae Poche, 1907; from the Red Sea fishes based on scanning electron microscopy. *Journal of Union of Arab Biologists* 5, 1–19.

SHEN, J. (1985a). Digenetic trematodes of fishes from the Xisha Islands, II. *Studia Marina Sinica* 24, 167–180.

SHEN, J. (1985b). Digenetic trematodes of fishes from the Xisha Islands, III (larval forms). *Studia Marina Sinica* 24, 181–188.

SHEN, J. (1990a). [Didymozoidae trematodes from marine fishes offshore of China]. *Marine Science Bulletin* 9, 46–54.

SHEN, J. (1990b). *Digenetic Trematodes of Marine Fishes from Hainan Island*. Beijing, Science Publications.

SHIMAZU, T. & MACHIDA, M. (1995). Some species of the genus *Bivesicula* (Digenea: Bivesiculidae), including three new species, from marine fishes of Japan and Palau. *Bulletin of the National Science Museum, Tokyo, Series A, Zoology* 21, 127–141.

SIDDIQI, A. H. & CABLE, R. M. (1960). Digenetic trematodes of marine fishes of Puerto Rico. *Scientific Survey of Porto Rico and the Virgin Islands* 17, 257–369.

SOGANDARES-BERNAL, F. (1959). Digenetic trematodes of marine fishes from the Gulf of Panama and Bimini, British West Indies. *Tulane Studies in Zoology* 7, 71–117.

SOGANDARES-BERNAL, F. & HUTTON, R. F. (1959). Studies on helminth parasites from the coast of Florida III. Digenetic trematodes of marine fishes from Tampa and Boca Ciega Bays. *Journal of Parasitology* 45, 337–346.

SPARKS, A. K. (1957). Some digenetic trematodes of marine fishes of the Bahama Islands. *Bulletin of Marine Science of the Gulf and Caribbean* 7, 255–265.

SWOFFORD, D. L. (1998). PAUP*. Phylogenetic Analysis Using Parsimony * and other methods. Sunderland, MA., Sinauer Associates Inc.

szuks, h. (1981). Bucephaliden (Trematoda: Digenea) aus Fischen der Küstengewässer Nordwestafrikas. *Wissenschaftliche Zeitschrift der Pädagogischen Hochschule 'Liselotte Herrmann' Güstrow Aus der Mathematisch-Naturwissenschaftlichen Fakultät* 2, 167–178.

tang, c. & xu, z. (1979). The 'black root' disease of the razor clam in estuary of Jiulong river, Fujian. *Acta Zoologica Sinica* 25, 336–346.

toman, g. (1992*a*). Digenetic trematodes of marine teleost fishes from the Seychelles, Indian Ocean. III. *Acta Parasitologica* 37, 119–126.

toman, g. (1992*b*). Digenetic trematodes of marine teleost fishes from the Seychelles, Indian Ocean. IV. *Acta Parasitologica Polonica* 37, 127–130.

vassiliadés, g. (1982). Helminthes parasites des Poissons de mer des côtes du Sénégal. *Bulletin de l'Institute Fondamental d'Afrique Noire* 44, 78–99.

velasquez, c. c. (1959). Studies on the family Bucephalidae Poche 1907 (Trematoda) from Philippine food fishes. *Journal of Parasitology* 45, 135–147.

velasquez, c. c. (1966). Some parasitic helminths of Philippine fishes. *The U. P. Research Digest* 5, 23–29.

velasquez, c. c. (1975). *Digenetic Trematodes of Philippine Fishes*. Quezon, University of Philippines Press.

vélez, i. (1978). Algunos trematodes (Diginea [*sic*]) de peces marinos del norte de Colombia. *Anales del Instituto de Investigaciones Marinas de Punta de Betin* 10, 223–243.

wang, p.-q. (1982*a*). Some digenetic trematodes of marine fishes from Fujian Province, China. *Oceanologia et Limnologia Sinica* 13, 179–194.

wang, p.-q. (1982*b*). Hemiuroid trematodes of marine fishes from Fujian Province, China. *Journal. Fujian Teacher's University. Natural Science Edition* 2, 67–80.

wang, y. y., wang, p.-q. & zhang, w. h. (1992). [Opecoelid trematodes of marine fishes from Fujian Province]. *Wuyi Science Journal* 9, 67–89.

williams, e. h. & bunkley-williams, l. (1996). *Parasites of Offshore Big Game Fishes of Puerto Rico and the Western Atlantic*. Mayaguez, Antillean College Press.

winter, h. a. (1960). Algunos trematodos digeneos de peces marinos de aguas del oceano pacifico del sur de California, U.S.A. y del litoral Mexicano. *Anales del Instituto de Biología. Universidad Nacional Autonóma de México* 30, 183–208.

yamaguti, s. (1934). Studies on the helminth fauna of Japan. Part 2. Trematodes of fishes. I. *Japanese Journal of Zoology* 5, 249–541.

yamaguti, s. (1938*a*). *Studies on the Helminth Fauna of Japan. Part 21. Trematodes of Fishes, IV*. Kyöto, Japan, Yamaguti, S.

yamaguti, s. (1938*b*). Studies on the helminth fauna of Japan Part 24. Trematodes of fishes, V. *Japanese Journal of Zoology* 8, 15–74.

yamaguti, s. (1939). Studies on the helminth fauna of Japan. Part 26. Trematodes of fishes, VI. *Japanese Journal of Zoology* 8, 211–230.

yamaguti, s. (1940). Studies on the helminth fauna of Japan. Part 31. Trematodes of fishes, VII. *Japanese Journal of Zoology* 9, 35–108.

yamaguti, s. (1942). Studies on the helminth fauna of Japan. Part 39. Trematodes of fishes mainly from Naha. *Transactions of the Biogeographical Society of Japan* 3, 329–398.

yamaguti, s. (1953). Parasitic worms mainly from Celebes. Part 3. Digenetic trematodes of fishes II. *Acta Medicinae Okayama* 8, 257–295.

yamaguti, s. (1958). Studies on the helminth fauna of Japan. Part 52. Trematodes of fishes, XI. *Publications of the Seto Marine Biological Laboratory* 6, 369–384.

yamaguti, s. (1965). New digenetic trematodes from Hawaiian fishes, I. *Pacific Science* 19, 458–481.

yamaguti, s. (1970). *Digenetic Trematodes of Hawaiian Fishes*. Tokyo, Keigaku Publishing.

yamaguti, s. (1971). *Synopsis of Digenetic Trematodes of Vertebrates*. Tokyo, Keigaku.

zaidi, d. a. & khan, d. (1977). Digenetic trematodes of fishes from Pakistan. *Bulletin of the Department of Zoology University of the Panjab* 9, 1–56.

Ecology of larval trematodes in three marine gastropods

L. A. CURTIS*

University Parallel Program, Department of Biological Sciences and College of Marine Studies, University of Delaware, Newark, DE 19711 U.S.A.

SUMMARY

To comprehend natural host–parasite systems, ecological knowledge of both hosts and parasites is critical. Here I present a view of marine systems based on the snail *Ilyanassa obsoleta* and its trematodes. This system is reviewed and two others, those of the snails *Cerithidea californica* and *Littorina littorea*, are then summarized and compared. Trematodes can profoundly affect the physiology, behaviour and spatial distribution of hosts. Studying these systems is challenging because trematodes are often embedded in host populations in unappreciated ways. Trematode prevalence is variable, but can be high in populations of all three hosts. Conditions under which single- and multiple-species infections can accumulate are considered. Adaptive relations between species are likely the most important and potentials for adaptation of parasites to hosts, hosts to parasites, and parasites to other parasites are also considered. Even if colonization rate is low, a snail population can develop high trematode prevalence, if infections persist long and the host is long-lived and abundant. Trematodes must be adapted to use their snail hosts. However, both *I. obsoleta* and *L. littorea* possess highly dispersed planktonic larvae and trematode prevalence is variable among snail populations. Host adaptation to specific infections, or even to trematodes in general, is unlikely because routine exposure to trematodes is improbable. Crawl-away juveniles of *C. californica* make adaptation to trematodes in that system a possibility. Trematode species in all three systems are not likely adapted to each other. Multiple-species infections are rare and definitive hosts scatter parasite eggs among snail populations with variable prevalences. Routine co-occurrence of trematodes in snails is thus unlikely. Adaptations of these larval trematodes to inhabit the snail host must, then, be the basis for what happens when they do co-occur.

Key words: Gastropod, trematode, *Ilyanassa obsoleta*, *Cerithidea californica*, *Littorina littorea*.

INTRODUCTION

Studies from an ecological or evolutionary perspective on larval trematodes in marine molluscs have become numerous in recent years. The attention to marine systems is useful. Molluscs contribute much and are central in trematode life cycles. The impact of trematodes on the ecology of molluscan hosts can be very great and must not be underestimated (Curtis & Hurd, 1983; Lauckner, 1987; Sousa, 1991; Thomas *et al.* 1997; Poulin 1999).

Ecological investigations of these systems are often effort-intensive, yet still may not provide explanations for the phenomena observed. While released cercariae can be used to assess a snail's infection status, the only sure way (external metacercarial infections excepted) is to look for parasite stages in host tissues. An assessment of the trematodes present in a host population therefore often requires many dissections. Popiel (as cited by Irwin, 1983) must hold the record for the largest number. Over 250000 *Littorina saxatilis* from the Welsh coast were examined and only 1 was infected with *Cercaria littorinae saxatilis* V. In Ireland, Irwin (1983) found this cercaria in 31 of 350 (9 %) snails (including in 7 double-species infections). This illustrates the degree of spatial heterogeneity in prevalence and variety of infections often involved in marine snail–trematode systems. There is seldom a tested, valid explanation for such differences.

My own work has been almost exclusively on the *Ilyanassa obsoleta* – trematode system in Delaware estuaries and I therefore tend to view mollusc–trematode dynamics through this eastern North American system. Commonalities among scientific phenomena are prized. It would be simpler, for example, if we could assume that the ecological determinants of freshwater and marine mollusc–trematode assemblages were the same, or that long-lived and short-lived snails are the same as resources for trematodes. System similarities are important to our understanding, but differences matter too. A comparative approach stands to inform our perspective on what drives the ecological phenomena associated with marine snails and trematodes. I will first rather extensively summarize what is known about *I. obsoleta* and its trematodes. Information is widely scattered in a literature spanning a century and a current summary of knowledge about this frequently trematode-infected and much-studied snail is warranted. Two other well-studied marine systems, those of the snails *Cerithidea californica* and *Littorina littorea*, will then be more briefly summarized and a comparison made.

* Correspondence address: Dr Lawrence A. Curtis, Cape Henlopen Laboratory, College of Marine Studies, University of Delaware, Lewes, DE 19958 U.S.A.
E-mail: lcurtis@udel.edu

THE *ILYANASSA OBSOLETA* SYSTEM

In a review of snails and trematodes, Esch & Fernandez (1994) wrote 'that knowledge of the biology of both the host and the parasite is imperative in order to understand the ecology of either organism.' The current comprehension of the *I. obsoleta*–trematode system is the product of such an integration. Considerable knowledge of the biology of this snail exists. Of particular note in the present context is the possession of a crystalline style (Jenner, 1956). This is a rare structure in neogastropods (Yonge, 1930) and probably accounts for *I. obsoleta's* abundance, which can be over 1000 m^{-2}. It allows the dietary utilization of an abundant food source, plant materials (Curtis & Hurd, 1979, 1981; Brenchley, 1987). Though snails consume carrion, most of the dietary intake comes from algal sources (Wetzel, 1977). The main habitats for *I. obsoleta* are mudflats, sandflats and saltmarshes on the Atlantic coast of North America. Growth and age structure were first studied in a Woods Hole population by Scheltema (1964). Egg cases are produced in the spring, from which veliger larvae emerge. After a tenure of weeks in the plankton they settle as juveniles at ∼1 mm shell height. By the third summer, snails reach ∼14 mm and the age of first reproduction. Growth slows after summer 3 and with yearly recruitment a local population will contain 3 size-classes, first and second summer snails plus a composite group (⩾ 18 mm) containing older individuals. Planktonic larval dispersal results in genetically homogeneous populations along the eastern seaboard of North America (Gooch, Smith & Knupp, 1972).

Miller & Northup (1926) provide early descriptions of many of the trematodes found in this snail around Woods Hole, Massachusetts. Further descriptions and elucidations of life cycles are given by later authors (Table 1). McDermott (1951) describes and figures the trematodes of *I. obsoleta* in New Jersey. Stunkard (1983*a*) lists the infecting species around Woods Hole as well as other hosts in the life cycles.

Physiological interactions of *I. obsoleta* and its trematodes are complex. They are important to consider, as they could dictate parasite transmission dynamics and changes in host growth, host or parasite survival, and spatial distribution of infected hosts. The physiological response to temperature of snails was found to be altered when infected (Vernberg, 1969; Vernberg & Vernberg, 1974). In the laboratory at 37 °C, uninfected and infected hosts survived about equally well, but at 39 and 41 °C survival of infected snails was reduced. The inference was that infection could jeopardize host survival in the intertidal zone. Riel (1975) found contrasting results wherein infected snails survived high temperatures better. Later work (Barber &

Caira, 1995; Curtis, Kinley & Tanner, 2000) showed that infected snails survive in the field for years. The upper temperature tolerance of parasite stages in snail hosts was deemed an adaptation to the definitive host environment (Vernberg & Vernberg, 1974). For example, the larvae of *Himasthla quissetensis*, a bird parasite, could survive 41 °C, whereas the larvae of *Zoogonus rubellus*, a fish parasite, were killed at 36 °C. Cercariae were better able to handle salinity extremes than the snail host (Vernberg, 1969). For example, *H. quissetensis* cercariae had about the same respiratory rate from 10 to 35, while the snail's rate declined below 15 psu (= g/kg). Kasschau (1975) found no difference in reaction to salinity between uninfected and infected hosts. Adult trematodes could tolerate anaerobic conditions for a day or more, but cercariae for no more than a few hours. This could negatively affect host-to-host transmission in the sometimes reduced O$_2$ conditions of saltmarshes (Vernberg & Vernberg, 1974). Sindermann, Rosenfield & Strom (1957) and Sindermann (1960) showed that *Austrobilharzia variglandis* cercarial output was reduced by withholding food from hosts, salinity extremes, low temperatures, and reduced oxygen tension. The question of gigantism, i.e. enhanced growth of parasitized *I. obsoleta*, which would affect host size-infection prevalence relationships, was examined in the laboratory by Cheng *et al.* (1983). They found that enhanced growth due to infection did not occur. Neither does gigantism occur in the wild; rates of shell height change in infected individuals were much reduced (Curtis, 1995). Infected *I. obsoleta* are sterile. Among thousands of dissections of infected snails, only 2 individuals still had gametes (Curtis, 1997). Cheng, Sullivan & Harris (1973) and Pearson & Cheng (1985) showed that castration by *Z. rubellus* infections results from a parasite secretion.

Trematodes affect snail movement and spatial distribution. There is a tendency for *I. obsoleta* to move to subtidal areas as winter approaches (Batchelder, 1915), but Sindermann (1960) noted that this was inhibited by infecting *A. variglandis*, *H. quissetensis* and *Z. rubellus*. Emphasizing the effects of *A. variglandis*, the following sequence was reported. In spring, snails move to the upper shore where migratory birds (the source of snail infections) congregate. Snails gain infections and infected snails shed cercariae, which infect bird hosts. Snails remain in the higher intertidal zone in summer and prevalence there increases to ∼ 25%. In autumn, uninfected snails move into subtidal winter aggregations, but infected snails tend to remain on the upper shore. The snail population overwinters essentially in this state. Partly to assess the findings of Sindermann (1960), Stambaugh & McDermott (1969) studied the effect of trematode infections on locomotion of *I. obsoleta*. In the laboratory, they studied 6 species of trematodes, but mainly *A.*

Table 1. *The trematodes of the estuarine snail* Ilyanassa obsoleta *encountered in Delaware, USA and other hosts in the life cycles*

Trematode	Cercariae produced in	2nd host (example)	Definitive host	Reference(s)
Himasthla quissetensis	rediae	bivalve	bird	Stunkard (1938*a*)
Lepocreadium setiferoides	rediae	flatworm	fish	Martin (1938); Stunkard (1972)
Zoogonus rubellus	sporocysts	polychaete	fish	Stunkard (1938*b*)
Austrobilharzia variglandis	sporocysts	none	bird	Stunkard & Hinchliffe (1952)
Gynaecotyla adunca	sporocysts	crustacean	bird, fish	Rankin (1940); Hunter (1952)
Stephanostomum dentatum	rediae	fish	fish	Stunkard (1961)
Stephanostomum tenue	rediae	fish	fish	Martin (1939)
Diplostomum nassa	sporocysts	unknown	unknown	Martin (1945); Stunkard (1973)
Pleurogonius malaclemys	rediae	*I. obsoleta*	terrapin	Hunter (1961; 1967)

variglandis, *H. quissetensis*, and *Z. rubellus*. There was much individual variation in locomotion, but Sindermann's results were supported, as tendency to move and rate of locomotion were reduced in infected snails. In Delaware, *A. variglandis* is infrequent (Curtis, 1997) and these migration patterns have not been seen. Nevertheless, the vertical position of the collection site and the particular trematodes present can greatly influence observed prevalence.

An *I. obsoleta* population on an apparently homogeneous sandflat, Cape Henlopen in Delaware Bay, was investigated by Curtis & Hurd (1983) and much unanticipated spatial heterogeneity was found. Infection prevalence increased exponentially with snail size, suggesting that snails do not lose infections. That larger snails were more likely to harbour trematodes added to infection prevalence heterogeneity because size classes were patchily distributed. Two trematode zones on the sandflat were identified, though no obvious physical feature was involved. There was much variation within zones, but in the southwest area a smaller proportion of snails was infected than in the northeast, where prevalence was generally 50–100%. An overall analysis showed that snails $\leqslant 18$ mm shell height were seldom infected in either zone; snails $\geqslant 24$ mm were infected wherever found; however, for unknown reasons, snails 19–23 mm showed greater trematode prevalence in the northeast section. It was determined that samples representative of the populations of snails and/or trematodes on the sandflat would be difficult to obtain.

Certain aspects of the spatial heterogeneity were later explained. Feeding aggregations on carrion promote snail growth and survival (Curtis & Hurd, 1979) and reproduction (Hurd, 1985). Curtis (1985) investigated the roles of snail sex and parasitism in forming these aggregations. In carrion-response experiments, control samples from enclosures estimated the proportions of resident snails of each sex that were infected with different trematodes. Responding snails crawling to carrion at the upstream end of enclosures were examined. Uninfected females, in reproductive condition, responded to carrion more often than males and carrion aggregations are often largely composed of females. Parasite influence could be detected only after the breeding season; e.g. *L. setiferoides*-infected snails of both sexes responded preferentially to carrion. Thus, trematodes can affect snail distribution other than by merely inhibiting locomotion.

Another behaviour alteration affecting snail spatial distribution was discovered. Curtis (1987) found that most snails stranded above the waterline by falling tides on beaches and sandbars had *Gynaecotyla adunca* infections. To investigate this pattern, a plot on a sandbar peak was cleared of snails on an initial low tide and subsequently on 29 low tides over a 2 week period. Snails recovered in the plot following high-waters were almost always *G. adunca*-infected; more were found on night low tides, and many harboured multiple-species infections. Usually, transmission-enhancing adaptations work through predation. The next hosts of *G. adunca* are semi-terrestrial amphipod or decapod crustaceans and they must be reached by cercariae (Rankin, 1940). Since hosts are castrated when infected, and since only *G. adunca*-infected snails frequented the upper shore, it was deduced that this altered behaviour was a parasite adaptation resulting in enhanced cercarial transmission to the second intermediate host.

Curtis (1990) followed the movements of individual *G. adunca*-infected snails. Two groups of 250 snails were collected: in one group, collected in the high intertidal zone, snails were likely to be infected with *G. adunca*; snails in the other group, collected lower down, were not. Snails were individually marked, released around a sandbar peak, and their locations noted on 16 subsequent low tides. Only *G. adunca*-infected snails were frequently seen near the sandbar peak and their preference for visiting on dark low tides was confirmed. Some individuals bearing this parasite visited the sandbar repeatedly, suggesting a regular migration. Curtis

(1993) investigated when and where *G. adunca* cercariae were released by infected snails. Hosts, harbouring active cercariae, began their upward migrations mostly on afternoon flood tides. As they were later left emerged by receding tides at night, they produced mucus trails containing perhaps thousands of cercariae. Oddly, trails left on daytime ebbing tides lacked cercariae. Cercariae are thus left where probability of reaching a nocturnal, semi-terrestrial, crustacean second intermediate host is greatest. How these migrations are cued and controlled is unknown.

Recently, McCurdy, Boates & Forbes (2000) described a different spatial distribution of *G. adunca*-infected snails in the Bay of Fundy. There the prevalence of *G. adunca* was greater in snails vertically lower down, associated with the distribution of the amphipod second host in that habitat. What might explain this curious geographical and distributional difference? Possibly the parasite is universally adapted to adjust the zonation of its first host according to which second intermediate host is available. Or, there may be differently evolved parasite ecotypes.

Longevity of individuals in a population is important from many perspectives, not least the accumulation of parasites. This snail has been cited as living 3 (Hyman, 1967), 5 (Scheltema, 1964) and 7 years or more (Jenner & Jenner, 1977). Earlier estimates of longevity were based on population-level studies. Curtis (1995) followed the growth histories of individual snails (most $\geqslant$ 18 mm shell height) in the field. In 1991, 1200 snails from an infrequently parasitized population that had further tested uninfected by failing to release cercariae, were individually marked, measured, and released onto Cape Henlopen. In 1993, another 200 such transplanted snails plus 300 native Henlopen snails were released. Marked snails were located and shell height changes noted through summers and autumns of both release years. Uninfected (n = 173) and infected (n = 49) snails were recovered. For uninfected snails, those initially smaller grew a little faster than those initially larger. Mean growth was 1·2 mm y^{-1} with considerable variation (95 % confidence interval ±1 mm). Growth of infected snails was much slower, estimated at 0·2 mm y^{-1}. Snails are longer lived than previously thought. A 27 mm snail, infected or not, was estimated to be 30–40 years old. Further, in the area where this study was done 98 % of 23 mm snails were already infected (Curtis & Hurd, 1983). Since many $\geqslant$ 25 mm infected snails were observed, which would require at least an additional decade of growth, it was inferred that infections persist for at least 10 years. Thus, hosts and infections live long and the *I. obsoleta*–trematode system is slowly paced.

A study done in the Savages Ditch area of Rehoboth Bay, Delaware, where many infected individuals occurred, confirmed this (Curtis *et al.* 2000). On Cape Henlopen 27 mm snails are scarce, while some Savages Ditch snails reach 37–39 mm. Is this the result of locally faster growth or greater age? Transplanted snails (n = 249), which had tested as uninfected, and native snails (n = 231), which were mostly already infected, were marked and released in 1996. Growth of uninfected snails was about the same as on Cape Henlopen (mean = 1·5 mm y^{-1} ± 0·7, n = 86). Some infected snails provided 4 year-long field growth histories. Growth rate, based on 94 infected snails, was again estimated at ~ 0·2 mm y^{-1}. Probable size at infection, the large size attained, growth rates, and habitat history, suggested that some snails stood to be as old as 60–70 years.

The long lifespan of the host must be coupled with the slow rate at which trematodes colonize snails. The 1991 and 1993 releases (above), totaling 1400 putatively uninfected snails (sentinels), assessed the probability of a snail becoming infected while in the field. These sentinels were released into an area where trematode prevalence was high (80–90 %) and it was expected that sentinels would quickly become infected. However, 185 sentinels were recovered after being free a mean of 87 d (range = 17–793) and only 1·6 % carried new infections (Curtis, 1996). The 249 sentinels released in 1996 at Savages Ditch, another high prevalence site, mostly disappeared from the release area and only 16 were recovered after being free for 104–776 days. Two became infected (over 2 years) and the infection probability estimate was 6·3 % y^{-1} (Curtis & Tanner, 1999). Juveniles are uninfected at settlement and support for a low colonization rate can be gained from the low infection prevalence in snails 1–2 (1·2 %) and 3–4 years old (4·0 %) (Curtis, 1997).

Curtis & Tanner (1999) also considered native snails from both Cape Henlopen and Savages Ditch that had certain infections when released. These served to gauge the probability of changes in infecting species composition over time. Apparent changes, however, must be interpreted wisely because if snails fail to release cercariae of all species harboured when tested, which happens (Curtis & Hubbard, 1990), infections will be incorrectly assessed. Most hosts demonstrated the same infecting species for their whole time free [currently up to 6 summers at Savages Ditch (Curtis, unpublished data)]. In total, 6 of 123 (4·9 %) native snails recovered over multiple years had clearly changed infecting species composition. The probability of a host changing infection status is therefore low, clearly under 5 % y^{-1}. Thus, a snail might easily go a decade and more without adding or losing infecting species. Host and infection longevity, not rapid colonization, are the keys to accumulation of single and multiple infections in populations.

Assemblages of trematode species infecting individual snails (infracommunities) were recently reviewed by Esch, Curtis & Barger (2001). De-Coursey & Vernberg (1974) first studied these assemblages in *I. obsoleta*. They had only a few multiply-infected snails to work with, but determined that in double infections, there was some displacement of stages from the usual sites occupied in single infections. They also counted *L. setiferoides* and *Z. rubellus* cercariae produced in a 24 h period by singly- (n = 5 and 3, respectively) and doubly-infected (n = 2) snails. The double infections produced fewer cercariae.

Curtis (1985, 1987, 1990) noted that multiple infections are relatively common in *I. obsoleta*, particularly if infected by *G. adunca*. Curtis & Hubbard (1993) took advantage of this to collect snails harbouring a variety of multiple infections from a sandbar where *G. adunca* was common. They collected 18 uninfected snails, 162 with single infections (5 species), 134 with double infections (11 combinations), and 65 with triple infections (5 combinations). Using these snails, they considered 4 possible evolutionary models for trematodes using this host: (1) parasites are adapted to the host; (2) the host is adapted to the parasites; (3) parasites are adapted to the host and to each other; and (4) there is a combination of (2) and (3), where all players are adapted to each other. With regard to (1), it was taken for granted that the trematodes are adapted to this obligatory host since they must encounter it in each life cycle. Number (2) would require that hosts encounter trematodes in succeeding generations. If so, hosts might be able to resist infection by the common species. Number (3) would require that succeeding generations of trematodes co-occur in infracommunities. Adaptations might be expected such that one species could counter (or possibly facilitate) co-occurrence with another species.

With regard to (2), the ability of the host to resist infections was not tested, but trematode prevalence is variable within and among *I. obsoleta* populations (Curtis & Hurd, 1983; Curtis, 1997) and the host life cycle includes an unpredictably dispersed planktonic larva. A lineage of snails bearing a mutation that, for example, defended against a particular trematode, or even trematodes in general, would not necessarily encounter the required selective agent. There is too much opportunity to exist free of trematodes. Thus, selective pressures would probably not be pervasive and consistent enough to generate host adaptations specific to trematodes. With regard to (3), adaptive relationships between trematode species, Curtis & Hubbard (1993) tested for changes in spatial distribution of trematodes within snails, and for complete suppression or reduction of cercarial output when species co-occurred. If adaptations had developed, they stood to be manifested in one or more of these ways. Much variability was encountered, but no consistent displacements of trematode larvae

within snails due to co-occurrence were detected. Mostly, parasite stages occurred throughout snails in all species combinations tested. No species' production of cercariae was suppressed, nor was it significantly reduced, by co-occurrence with other species. Trematode species seemed uninfluenced by other species present in the same host. Further, in the global *I. obsoleta* population, species of trematodes only seldom and irregularly co-occur (Curtis, 1997). Definitive hosts disperse eggs of trematodes unpredictably and they may easily disperse them to sites where, even if miracidia successfully colonize a snail, the selective force (another trematode) is likely rare or absent. Therefore, the only likely adaptive relationship in this system is parasite to host.

Studies of trematode species present in *I. obsoleta* populations (component communities) usually reveal low prevalence and few multiple-species infections (reviewed in Curtis, 1997). Given problems associated with obtaining representative samples, different methods of collection and analysis, as well as actually varying assemblages across landscapes (Esch *et al.* 2001), variable results may be expected. Major works on *I. obsoleta* include Gambino (1959), who examined snails from a Rhode Island population by dissection, Vernberg, Vernberg & Beckerdite (1969), who examined snails from North Carolina, first by cercarial release and then dissection of those shedding, and McDaniel & Coggins (1972), who assessed snails from North Carolina by cercarial release. Curtis (1997) presented data from 11 774 dissections of snails from 9 Delaware sites collected between 1981 and 1995. Samples were not usually meant to be representative of the populations sampled, and because of this calculations were not made to compare observed and expected frequencies of multiple infections, as is sometimes done (e.g. Vernberg *et al.* 1969). Results would probably have given some erroneous impressions. Overall, 51·04 % of snails were infected (range across 9 sites = 8·7–100 %) with 1 or more species. Five core species (occurring in ⩾ 1 % of snails) and 4 satellite species (occurring in < 1 %) were noted. Of snails examined, 12·57 % carried multiple-species infections, usually involving core species. There were 16 double, 7 triple, and 1 quadruple combinations. This large *I. obsoleta* sample, admittedly biased toward older snails and a high prevalence site (Cape Henlopen), had greater prevalence of single and multiple infections than any so far collected. Gambino's (1959) sample was most similar; 25·66 % of 5717 snails were infected and 0·91 % had unspecified double infections.

SYSTEMS FOR COMPARISON

Cerithidea californica and its trematodes bear comparison with the *I. obsoleta* system. The natural history of this snail, which occurs from just north of

San Francisco to central Baha California, is treated by Race (1981). This intertidal saltmarsh inhabitant is a mesogastropod (herbivorous) deposit feeder. It sometimes reaches densities of $> 1000\,\text{m}^{-2}$. Juveniles crawl away from the egg case and adults are mostly sedentary, leading to limited gene flow among populations. Young snails are about $0.25\,\text{mm}$ at release in June, reach $3\,\text{mm}$ by August, cease growing in winter, and by the end of the second summer are $\sim 15\,\text{mm}$. Size at first reproduction is $\sim 20\text{--}24\,\text{mm}$, probably when 3 years old. The specific habitat occupied and snail size can affect growth rates. Growth of snails in submerged pans is faster ($\sim 4\,\text{mm mo}^{-1}$ for $15\text{--}20\,\text{mm}$ snails) than those in emerged pans or on mudflats ($\sim 1\,\text{mm mo}^{-1}$). Snails $20\text{--}25\,\text{mm}$ grow only about $1\,\text{mm mo}^{-1}$ and snails $> 25\,\text{mm}$ slower still. Individuals $> 30\,\text{mm}$ are often observed and a longevity of $8\text{--}10$ years is given by Sousa (1983), 20 years by Byers & Goldwasser (2001).

Populations of this species can have substantial trematode prevalence, but there is much spatial variability (Lafferty, 1993*a*). Birds are the definitive hosts of all trematodes in this snail (Sousa, 1983). Martin (1955) examined 12 995 snails $\geqslant 20\,\text{mm}$, that were collected over a year at a bird congregation site in Upper Newport Bay, California. Overall, trematode prevalence was 66.79% and double (38 combinations) and triple (9 combinations) infections occurred in 5.31%. At other California sites prevalence was lower and multiple infections fewer: in Gloleta Slough prevalence was 15.4% ($n = 2910$), with 0.45% doubles (5 combinations) (Yoshino, 1975); in Bolinas Lagoon overall prevalence was 16.90% ($n = 25\,859$, over 7 years) and 0.35% had multiple infections (21 double combinations plus a single occurrence of a triple) (Sousa, 1990).

Trematode-induced behaviour alterations affecting aggregation and movement of *C. californica* have not been reported. Sousa (1983) noted that infected snails did not exhibit copulatory behaviour. Growth of infected snails is stunted (Sousa, 1983; Lafferty, 1993*b*). Sousa (1983) illustrates increasing prevalence with size, all snails $\geqslant 33\,\text{mm}$ being infected. Infections appear to be permanent and greater prevalence in older snails might be predicted because they have been longer exposed to infection. At sites regularly visited by definitive hosts, high trematode prevalence can occur, e.g. Martin's (1955) collection site. Data seem not to exist on the probability that an uninfected *C. californica* will become infected. However, Sousa (1993) reports that, of 1170 snails previously infected, 8.3% changed infections during field release times up to 4.1 years. This shows that infected snails can live for years. It also suggests a low colonization rate, as does the few infections in $1\text{--}2$ years old snails (Sousa, 1983).

Littorina littorea and its trematodes may also be compared with the *I. obsoleta* system. This abundant mesogastropod is native to European waters, but occurs in eastern North America and on the Pacific coast as well (Carlton, 1969). Larvae are planktonic, gene flow is substantial, and populations are genetically homogeneous over broad regions (Berger, 1977). Distribution on shores is variable. Around England and the North Sea coast of Germany it occurs mostly on rocky intertidal shores. In the Baltic Sea, with its reduced salinity and tidal range, its vertical distribution is displaced downward into the subtidal zone (Lauckner, 1984). On a Wadden Sea tidal flat, Saier (2000) found that all but a few large adults occur intertidally and recruitment occurs there. In North America, this snail was first reported in Nova Scotia in 1840 and has subsequently spread southward (Berger, 1977). It displaces co-occurring *I. obsoleta* (Brenchley & Carlton, 1983). In Nova Scotia, juveniles recruit subtidally and then move shoreward (Lambert & Farley, 1968). A migration from the high to low intertidal zone occurs in autumn, but this is not seen in Europe (Sindermann & Farrin, 1962; Lambert & Farley, 1968).

On age and growth of *L. littorea*, Moore (1966) notes that sexual maturity is attained when 2 or 3 years old. Hyman (1967) gives a longevity of $4\text{--}10$ years, 20 years in captivity, with snails reaching $\sim 27\,\text{mm}$ shell height in 4.5 years. Hughes & Answer (1982) indicate that snails attain $\sim 12\text{--}15\,\text{mm}$ in 1 year, $\sim 18\text{--}20\,\text{mm}$ by 2 years, and with subsequently slower growth they reach an asymptotic size at $30\,\text{mm}$. Robson & Williams (1971) note that some reach larger than $30\,\text{mm}$. In Nova Scotia, growth rates are similar and the largest snails are $\geqslant 4$ years old (Lambert & Farley, 1968).

The extensive data on prevalence and diversity of trematode infections in *L. littorea* show variability within and among regions (Table 2). In England, James (1971) lists 11 trematode species; on North Sea coasts, Lauckner (1980) lists 6. *Podocotyle atomon* uses a fish definitive host, but all other species use birds and 4 of these occur in Baltic Sea *L. littorea* (Lauckner, 1984). In North America, trematode diversity in this snail is lower. Stunkard (1983) lists 2 species and Pohley (1976) 3, one observed only once. An important trematode, common to both sides of the Atlantic and the Baltic Sea, is *Cryptocotyle lingua*. Double infections are most common on the North Sea coast of Germany (Werding, 1969; Lauckner, 1980), where 7 different combinations are recorded (Table 2). Triple infections are not reported in *L. littorea*.

Effects on migratory behaviour can be pronounced. In Maine (Sindermann & Farrin, 1962) and Nova Scotia (Lambert & Farley, 1968) the down-shore movement of snails for winter is inhibited by infections with *C. lingua*. Williams & Ellis (1975) followed the vertical movements of marked uninfected and infected snails for 8 weeks in Yorkshire, England. Uninfected snails moved down-

Table 2. Trematode prevalence observed in *Littorina littorea* populations. Number of species combinations is given in parentheses

Locality	Number of snails examined	% infected	Number of trematode species	% doubles	Reference
West England	2500	3–4	7	—	Rees (1936)
West England	6165	4·8	6	—	James (1968)
West England					Hughes & Answer (1982)*
Orme area	60	31·7	?		
Porth Cwyfan	299	3·0	?		
Trwyn-Penrhyn	289	7·6	?		
Gorad-y-Gyt	230	17·0	?		
Foryd Bay	377	18·6	?		
South England (Plymouth)					Rees (1936)
Drake Island	220	10·0	2	—	
Trevol	550	2·0	2	—	
South England	800+	13·3	3	—	Watts (1971)
East England					Robson & Williams (1970)
Scalby Rocks	5878	39·9	4	2·5 (2)	
South Bay	1009	3·2	?	—	
Scalby Pipe	1194	20·7	?	—	
Burniston Bay	434	2·5	?	—	
East Scotland (Ythan Estuary)					Huxham *et al.* (1993)
Newburgh Quay	200	11·5	3	—	
Quay Inches	200	14·5	4	—	
Burnmouth Scalp	200	2·5	1	—	
Sheepfold Burn	200	6·0	2	—	
Northern Ireland (Belfast)					Matthews *et al.* (1985)
Portavogie shore	101	52·5	3	—	
Portavogie offshore	300	65·3	3	1·0 (2)	
German North Sea coast					Lauckner (1980)
North Sea coast	2691	42·8	6	3·3 (7)	
Isle of Sylt	30811	19·4	6	—	
Schleswig-Holstein	1496	5·7	6	—	
Baltic Sea/Kattegat	11571	15·0	5	—	
Baltic Sea/Kattegat					Lauckner (1984)
Neustadt Bay coast	471	1·6	4	—	
Neustadt Bay 16 m	577	18·7	2	—	
Sletterhage 20–26 m	334	12·6	4	—	
New England (Isles of Shoales)					Hoff (1941)
gull feeding area	501	6·2	1	—	
gull roosting area	86	19·8	1	—	
gull nesting area	134	20·9	1	—	
around docks	96	0·0	—	—	
New England (Maine)					Sindermann & Farrin (1962)
high-tide zone	3000	65·0	1	—	
mid-tide zone	3000	45·0	1	—	
low-tide zone	3000	46·0	1	—	
New England (Maine and Rhode Island)					Pohley (1976)
Eastport	464	1·5	1	—	
Rogue Bluffs	651	10·5	3	—	
Watch Hill	925	2·4	2	—	

* These authors wrote generally of 4 species of trematodes and found 2 double infections in 2 species combinations.

ward a few metres in winter; snails infected with *C. lingua* rediae or *Renicola roscovita* sporocysts moved downward less than uninfected snails. Here, similar to *I. obsoleta*, the effects on hosts of trematodes themselves can bring about spatial heterogeneity in prevalence.

Trematodes have extensive detrimental effects on *L. littorea* (Lauckner, 1980). Infection causes, if not outright castration, a severe reduction in gonadal output (Huxham, Raffaelli & Pike, 1993). Probability

of bearing an infection increases with size (age) (e.g. Lauckner, 1980; Hughes & Answer, 1982). *Renicola roscovita* is found only in smaller adults, but the general notion is that only adult *L. littorea* can be infected (James, 1968; Robson & Williams, 1970; Lauckner, 1980). However, snails < 1 year old can be infected. Lambert & Farley (1968) make note of a few snails < 11 mm infected with *C. lingua* in Nova Scotia. In England, Robson & Williams (1970) recorded 2 species infecting <10 mm snails and 3 infecting 10·1–15·0 mm snails (sometimes doubly). Infections in immature snails do occur, but are seldom recorded, possibly because prevalence is low and such hosts are less frequently examined.

Several authors have considered enhanced growth (gigantism) in infected *L. littorea* and it seems not to occur. Robson & Williams (1971) could not confirm or deny it and Hughes & Answer (1982) found growth unchanged. Huxham *et al.* (1993) noted that growth of caged, infected snails in Scotland was very slow or stopped. Mouritsen, Gorbushin & Jensen (1999) found that, depending on habitat, growth is either stunted or unaffected.

Longevity of infections in *L. littorea* has been studied. Rothschild (1942) observed an active *C. lingua* infection in the laboratory for 7 years. This alone makes the case for potentially prolonged viability. Robson & Williams (1970) kept snails in the laboratory infected with *C. lingua* (n = 49), or *Cercaria A* (*Renicola roscovita*) (n = 35), or both (n = 3) for 6·5 months. At the end of that time, 3 initially shedding *C. lingua* and 11 *R. roscovita* had died, some (n = 4) had lost their infections, and 8 snails were unaccountably lost. All 3 double infections survived. The frequent death of *R. roscovita*-infected snails suggested that this trematode is ultimately lethal to the snail. Huxham *et al.* (1993) caged a total of 800 uninfected and infected snails in the field over 530 days, 96 died, 16 were lost, and infected snails were somewhat more likely to die. Large *L. littorea* are not heavily preyed upon by birds and mortality among them was attributed to trematodes.

There is little information on the colonization rate of *L. littorea* by trematodes, but it appears to be low. Hughes & Answer (1982) did not estimate a rate but hypothesized that the total risk of a snail becoming infected is constant throughout life, though for a particular trematode it could vary in time. As described above, immature snails can be infected and prevalence increases with size (age). In their study of snail movement, Williams & Ellis (1975) noted that none of their uninfected snails became infected during 2 months of shore exposure.

A COMPARISON

In their native areas, one feature the *I. obsoleta*, *C. californica* and *L. littorea* systems have in common is a widespread, abundant host and a diversity of infecting trematodes. That long-standing residence within a geographic region matters to trematode diversity seems clear. The native area for *C. californica* is the southern west coast of North America where 18 species infect the snail (Ching, 1991). The native area for *I. obsoleta* is the eastern seaboard of North America where at least 9 species infect it (Table 1). It has been introduced on the west coast (Race, 1981), but only 1 species, *A. variglandis*,

has been reported there (Grodhaus & Keh, 1958; Ching, 1991). The native area for *L. littorea* is Europe, where up to 11 trematodes infect it (James, 1971). Berger (1977) reported fossil *L. littorea* shells from Nova Scotia estimated to be ∼ 700 years old. This suggests a resident population prior to European colonization of North America. He posits that the modern snail population passed through a genetic bottleneck from this older population. If so, it would be instructive to have an indication of trematode diversity in the ancestral population. There may have been a parasite bottleneck as well. Currently, only *C. lingua* is common (Pohley, 1976; Stunkard, 1983). In western North America, Ching (1991) lists no infecting species. In non-native areas then, because of the few available species, multiple infections do not occur in *I. obsoleta* or *L. littorea*. This historical factor largely determines the species richness and variety of infra- and component trematode communities.

The difficulty of obtaining samples that, even if taken at random, are representative of populations is an important consideration. This is analogous to the problem of unwittingly including parasitized hosts in experiments on supposedly unparasitized hosts (Curtis & Hurd, 1983; Lauckner, 1987) because much unexplainable or misinterpreted variability may result. The point in emphasizing the general spatial heterogeneity inherent in these systems, and particularly the effects of trematodes on the spatial distribution of their hosts, is that species prevalence estimates stand to be affected. Prevalence estimates may be incorrect with respect to the general conditions under which snails become colonized, if such general conditions exist. If so, predictions about probability of species co-occurrences by way of a null model, basically making predictions by multiplying trematode prevalence estimates together, will be in error.

Examples of infected snails occupying restricted portions of the general host distribution can be noted in *I. obsoleta*. Unanticipated spatial heterogeneity, involving both parasites and host, was discovered in a sandflat population (Curtis & Hurd, 1983). The basis for some of it was later explained (Curtis, 1985, 1987) but if samples had been unknowingly taken where certain trematodes are common, such as from the northeast section of the sandflat, carrion aggregations after the breeding season, or beaches and sandbars during summer, invalid interpretations might have emerged. In the global population, these are special, unrepresentative situations. The abundance of certain infections might have been considered a consequence of an influx of definitive hosts and a matter of seasonal variation. However, this host and its infections live long and seasonality has little to do with the infections found in a sample (Esch *et al.* 2001). If a sample were collected from a sandbar, where *G. adunca* (with its attendant multiple-species infections) is common, it might have been concluded that species interactions within snails are much more frequent and important than is actually the case. On the obverse of this point, as extensive as the sampling of Curtis & Hurd (1983) was, they did not even detect *G. adunca* infecting snails because they did not happen to take samples from sandbars. Had probabilities of co-occurrences been calculated, an important species in this regard would have been utterly missed. Sindermann & Farrin (1962), Lambert & Farley (1968), and Williams & Ellis (1975) signal similar concerns with *L. littorea*. This style of analysis, with the trematode species considered here and others, has often been done (e.g. Vernberg *et al.* 1969; Lauckner, 1980; Kuris & Lafferty, 1994) and species interactions are sometimes concluded to be important in structuring trematode component communities. Knowledge of both host and parasite ecology, sampling wisdom, and caution would be wisely exercised before drawing inferences.

Complex trematode assemblages do not occur everywhere snail populations do. All three hosts considered are abundant and trematode prevalence in all three can be locally high, with the consequent presence of relatively frequent multiple-species infra-communities. High prevalences are often associated with substantial input of infective propagules from definitive hosts (e.g. Hoff, 1941; Martin, 1955; Robson & Williams, 1970; Curtis, 1997). A recent field experiment with an estuarine snail, *Cerithidea scalariformis*, confirms this connection (Smith, 2001). Host longevity is another factor in common. *Ilyanassa obsoleta* longevity is extraordinary, with individuals living for several decades (Curtis, 1995; Curtis *et al.* 2000). Longevity of *C. californica* (Sousa, 1983) and *L. littorea* (Hyman, 1967) is 1 or 2 decades. Single and multiple infections also seem to live long in all 3 hosts. In *I. obsoleta*, Curtis (1995) inferred that infections persist for 10 years or more; Curtis & Tanner (1999) observed only a few infracommunity species composition changes in the field over 3 years and Curtis *et al.* (2000) observed the same snail infected with the same trematode species for up to 5 years. In *C. californica*, infections seem permanent and long-lived (Sousa, 1983). Infracommunity composition does change (Sousa, 1993), but relatively infrequently. No data on probability of *L. littorea* changing infection status in the field seem to exist, except for the lack of new infections among the marked snails of Williams & Ellis (1975). *Renicola roscovita* (Robson & Williams, 1970) apparently causes the death of this snail and, in the laboratory, individuals sometimes lose infections (Robson & Williams, 1971). In any event, overall prevalence increases with snail age (Lauckner, 1980) and most infections persist throughout the host's life. In the trematode component communities harboured by

certain populations of these snails, gradual colonization of long-lived hosts by long-lived infections seems of importance in the accumulation of high prevalence and a variety of multiple infections. If colonization is disrupted naturally, or by human impact (Lafferty & Kuris, 1999), high prevalence will not develop.

If the longevity of hosts and infections is substantial, even a low colonization rate will lead to high prevalence, and a low rate seems to be the case in all three systems. This is most clear in *I. obsoleta* where direct assessments of field colonization rates exist (Curtis, 1996; Curtis & Tanner, 1999). If prevalence of single and multiple infections were to be high in cases of shorter-lived snails and infections, it would likely stem from a higher colonization rate.

With regard to interspecific adaptive relationships in these three systems, Williams' (1966) admonition is relevant: 'adaptation is a special and onerous concept that should be used only where it is really necessary.' If there are evolutionary (adaptive) relationships between species in snail–trematode systems, it is important to recognize them because these relationships and their consequences would be pervasive and ecologically important. For adaptions to develop, all conditions must be in place for natural selection to work. The view presented here is that, in the global snail or trematode populations considered, selection pressures emanating from other species can be rarely encountered and certain adaptive relations are unlikely to evolve.

In all three systems, trematodes must be adapted to use their snail hosts, as the snail environment is a necessary part of each life cycle. However, the possibility of the host being adapted to the presence of trematodes is less certain. The *C. californica* system is interesting in this regard. Because of crawl-away juveniles, gene flow is limited and local snail populations could possibly become differentially adapted to trematodes. Lafferty (1993*a*) may have identified such a situation. This snail matures at smaller sizes where trematodes are highly prevalent. In a reciprocal transplant experiment, he collected 550 immature snails from a site with 52% prevalence and 660 immature snails from a site with 0% prevalence. Snails were marked and some were released into their native environments as controls, others were transplanted either from the high to low, or low to high, prevalence sites. In essence, snails from the high prevalence site retained their smaller maturation size in both their own environment and the low prevalence environment; larger maturation size was retained by low prevalence site snails in both situations. Are these different adaptive responses of local host populations to parasite prevalence, or are all populations adapted similarly and this is a local environmentally induced response, perhaps parasite-mediated (e.g. miracidial presence signals snails to mature earlier)? The question could not be definitively resolved, but a genetic difference between populations is suggested. If local populations of this snail can evolve adaptive responses to trematodes, this would be unique among the three systems considered because both *I. obsoleta* and *L. littorea* have planktonic larvae and gene flow among populations is extensive (Gooch *et al.* 1972; Berger, 1977). If a host has a widely dispersed planktonic larva, it would be difficult or impossible to develop adaptations to trematodes because of spatial heterogeneity in trematode occurrence (Curtis & Hurd, 1983; Curtis & Hubbard, 1993; Curtis, 1997; Table 2). A veliger larva bearing a mutation that would help it resist a future trematode infection would not necessarily encounter the selective pressure at the site where it metamorphosed into a benthic juvenile. Whatever fitness benefit there might have been would seldom be gained. Until we can follow host genotype dispersal by planktonic larvae [see DiBacco & Levin (2000) for an advance in this area] and couple it with spatial heterogeneity in trematode prevalence, this argument will depend on what we do not know about planktonic larval dispersal rather than what we do.

The potential for trematode species to be adapted to one another seems remote in all three systems. Multiple infections are generally scarce and patchy in spatial distribution. A trematode genetic strain with a heritable advantage, in dealing with another trematode species on co-occurrence in a snail host, would have it eggs dispersed by a vertebrate definitive host, which may well disperse the eggs to sites where the other trematode, the supposed selective pressure, is rare or absent. On this reasoning, trematode species co-occurring in the same snail is an infrequent ecological accident. How they get along would be based on the one adaptational constant in these systems, how each species is adapted to utilize the host. So, for example, Esch *et al.* (2001) noted that rediae have a mouth and are adapted to 'chew' on host tissues. If rediae eat the stages of another trematode, which they sometimes do (e.g. Sousa, 1993), it results from the way rediae are adapted to utilize the host, not an adaptation to counter the presence of another trematode. In Williams' (1966) terms, chewing on host tissue would be a 'function' (directly selected), chewing on other parasites would be an 'effect' (not directly selected).

To support or refute the existence of adaptive relationships among species in snail-trematode systems, would require specific data on some very large-scale systems. For starters, answers to these questions would be needed: (1) how patchily are snail populations with intense trematode presence distributed? (2) what is the probability that planktonic larvae will be dispersed to these patches? and (3) what is the probability that definitive (and possibly other) hosts will disperse trematode propagules to

these patches? The answer to the first seems to be that high prevalence sites are quite localized. The second and third are not answerable at present, but there must be substantial probability that veliger larvae and trematode eggs will be scattered among various host populations, many of which lack high trematode prevalence. Natural selection leading to interspecific adaptations under these circumstances is reasonably questioned.

Another factor diluting chances of larval trematodes being adapted to each other is that some trematodes can use alternate snail species. In *I. obsoleta*, taxonomic uncertainties could be involved, but *H. quissetensis* was reported in *Nassarius vibex* by Holliman (1961) and Rohde (1977) suggested that *Austrobilharzia terregalensis* is probably synonymous with *A. variglandis* and has a worldwide distribution in a variety of snail species. In *C. californica*, only *Parorchis acanthus* also occurs in other snail species (n = 2) (Ching, 1991). For trematodes in these two systems, then, there is limited latitude as to host snail selection. However, *L. littorea* co-occurs with several congeners and trematodes may infect any of several hosts. In North America, Stunkard (1983) lists the trematodes of 3 littorinid species, *L. littorea*, *L. obtusata*, and *L. saxatilis*. *Cryptocotyle lingua* occurs in all 3, as does *Cercaria parvicaudata*. In Europe, there are more parasite species and a greater diversity of littorinids. James (1971), working mainly on the coast of Wales in England, lists *L. littorea*, *L. littoralis*, *L. neritoides* and *L. saxatilis* (with 4 subspecies). In *L. littorea* 11 trematodes are listed, 6 as occurring exclusively in *L. littorea*. However, 1 of the 6 is *C. lingua*, which has been observed in other littorinids (Lauckner, 1980; Stunkard, 1983). Thus, 6 of 11 species may be listed as infecting multiple littorinid hosts. If trematodes can use alternate snail hosts, the host–parasite system becomes increasingly diffuse and likelihood of establishing adaptations between trematodes in any 1 snail host is even less.

CODA

Questions of how frequently multiple infections occur in marine snails, and how they form and persist, have been addressed recently by several authors (Lauckner, 1980; Sousa, 1990, 1993, 1994; Kuris, 1990; Kuris & Lafferty, 1994; Lafferty, Sammond & Kuris, 1994; Curtis, 1995, 1997; Esch *et al.* 2001). Often these questions have been addressed from a statistical point of view, using component community patterns to infer infracommunity dynamics. Relatively frequent multiple trematode infections have been observed in certain populations of all three snails considered here. When species co-occur in individual snails, consistent patterns of association may emerge, and these dictate to some extent the variety of multiple infections observed in component communities. Examples are:

in *I. obsoleta*, *H. quissetensis* and *L. setiferoides* seldom occur in the same snail, but these species variously co-occur with *Z. rubellus*, *G. adunca* and *A. variglandis* (Vernberg *et al.* 1969; Curtis, 1997); in *L. littorea*, *Himasthla elongata* cohabits with *R. roscovita*, but apparently never with the common *C. lingua* (Lauckner, 1980); and in *C. californica*, *Parorchis acanthus* and *Himasthla rhigedana* do not often coexist with other species, except the schistosome *Austrobilharzia* sp. (Kuris, 1990; Sousa, 1993). These biological relationships clearly exist in infracommunities, but they are infrequent in the global populations of these marine snails. The contention here is that these relationships result from trematode adaptations to the host snail, not adaptation of one parasite to another. For parasites to be adapted to one another would require interactive infracommunities, which require frequent additions (colonizations) and eliminations of species (Holmes, 1987). In these marine snails, most clearly in *I. obsoleta*, these appear to be infrequent events and infracommunities would therefore be isolationist rather than interactive.

Valuable clues about infracommunity dynamics may be gained by using co-occurrence patterns in component communities. However, single and multiple-species infections are often embedded in host snail populations in unappreciated ways, leading to difficulties in obtaining representative samples. This is a major concern. If samples are not representative, then the inferences drawn from them may mislead. Statistical concerns aside, the connection between infracommunity dynamics and component community structure is tenuous, if not absent (Esch *et al.* 2001). Esch & Fernandez (1994) note factors internal and external to the host snail that are involved in determining the structure of trematode component communities. These three marine systems, suggest that external factors involving history and the ecological characteristics of parasites and hosts are the key. Far more important than infracommunity dynamics, are (1) frequency of definitive host visitation, (2) the availability of several trematode species, even infecting at a low rate, (3) an abundant, long-lived host snail, and (4) long-lived infections.

ACKNOWLEDGEMENTS

Thanks to G. W. Esch and also K. Hubbard for reading and commenting on previous drafts. Useful discussion was provided by H. V. Cornell. I am solely responsible for the content.

REFERENCES

BARBER, K. E. & CAIRA, J. N. (1995). Investigation of the life cycle and adult morphology of the avian blood fluke *Austrobilharzia variglandis* (Trematoda: Schistosomatidae) from Connecticut. *Journal of Parasitology* **81**, 584–592.

BATCHELDER, C. H. (1915). Migration of *Ilyanassa obsoleta*, *Littorina littorea*, and *Littorina rudis*. *The Nautilus* **29**, 43–46.

BERGER, E. (1977). Gene-enzyme variation in three sympatric species of *Littorina*. II. The Roscoff population, with a note on the origin of North American *L. littorea*. *Biological Bulletin* **153**, 255–264.

BRENCHLEY, G. A. (1987). Herbivory in juvenile *Ilyanassa obsoleta*. *The Veliger* **30**, 167–172.

BRENCHLEY, G. A. & CARLTON, J. T. (1983). Competitive displacement of native mudsnails by introduced periwinkles in the New England intertidal zone. *Biological Bulletin* **165**, 543–558.

BYERS, J. E. & GOLDWASSER, L. (2001). Exposing the mechanism and timing impact of nonindigenous species on native species. *Ecology* **82**, 1330–1343.

CARLTON, J. T. (1969). *Littorina littorea* in California (San Francisco and Trinadad Bays). *The Veliger* **11**, 283–284.

CHENG, T. C., SULLIVAN, J. T. & HARRIS, K. R. (1973). Parasitic castration of the marine prosobranch gastropod *Nassarius obsoletus* by sporocysts of *Zoogonus rubellus* (Trematoda): histopathology. *Journal of Invertebrate Pathology* **21**, 183–190.

CHENG, T. C., SULLIVAN, J. T., HOWLAND, K. H., JONES, T. F. & MORAN, H. J. (1983). Studies on parasitic castration: soft tissue and shell weights of *Ilyanassa obsoleta* (Mollusca) parasitized by larval trematodes. *Journal of Invertebrate Pathology* **42**, 143–150.

CHING, H. L. (1991). Lists of larval worms from marine invertebrates of the Pacific Coast of North-America. *Journal of the Helminthological Society of Washington* **58**, 57–68.

CURTIS, L. A. (1985). The influence of sex and trematode parasites on carrion response of the estuarine snail *Ilyanassa obsoleta*. *Biological Bulletin* **169**, 377–390.

CURTIS, L. A. (1987). Vertical distribution of an estuarine snail altered by a parasite. *Science* **235**, 1509–1511.

CURTIS, L. A. (1990). Parasitism and the movement of intertidal gastropod individuals. *Biological Bulletin* **179**, 105–112.

CURTIS, L. A. (1993). Parasite transmission in the intertidal zone: vertical migrations, infective stages, and snail trails. *Journal of Experimental Marine Biology and Ecology* **173**, 197–209.

CURTIS, L. A. (1995). Growth, trematode parasitism, and longevity of a long-lived marine gastropod (*Ilyanassa obsoleta*). *Journal of the Marine Biological Association of the United Kingdom* **75**, 913–925.

CURTIS, L. A. (1996). The probability of a marine gastropod being infected by a trematode. *Journal of Parasitology* **82**, 830–833.

CURTIS, L. A. (1997). *Ilyanassa obsoleta* (Gastropoda) as a host for trematodes in Delaware estuaries. *Journal of Parasitology* **83**, 793–803.

CURTIS, L. A. & HUBBARD, K. M. (1990). Trematode infection in a gastropod host misrepresented by observing shed cercariae. *Journal of Experimental Marine Biology and Ecology* **143**, 131–137.

CURTIS, L. A. & HUBBARD, K. M. (1993). Species relationships in a marine gastropod-trematode ecological system. *Biological Bulletin* **184**, 25–35.

CURTIS, L. A. & HURD, L. E. (1979). On the broad nutritional requirements of the mud snail, *Ilyanassa (Nassarius) obsoleta* (Say) and its polytrophic role in the food web. *Journal of Experimental Marine Biology and Ecology* **41**, 289–297.

CURTIS, L. A. & HURD, L. E. (1981). Crystalline style cycling in *Ilyanassa obsoleta* (Say) (Mollusca: Neogastropoda): further studies. *The Veliger* **24**, 91–96.

CURTIS, L. A. & HURD, L. E. (1983). Age, sex and parasites: spatial heterogeneity in a sandflat population of *Ilyanassa obsoleta*. *Ecology* **64**, 819–828.

CURTIS, L. A., KINLEY, J. L. & TANNER, N. L. (2000). Longevity of oversized individuals: growth, parasitism, and history in an estuarine snail population. *Journal of the Marine Biological Association of the United Kingdom* **80**, 811–820.

CURTIS, L. A. & TANNER, N. L. (1999). Trematode accumulation by the estuarine gastropod *Ilyanassa obsoleta*. *Journal of Parasitology* **85**, 419–425.

DECOURSEY, P. J. & VERNBERG, W. B. (1974). Double infections of larval trematodes: competitive interactions. In *Symbiosis in the Sea* (ed. Vernberg, W. B.), pp. 93–109. Columbia, USA: University of South Carolina Press.

DIBACCO, C. & LEVIN, L. A. (2000). Development and application of elemental fingerprinting to track the dispersal of marine invertebrate larvae. *Limnology and Oceanography* **45**, 871–880.

ESCH, G. W., CURTIS, L. A. & BARGER, M. A. (2001). A perspective on the ecology of trematode communities in snails. *Parasitology* **123**, S57–S75.

ESCH, G. W. & FERNANDEZ, J. C. (1994). Snail trematode interactions and parasite community dynamics in aquatic systems: a review. *American Midland Naturalist* **131**, 209–237.

GAMBINO, J. J. (1959). The seasonal incidence of infection of the snail *Nassarius obsoletus* (Say) with larval trematodes. *Journal of Parasitology* **45**, 440, 456.

GOOCH, J. L., SMITH, B. S. & KNUPP, D. (1972). Regional surveys of gene frequencies in the mud snail *Nassarius obsoletus*. *Biological Bulletin* **142**, 36–48.

GRODHAUS, G. & KEH, B. (1958). The marine dermatitis-producing cercaria of *Austrobilharzia variglandis* in California (Trematoda: Schistosomatidae). *Journal of Parasitology* **44**, 633–638.

HOFF, C. C. (1941). A case of correlation between infection of snail hosts with *Cryptocotyle lingua* and the habits of gulls. *Journal of Parasitology* **27**, 539.

HOLLIMAN, R. B. (1961). Larval trematodes from Apalachee Bay area, Florida with a checklist of known marine cercariae arranged in a key to their superfamilies. *Tulane Studies in Zoology* **9**, 1–74.

HOLMES, J. C. (1987). The structure of helminth communities. *International Journal for Parasitology* **17**, 203–208.

HUGHES, R. N. & ANSWER, P. (1982). Growth, spawning and trematode infection of *Littorina littorea* from an exposed shore in North Wales. *Journal of Molluscan Studies* **48**, 321–330.

HUNTER, W. S. (1952). Contributions to the morphology and life-history of *Gynaecotyla adunca* (Linton, 1905) (Trematoda: Microphallidae). *Journal of Parasitology* **38**, 308–314.

HUNTER, W. S. (1961). A new monostome, *Pleurogonius malaclemys*, n. sp. (Trematoda: Pronocephalidae) from Beaufort, North Carolina. *Proceedings of the Helminthological Society of Washington* **28**, 111–114.

HUNTER, W. S. (1967). Notes on the life history of *Pleurogonius malaclemys* Hunter, 1961 (Trematoda: Pronocephalidae) from Beaufort, North Carolina, with a description of the cercaria. *Proceedings of the Helminthological Society of Washington* **34**, 33–40.

HURD, L. E. (1985). On the importance of carrion to reproduction in an omnivorous estuarine neogastropod, *Ilyanassa obsoleta* (Say). *Oecologia* **65**, 513–515.

HUXHAM, M., RAFFAELLI, D. & PIKE, A. (1993). The influence of *Cryptocotyle lingua* (Digenea, Platyhelminthes) infections on the survival and fecundity of *Littorina littorea* (Gastropoda, Prosobranchia): an ecological approach. *Journal of Experimental Marine Biology and Ecology* **168**, 223–238.

HYMAN, L. H. (1967). *The Invertebrates: Volume VI Mollusca I*. New York: McGraw-Hill Book Company.

IRWIN, S. W. B. (1983). Incidence of trematode parasites in two populations of *Littorina saxatilis* (Olivi) from the North Shore of Belfast Lough. *Irish Naturalist Journal* **21**, 26–29.

JAMES, B. L. (1968). The distribution and keys of species in the family Littorinidae and of their digenean parasites, in the region of Dale, Pembrokeshire. *Field Studies* **2**, 615–650.

JAMES, B. L. (1971). Host selection and ecology of marine digenean larvae. In *Fourth European Marine Biology Symposium* (ed. Crisp, D. J.), pp. 179–196. Cambridge: Cambridge University Press.

JENNER, C. E. (1956). The occurrence of a crystalline style in the marine snail, *Nassarius obsoletus*. *Biological Bulletin* **111**, 304.

JENNER, C. E. & JENNER, M. G. (1977). Experimental displacement of an *Ilyanassa obsoleta* aggregation. *Journal of the Elisha Mitchell Scientific Society* **93**, 103.

KASSCHAU, M. R. (1975). Changes in concentration of free amino acids in larval stages of the trematode *Himasthla quissetensis* and its intermediate host *Nassarius obsoletus*. *Comparative Biochemistry and Physiology* **51B**, 273–280.

KURIS, A. M. (1990). Guild structure of larval trematodes in molluscan hosts: prevalence, dominance and significance of competition. In *Parasite Communities: Patterns and Processes* (ed. Esch, G. W., Bush, A. O. & Aho, J.), pp. 69–100. London: Chapman and Hall.

KURIS, A. M. & LAFFERTY, K. D. (1994). Community structure: larval trematodes in snail hosts. *Annual Review of Ecology and Systematics* **25**, 189–217.

LAFFERTY, K. D. (1993 a). The marine snail, *Cerithidea californica*, matures at smaller sizes where parasitism is high. *Oikos* **68**, 3–11.

LAFFERTY, K. D. (1993 b). Effects of parasitic castration on growth, reproduction and population dynamics of the marine snail *Cerithidea californica*. *Marine Ecology-Progress Series* **96**, 229–237.

LAFFERTY, K. D. & KURIS, A. M. (1999). How environmental stress affects the impacts of parasites. *Limnology and Oceanography* **44**, 925–931.

LAFFERTY, K. D., SAMMOND, D. T. & KURIS, A. M. (1994). Analysis of larval trematode communities. *Ecology* **75**, 2275–2285.

LAMBERT, T. C. & FARLEY, J. (1968). The effect of parasitism by the trematode *Cryptocotyle lingua* (Creplin) on zonation and winter migration of the common periwinkle *Littorina littorea* (L.). *Canadian Journal of Zoology* **46**, 1139–1147.

LAUCKNER, G. (1980). Diseases of Mollusca: Gastropoda. In *Diseases of Marine Animals, Vol. I* (ed. Kinne, O.), pp. 311–424. New York: John Wiley & Sons.

LAUCKNER, G. (1984). Brackish-water submergence of the common periwinkle, *Littorina littorea*, and its digenean parasites in the Baltic Sea and in the Kattegat. *Helgolander Meeresuntersuchungen* **37**, 177–184.

LAUCKNER, G. (1987). Ecological effects of larval trematode infestation on littoral marine invertebrate populations. *International Journal for Parasitology* **17**, 391–398.

MARTIN, W. E. (1938). Studies on trematodes of Woods Hole: the life cycle of *Lepocreadium setiferoides* (Miller and Northup) Allocreadiidae, and the description of *Cercaria cumingiae* n. sp. *Biological Bulletin* **75**, 463–472.

MARTIN, W. E. (1939). Studies on the trematodes of Woods Hole. II. The life cycle of *Stephanostomum tenue* (Linton). *Biological Bulletin* **77**, 65–73.

MARTIN, W. E. (1945). Two new species of marine cercariae. *Transactions of the American Microscopical Society* **64**, 203–212.

MARTIN, W. E. (1955). Seasonal infections of the snail, *Cerithidea californica* Haldeman, with larval trematodes. In *Essays in the Natural Sciences in honor of Captain Allan Hancock, Allan Hancock Foundation*, pp. 203–210. Los Angeles: University of Southern Californica Press.

MATTHEWS, P. M., MONTGOMERY, W. I. & HANNA, R. E. B. (1985). Infestation of littorinids by larval Digenea around a small fishing port. *Parasitology* **90**, 277–287.

McCURDY, D. G., BOATES, J. S. & FORBES, M. R. (2000). Spatial distribution of the intertidal snail *Ilyanassa obsoleta* in relation to parasitism by two species of trematodes. *Canadian Journal of Zoology* **78**, 1137–1143.

McDANIEL, J. S. & COGGINS, J. R. (1972). Seasonal larval trematode infection dynamics in *Nassarius obsoletus* (Say). *Journal of the Elisha Mitchell Scientific Society* **88**, 55–57.

McDERMOTT, J. J. (1951). Larval trematode infection in *Nassa obsoleta* from New Jersey waters. M.S. Thesis, Rutgers University, New Brunswick, New Jersey, 72 pp.

MILLER, H. M. Jr. & NORTHUP, F. E. (1926). The seasonal infestation of *Nassa obsoleta* (Say) with larval trematodes. *Biological Bulletin* **50**, 490–508.

MOORE, H. B. (1966). *Marine Ecology*. New York: John Wiley & Sons.

MOURITSEN, K. M., GORBUSHIN, A. M. & JENSEN, K. T. (1999). Influence of trematode infections on *in situ* growth rates of *Littorina littorea*. *Journal of the Marine Biological Association of the United Kingdom* **79**, 425–430.

PEARSON, E. J. & CHENG, T. C. (1985). Studies on parasitic castration: occurrence of a gametogenesis inhibiting factor in extract of *Zoogonus lasius* (Trematoda). *Journal of Invertebrate Pathology* **46**, 239–246.

POHLEY, W. J. (1976). Relationships among three species of *Littorina* and their larval Digenea. *Marine Biology* **37**, 179–186.

POULIN, R. (1999). The functional importance of parasites in animal communities: many roles at many levels. *International Journal for Parasitology* **29**, 903–914.

RACE, M. S. (1981). Field ecology and natural history of *Cerithidea californica* (Gastropoda: Prosobranchia) in San Francisco Bay. *The Veliger* **24**, 18–27.

RANKIN, J. S. Jr. (1940). Studies on the trematode family Microphallidae Travassos, 1921. IV. The life cycle and ecology of *Gynaecotyla nassicola* (Cable and Hunninen, 1938) Yamaguti, 1939. *Biological Bulletin* **79**, 439–451.

REES, W. J. (1936). The effect of parasitism by larval trematodes on the tissues of *Littorina littorea* (Linne'). *Proceedings of the Zoological Society of London* **1936**, 357–368.

RIEL, A. (1975). Effect of trematodes on survival of *Nassarius obsoletus* (Say). *Proceedings of the Malacological Society of London* **41**, 527–528.

ROBSON, E. M. & WILLIAMS, I. C. (1970). Relationships of some species of Digenea with the marine prosobranch *Littorina littorea* (L.). I. The occurrence of larval Digenea in *L. littorea* on the north Yorkshire coast. *Journal of Helminthology* **44**, 153–168.

ROBSON, E. M. & WILLIAMS, I. C. (1971). Relationships of some species of Digenea with the marine prosobranch *Littorina littorea* (L.). II. The effect of larval digenea on the reproductive biology of *L. littorea*. *Journal of Helminthology* **45**, 145–159.

ROHDE, K. (1977). The bird schistosome *Austrobilharzia terrigalensis* from the Great Barrier Reef, Australia. *Zeitschrift für Parasitenkunde* **52**, 39–51.

ROTHSCHILD, M. (1942). A seven year old infection of *Cryptocotyle lingua* (Creplin) in the winkle *Littorina littorea* (L.). *Journal of Parasitology* **28**, 350.

SAIER, B. (2000). Age-dependent zonation of the periwinkle *Littorina littorea* (L.) in the Wadden Sea. *Helgoland Marine Research* **54**, 224–229.

SCHELTEMA, R. S. (1964). Feeding and growth in the mud-snail *Nassarius obsoletus*. *Chesapeake Science* **5**, 161–166.

SINDERMANN, C. J. (1960). Ecological studies of marine dermatitis-producing schistosome larvae in northern New England. *Ecology* **41**, 678–684.

SINDERMANN, C. J. & FARRIN, A. E. (1962). Ecological studies of *Cryptocotyle lingua* (Trematoda: Heterophyidae) whose larvae cause "pigment spots" of marine fish. *Ecology* **43**, 69–75.

SINDERMANN, C. J., ROSENFIELD, A. & STROM, L. (1957). The ecology of marine-dermatitis-producing schistosomes. II. Effects of certain environmental factors on emergence of cercariae of *Austrobilharzia variglandis*. *Journal of Parasitology* **43**, 382.

SMITH, N. F. (2001). Spatial heterogeneity in recruitment of larval trematodes to snail intermediate hosts. *Oecologia* **127**, 115–122.

SOUSA, W. P. (1983). Host life history and the effect of parasitic castration on growth: a field study of *Cerithidea californica* Haldeman (Gastropoda: Prosobranchia) and its trematode parasites. *Journal of Experimental Marine Biology and Ecology* **73**, 273–296.

SOUSA, W. P. (1990). Spatial scale and the processes structuring a guild of larval trematodes. In *Parasite Communities: patterns and processes* (ed. Esch, G. W., Bush, A. O. & Aho, J.), pp. 41–67. London: Chapman and Hall.

SOUSA, W. P. (1991). Can models of soft-sediment community structure be complete without parasites? *American Zoologist* **31**, 821–830.

SOUSA, W. P. (1993). Interspecific antagonism and species coexistence in a diverse guild of larval trematode parasites. *Ecological Monographs* **63**, 103–128.

SOUSA, W. P. (1994). Patterns and processes in communities of helminth parasites. *Trends in Ecology & Evolution* **9**, 52–57.

STAMBAUGH, J. E. & MCDERMOTT, J. J. (1969). The effects of trematode larvae on the locomotion of naturally infected *Nassarius obsoletus* (Gastropoda). *Proceedings of the Pennsylvania Academy of Sciences* **43**, 226–231.

STUNKARD, H. W. (1938 *a*). The morphology and life cycle of the trematode *Himasthla quissetensis* (Miller and Northup, 1926). *Biological Bulletin* **75**, 146–164.

STUNKARD, H. W. (1938 *b*). *Distomium lasium* Leidy, 1891 (Syn. *Cercariaeum lintoni* Miller and Northup, 1926), the larval stage of *Zoogonus rubellus* (Olson, 1898). *Biological Bulletin* **75**, 308–334.

STUNKARD, H. W. (1961). *Cercaria dipterocerca* Miller and Northup, 1926 and *Stephanostomum dentatum* (Linton, 1900) Manter, 1931. *Biological Bulletin* **120**, 221–237.

STUNKARD, H. W. (1972). Observations on the morphology and life-history of the digenetic trematode, *Lepocreadium setiferoides* (Miller and Northup, 1926) Martin 1938. *Biological Bulletin* **142**, 326–334.

STUNKARD, H. W. (1973). Studies on larvae of strigeoid trematodes from the Woods hole, Massachusetts region. *Biological Bulletin* **144**, 525–540.

STUNKARD, H. W. (1983). The marine cercariae of the Woods Hole, Massachusetts region, a review and a revision. *Biological Bulletin* **164**, 143–162.

STUNKARD, H. W. & HINCHLIFFE, M. C. (1952). The morphology and life-history of *Microbilharzia variglandis* (Miller and Northup, 1926) Stunkard and Hinchliffe, 1951, avian blood-flukes whose larvae cause "swimmer's itch" of ocean beaches. *Journal of Parasitology* **38**, 248–265.

THOMAS, F., CRIVELLI, A., CEZILLY, F., RENAUD, F. & DE MEEUS, T. (1997). Parasitism and ecology of wetlands: a review. *Estuaries* **20**, 646–654.

VERNBERG, W. B. (1969). Adaptations of host and symbionts in the intertidal zone. *American Zoologist* **9**, 357–365.

VERNBERG, W. B. & VERNBERG, F. J. (1974). Metabolic pattern of a trematode and its host: a study in the evolution of physiological responses. In *Symbiosis in the Sea* (ed. Vernberg, W. B.), pp. 161–172. Columbia, USA: University of South Carolina Press.

VERNBERG, W. B., VERNBERG, F. J. & BECKERDITE, F. W. Jr. (1969). Larval trematodes: double infections in common mud-flat snail. *Science* **164**, 1287–1288.

WATTS, S. D. M. (1971). Effects of larval Digenea on the free amino acid pool of *Littorina littorea* (L.). *Parasitology* **62**, 361–366.

WERDING, B. (1969). Morphologie, entwichlung und oikologie digener trematoden-larven der strandschnecke *Littorina littorea*. *Marine Biology* **3**, 306–333.

WETZEL, R. L. (1977). Carbon resources of a benthic salt marsh invertebrate *Nassarius obsoletus* Say (Mollusca: Nassariidae). In *Estuarine Processes, Vol. II* (ed. Wiley, M. L.), pp. 293–308. New York: Academic Press.

WILLIAMS, G. C. (1966). *Adaptation and Natural Selection. A Critique of Some Current Evolutionary Thought*. Princeton, New Jersey: Princeton University Press.

WILLIAMS, I. C. & ELLIS, C. (1975). Movements of the common periwinkle, *Littorina littorea* (L.), on the Yorkshire coast in winter and the influence of infection with larval Digenea. *Journal of Experimental Marine Biology and Ecology* **17**, 47–58.

YONGE, C. M. (1930). The crystalline style of the Mollusca and a carnivorous habit cannot normally coexist. *Nature* **125**, 444–445.

YOSHINO, T. P. (1975). A seasonal and histologic study of larval Digenea infecting *Cerithidea californica* from Goleta Slough, Santa Barbara County, California. *The Veliger* **18**, 156–161.

Order in ectoparasite communities of marine fish is explained by epidemiological processes

S. MORAND[1]*, K. ROHDE[2] *and* C. HAYWARD[2,3]

[1] *Laboratoire de Biologie Animale (UMR 5555 CNRS), Centre de Biologie et d'Ecologie Tropicale et Méditerranéenne, Université de Perpignan, 66860 Perpignan Cedex, France*
[2] *School of Biological Sciences, Division of Zoology, University of New England, Armidale NSW 2351, Australia*
[3] *Laboratory of Aquatic Animal Diseases, College of Veterinary Medicine, Chungbuk National University, Chongju, Chungbuk-do 361-763, Korea*

SUMMARY

Two kinds of community structure referred to, nestedness and bimodal distribution, have been observed or were searched for in parasite communities. We investigate here the relation between these two kinds of organisation, using marine fishes as a model, in order to show that parasite population dynamics may parsimoniously explain the patterns of ectoparasite species distribution and abundance. Thirty six assemblages of metazoan ectoparasites on the gills and heads of marine fish showed the following patterns: a positive relationship between abundance and the variance of abundance; a positive relationship between abundance and prevalence of infection; a bimodal pattern of the frequency distribution of prevalence of infection; nestedness as indicated by Atmar and Patterson's thermodynamic measure (a mean of 7·9°C); a unimodal distribution of prevalence in parasite assemblages with a temperature lower than the mean, and a bimodal distribution in assemblages with a temperature higher than the mean. We conclude that patterns are the result of characteristics of the parasite species themselves and that interspecific competition is not necessary to explain them. We emphasize that a holistic approach, taking all evidence jointly into account, is necessary to explain patterns of community structure. Ectoparasite assemblages of marine fish are among the animal groups that have been most thoroughly examined using many different methods, and all evidence supports the view that these animals live under non-equilibrium conditions, in largely non-saturated niche space in which interspecific competition occurs but is of little evolutionary importance.

Key words: Ectoparasites, fish parasites, community ecology, nested patterns, epidemiology, bimodal distribution, species coexistence, competition.

INTRODUCTION

The search of order is intricately linked with the development of ecological theories, and investigating patterns (of distribution and abundance of species, for example) is considered as one way to highlight the existence of order. Processes are inferred from observed patterns (May, 1976; Brown, 1995). Within this context, insular biogeography was and still is the paradigm in evolutionary and ecology research (at least in a macro-perspective approach). Patterns of species richness and distribution among islands, and adaptive radiation and character displacement through competition between species on islands have been investigated by many authors (e.g. Grant, 1968, 1975; Brown, 1995).

The concepts developed in insular biogeography were directly applied to parasites based on the simple analogy that hosts are islands for parasites (Dritschilo *et al.* 1975; Kuris, Blaustein & Alió 1980). Thus, arguments of insular biogeography were invoked to

predict that parasite species richness should be positively linked with host body size (Kuris *et al.* 1980; Guégan *et al.* 1992; Morand, 2000, and further references therein). However, the dynamics of species diversity, i.e. extinction and colonisation processes, which are the core of the theory of insular biogeography, were not taken into account. Moreover, geographic distance, which is an important feature of the theory, was also ignored, i.e. no analogy was proposed to adapt some of the hypotheses of insular biogeography to the case of host-parasite systems.

The analysis and interpretation of community structure were influenced by the hypothesis of competitive exclusion, according to which species may coexist only if they differ in certain characters, e.g. body size or trophic structures (Hutchinson, 1959; MacArthur, 1972; Simberloff & Boecklen, 1981). More specifically, the coexistence of insular free-living organisms seems to be favoured by divergence in morphology, especially the morphology of specialised organs such as the feeding apparatus (Grant & Schluter, 1984). Accordingly, it was proposed that parasite communities are also structured by competition (Holmes, 1990). However, although restriction of niche size by competing species is observed in the case of some endoparasites

* Corresponding Author: Serge Morand, Laboratoire de Biologie Animale (UMR 5555 CNRS). Centre de Biologie et d'Ecologie Tropicale et Méditerranéenne. Université de Perpignan, 66860 Perpignan Cedex France
E-mail: morand@univ-perp.fr

and mostly in species-rich communities, this is not the case for most ectoparasites, and there is no evidence for parasite species exclusion (Rohde, 1977, 1979, 1991).

Free-living coexisting species usually do not live in random assemblages. Nested subset structures in ecological communities are commonly found, that is, communities with successively lower species richness tend to be subsets of richer assemblages (Patterson & Atmar, 1986). Nested subset structure is often interpreted as the effect of highly predictable colonisation/extinction processes that determine the composition of communities (Brown, 1995). Hence, the search for nested patterns in parasite communities was the subject of several studies, and emphasis was placed on either the role of hosts and/or competition (Guégan & Hugueny, 1994; Hugueny & Guégan, 1997; Poulin & Guégan, 2000). The intrinsic character of parasites, i.e. their dynamics, was rarely taken into account when explaining nestedness patterns (but see Worthen & Rohde 1996; Rohde *et al.* 1998).

A different kind of community order investigated in free-living organisms is core-satellite organisation (Hanski, 1982). The core-satellite hypothesis is included in metapopulation theory (Hanksi & Gyllenberg 1993). The core-satellite hypothesis predicts a bimodal distribution of organisms in their environment, that is, the majority of species is present in most patches, or in only a small fraction of patches. The core-satellite hypothesis does not invoke competition but a rescue effect, i.e. the ability to re-colonize empty patches after extinction. It is based on the character of species, i.e. their population dynamics, and does not need an explanation based on competition and/or special attributes of the environment, nor the patch or the host.

The two kinds of community structure referred to, nestedness and bimodal distribution, have been observed or were searched for in parasite communities. However, nobody has investigated the relation between these two kinds of organisation. Here, using marine fishes as a model, we show that epidemiology, i.e. parasite population dynamics, parsimoniously explains the patterns of parasite species distribution and abundance.

METHODS

Core-satellite hypothesis and the distribution and abundance of parasites

Morand & Guégan (2000) have investigated the patterns of abundance and distribution of mammalian nematodes. They found a bimodal distribution of worm prevalence and a positive relationship between abundance and prevalence. They argued that these patterns are not the result of host specialisation but simply the results of demographic and stochastic processes.

Here we propose to reinvestigate this pattern for the case of fish ectoparasites. We use the data of Rohde, Hayward & Heap (1995 and unpublished records of Hayward) on 36 communities of gill and head ectoparasites of marine fish (only fish species with at least 5 ectoparasite species were included in the analyses, a further 97 fish species had fewer ectoparasite species).

Epidemiological modelling shows that the mean worm burden $M(t)$ is linked to the prevalence of infection $P(t)$ at time t according to:

$$P(t) = 1 - [1 + M(t)/k]^{-k}$$

where k is the parameter of the negative binomial distribution.

Perry & Taylor (1986) emphasised that k is linked to the mean worm burden M:

$$1/k = a\, M^{b-2} - 1/M$$

with a and b the two parameters of

$$s^2 = a \times M^b$$

where a represents a constant parameter, b an index of spatial heterogeneity, M the mean abundance and s^2 its variance (Taylor, Woiwod & Perry 1978, Taylor *et al.* 1983)

Measure of order

Several problems may arise when testing for nested patterns. For example, the lack of a significant nested pattern does not necessarily imply a random organisation (Poulin & Guégan, 2000). Atmar & Patterson (1993) have proposed to use a direct measure of order, which is based on entropy. As emphasised by Atmar & Patterson (1993): "statistical stochasticity is a concept closely related to heat, information, noise, order and disorder". They proposed the use of a metric that measures the *heat of disorder* inherent in the historical biogeography of an archipelago (i.e. historical organisation of communities). We use this method because it provides a simple thermodynamic measure of the order and disorder to describe nested patterns. A matrix temperature of perfect order assumes the attributes of a frozen liquid, where complete order exists only at 0 °C. As the temperature rises, turbulence is imposed on the system, at 100 °C no discernible extinction order remains, the presence-absence matrix has assumed the attributes of a free gas. The temperature of a matrix is inherent in the manner in which species are distributed throughout the matrix. Changes in temperature between 0 °C and 100 °C are assumed to be continuous.

RESULTS

Abundance–variance relationship

Treating each ectoparasite population as an independent observation, we found a positive re-

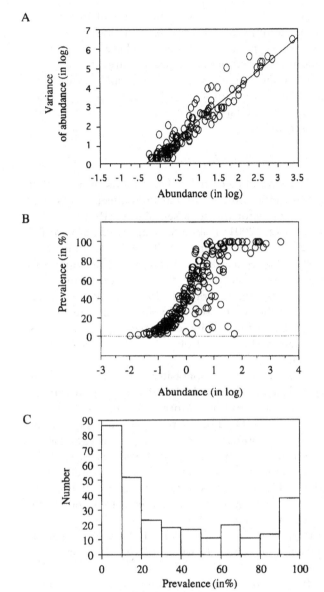

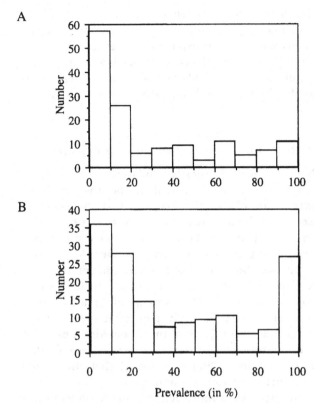

Fig. 1. (A) Relationship between the logarithms of the variance (s^2) and abundance (M), across populations of fish ectoparasites, fitted to a power-function with the intercept $a = 0.57 \pm 0.04$ and the slope, $b = 1.71 \pm 0.04$ ($r^2 = 0.92$; $P < 0.0001$; $n = 171$ populations). (B) Relationship between abundance (average parasite burden) and prevalence of fish ectoparasites (171 populations). (C) Bimodal distribution of prevalence of fish ectoparasites.

lationship between the mean abundance (in log) and the variance of abundance (in log) (Fig. 1A), with estimates of parameters $b = (1.71 \pm 0.04)$ and $a = (0.57 \pm 0.04)$. The values of these estimates are within the ranges typically observed in various assemblages of parasites (Morand & Guégan, 2000).

Bimodal distribution

A positive relationship between abundance and prevalence of ectoparasites was observed (Fig. 1B). The frequency distribution of ectoparasite preva-lence showed a bimodal pattern (Fig. 1C).

Fig. 2. (A) Unimodal distribution of prevalence of ectoparasite populations in assemblages characterized by a low temperature matrix (see Methods). (B) Bimodal distribution of prevalence of ectoparasite populations in assemblages characterized by a high temperature matrix (see Methods).

Temperature of ectoparasite assemblages

The temperatures of ectoparasite assemblages ranged from 4·9 °C to 43 °C with a mean value of 7·9 °C ± 0·5. The assemblages with a temperature lower than the mean value of 7·9 showed a unimodal distribution of prevalence (Fig. 2A), whereas the assemblages with a temperature higher than the mean value showed a bimodal distribution of prevalence (Fig. 2B). This suggests that increase of temperature, i.e. increase in disorder, leads to a core-satellite distribution of parasites within the assem-blage.

DISCUSSION

Epidemiological processes appear to be the most parsimonious explanation for the diversity, abun-dance and distribution of ectoparasite species infect-ing fish.

Distribution and abundance of ectoparasites on marine fish

As predicted by the core-satellite model, rare (satellite) ectoparasite species were observed to be more frequent in the environment (host) than locally abundant (core) species. According to Morand & Guégan (2000), a positive relationship between abundance and prevalence is purely the result of epidemiological processes. Demographic explanations may therefore explain the observed patterns of bimodality of prevalence when making Monte-Carlo simulations using epidemiological modelling frameworks (Anderson & May, 1985; Morand & Guégan, 2000).

Nestedness is the result of epidemiological processes

Nested patterns, mostly investigated in biogeographical studies, are said to be the result of extinction/colonisation events on archipelagos or isolated habitats (Atmar & Patterson, 1993). The observation of nested patterns in ectoparasite assemblages of tropical fish was explained by considering host body size as the determinant (Guégan & Hugueny, 1994), leading to the conclusion that hosts are a major determinant of parasite infracommunity structure (Guégan & Hugueny, 1994). But it was also suggested that competition between species may lead to non-randomness (see Poulin, 1996). Poulin (1997) emphasised that the nested structure of parasite communities may be the result of interspecific competition, but found no empirical support. In contrast, Rohde et al. (1998) emphasised that nestedness structure of parasite communities is not a proof for the existence of interspecific competition, since it may also (and more likely) result from different colonisation sequences, as suggested by the observation that nestedness in populations of a marine fish species was found only when juvenile and adult fish were tested jointly, but not when they were tested separately (Kleeman, 1996). Host specificity determines if a parasite is able to colonize a host and then may affect the nestedness structure parasite communities (Matejusova, Morand & Gelnar, 2000). However, there is not sufficient information on host specificity concerning marine ectoparasites to investigate this effect.

Nested patterns were observed in some studies (Guégan & Hugueny, 1994; Hugueny & Guégan, 1997), but in the majority of cases nested patterns were rarely observed and ectoparasite communities of fish seem to form random, unstructured assemblages (Rohde et al. 1994, 1995, 1998; Poulin, 1996; Worthen & Rohde, 1996).

Rather than using a dichotomic classification of nestedness (nested versus non-nested) as done in previous studies, we preferred here to use a measure of order following Atmar & Patterson (1993) that permits a continuous gradation from 0-nestedness to 100 % nestedness. We found that each assemblage of ectoparasite species is characterized by a temperature cooler than expected by chance. We also found that assemblages characterized by a low temperature form unimodal distributions and that assemblages characterized by a hot temperature form bimodal distributions (i.e. core satellite pattern) (Fig. 3). We

suggest that nestedness is the result of differential colonisation/extinction processes acting at the level of each parasite species. These differential colonisation/extinction processes are attributes of species and related to birth and death processes in population dynamics, i.e. they are not the consequence of interspecific competition, as further discussed below.

Competition is not important

A large number of field observations indicate that interspecific competition does not affect the "structure" of ectoparasite assemblages (Rohde, 1979, 1989, 1991, 2001; Rohde & Heap, 1998; Simkova et al. 2001; Lo & Morand, 2000, 2001). Morand et al. (1999) showed a lack of saturation of fish ectoparasite communities suggesting that infracommunities of parasites are not saturated by local parasite residents. This implies that resources provided by the fish are far from being totally exploited by parasites, as previously shown by Rohde (e.g. 1979, 1989, 1998, 2001).

Coexistence of ectoparasites is favoured by intraspecific aggregation, which is the common feature of parasite distribution (Morand et al. 1999; Lo & Morand, 2001). The increase of intraspecific aggregation compared with interspecific aggregation when total parasite species richness increases facilitates ectoparasite species coexistence (Morand et al. 1999). However, this does not suggest that interspecific aggregation is a response to interspecific competition, for which there is no evidence. Rather, interspecific aggregation is likely to be the result of behavioural characteristics of each ectoparasite species.

In conclusion, epidemiology is the most parsimonious explanation of the order that may be observed in ectoparasite assemblages. The observation of pattern (or the lack of pattern) is simply the result of demographic characteristics of each ectoparasite species in an assemblage, and competition does not need to be invoked. The assemblage pattern that can emerge may be the consequence of many different factors that affect the probability of a given host of being infected by a particular parasite.

A holistic approach to understanding parasite assemblages

We have shown above that ectoparasites of marine fish do not live in entirely random assemblages. Core and satellite species can be distinguished and some degree of nestedness occurs. However, these patterns can be explained by epidemiological processes, i.e. by characteristics of the various parasite species. It is not necessary to invoke interspecific interactions to explain them. Nevertheless, we wish to emphasize that different methods and different approaches may lead to different conclusions. Therefore, a holistic

Nested pattern   Unimodal distribution

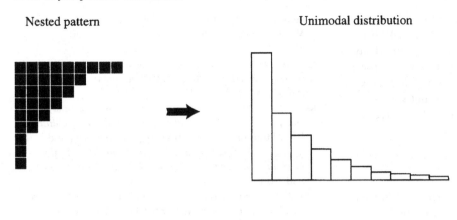

Non nested pattern   Bimodal distribution

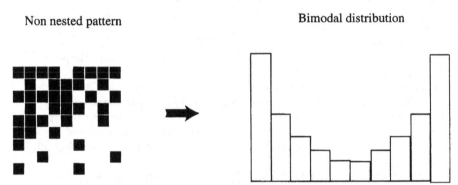

Fig. 3. Link between nested structure and prevalence distribution of parasites. (A) a nested structure, generally characterized by a low temperature, leads to a unimodal distribution of ectoparasite prevalence values. (B) a non nested structure, generally characterized by a high temperature, leads to a bimodal distribution of ectoparasite prevalence values.

approach is necessary, an approach that takes all evidence jointly into account, as suggested by Rohde (in press) when considering evidence for the mating hypothesis of niche restriction, that is, for the hypothesis that facilitation of mating is of great importance in restricting niches of parasites.

Assemblages of ectoparasites (and to a lesser degree of endoparasites) of marine fishes are now among the best known animal groups that have been studied with regard to the question of whether equilibrium or non-equilibrium conditions prevail. The evidence given by various authors for equilibrium conditions is complete or partial competitive exclusion or habitat shifts in the presence of other species, character displacement and particularly differences in the size of feeding organs of species using similar food resources (e.g. Krebs, 1997), as well as an asymptotic relationship between local and regional species richness (Cornell & Lawton, 1992). Many studies have shown that, for marine parasites, all evidence very strongly suggests that these animals live under non-equilibrium conditions. Evidence is as follows: (1) there is a high degree of non-saturation, i.e. many habitats are empty, as shown by

comparison of host species of similar size and from similar habitats with few and many parasite species (e.g. Rohde, 1979); (2) many species are little or not affected by the presence of other potentially competing species (e.g. Rohde, 1991 and further references therein); (3) differences in the size of feeding organs do also occur when resources are in unlimited supply, suggesting that such differences may be fortuitous (Rohde, 1991); (4) many examples of character displacement (of reproductive organs) can best be explained by reinforcement of reproductive barriers, and niche restriction may often be the result of selection to facilitate mating and not of competition (Rohde, in press); (5) an asymptotic relationship between local and regional diversity may be a consequence of differential likelihoods of species to appear in a community because of different colonization rates and life spans, and interactions between species are not necessary to explain the relationship (Rohde, 1998); (6) interspecific aggregation is reduced relative to intraspecific aggregation (Morand *et al.* 1999); (7) positive associations are much more common than negative ones (Rohde *et al.* 1994); (8) hyperparasites of various degrees are very rare, i.e. many habitats (hosts) are empty for future colonization (e.g. Rohde, 1989); (9) parasites do not conform to the packing rules derived from spatial scaling laws (fractional geometry) (Rohde, 2001); (10) nestedness, when it occurs, is not the result of interspecific competition but of characteristics of the various species themselves (this paper).

REFERENCES

ANDERSON, R. M. & MAY, R. M. (1985). Helminth infection of humans: mathematical models, population dynamics and control. *Advances in Parasitology* **24**, 1–101.

ATMAR, W. & PATTERSON, B. D. (1993). The measure of order and disorder in the distribution of species in fragmented habitat. *Oecologia* **96**, 373–382.

BROWN, J. H. (1995). *Macroecology*. Chicago, University Chicago Press,.

CORNELL, H. V. & LAWTON, J. H. (1992). Species interactions, local and regional processes, and limits in the richness of ecological communities: a theoretical perspective. *Journal of Animal Ecology* **61**, 1–12.

DOBSON, A. P. (1990). Models for multi-species-host-communities. In *Parasite Communities: Patterns and Processes* (ed. Esch, G. W., Bush A. O. & Aho, J. M.), pp. 260–288. London, U.K. Chapman & Hall Ltd.

DRITSCHILO, W., CORNELL, H., NAFUS, D. & O'CONNOR, B. (1975). Insular biogeography: of mice and mites. *Science* **190**, 467–469.

GRANT, P. R. (1968). Bill size, body size, and the ecological adaptations of bird species to competitive situations on islands. *Systematic Zoology* **14**, 319–333.

GRANT, P. R. (1975). The classical case of character displacement. *Evolutionary Biology* **8**, 237–337.

GRANT, P. & SCHLUTER, D. (1984). Interspecific competition inferred from patterns of guild structure. In *Ecological Communities: Conceptual Issues and the Evidence* (ed. Strong, D. R., Jr., Simberloff, D., Abele, L. G. & Thistle, A. B.), pp. 201–231. Princeton, New Jersey, Princeton University Press.

GUÉGAN, J. F. & HUGUENY, B. (1994). A nested parasite species subset pattern in tropical fish: host as major determinant of parasite infracommunity structure. *Oecologia* **100**, 184–189.

GUÉGAN, J. F., LAMBERT, A., LÉVEQUE, C., COMBES, C. & EUZET, L. (1992). Can host body size explain the parasite species richness in tropical freshwater fishes? *Oecologia* **90**, 197–204.

HANSKI, I. (1982). Dynamics of regional distribution: the core and satellite species hypothesis. *Oikos* **38**, 210–221.

HANSKI, I. & GYLLENBERG, M. (1993). Two general metapopulation models and the core-satellite species hypothesis. *The American Naturalist* **142**, 17–41.

HOLMES, J. C. (1990). Helminth communities in marine fishes. In *Parasite Communities: Patterns and Processes* (ed. Esch, G. W., Bush, A. O. & Aho, J. M.), pp. 101–130. London, New York, Chapman and Hall.

HUGUENY, B. & GUÉGAN, J. F. (1997). Community nestedness and the proper way to assess statistical significance by Monte-Carlo tests: some comments on Worthen and Rohde's (1996) paper. *Oikos* **80**, 572–574.

HUTCHINSON, G. E. (1959). Homage to Santa Rosalia, or why are there so many kinds of animals? *American Naturalist* **93**, 145–159.

KLEEMAN, S. (1996). *Community ecology of ecto- and endoparasites of a tropical fish species, Siganus doliatus*. BSc. Honours thesis, University of New England.

KREBS, C. J. (1997). *Ecology: The Experimental Analysis of Abundance and Distribution*. 4th ed. N. Y. Harper and Row

KURIS, A. M., BLAUSTEIN, A. R. & ALIÓ, J. J. (1980). Hosts as islands. *American Naturalist* **116**, 570–586.

LO, C. M. & MORAND, S. (2000). Spatial distribution and coexistence of monogenean gill parasites inhabiting two damselfishes from Moorea island (French Polynesia). *Journal of Helminthology* **74**, 329–336.

LO, C. M. & MORAND, S. (2001). Gill parasites of *Cephalopholis argus* (Teleostei: Serranidae) from Moorea island (French Polynesia): site selection and coexistence. *Folia Parasitologica* **48**, 30–36.

MACARTHUR, R. H. (1972). *Geographical Ecology: Patterns in the Distribution of Species*. New York, Harper & Row.

MATEJUSOVÁ, I., MORAND, S. & GELNAR, M. (2000). Nestedness in assemblages of gyrodactylids (Monogenea: Gyrodactylidae) parasitising two species of cyprinid – with reference to generalists and specialists. *International Journal for Parasitology* **30**, 1153–1158.

MAY, R. M. (1976). Patterns in multi-species communities. In *Theoretical Ecology. Principles and Applications* (ed. May, R. M.), pp. 142–162. Oxford, U.K, Blackwell.

MORAND, S. (2000). Wormy world: comparative tests of theoretical hypotheses on parasite species richness. In *Evolutionary Biology of Host-parasite Relationships: Theory Meets Reality*, pp. 63–79 (ed. Poulin, R., Morand, S. & Skorping, A.), pp. 63–79. Amsterdam, Elsevier.

MORAND, S. & GUÉGAN, J. F. (2000). Abundance and distribution of parasitic nematodes: ecological specialisation, phylogenetic constraints or simply epidemiology? *Oikos* **55**, 563–573.

MORAND, S., POULIN, R., ROHDE, K. & HAYWARD, C. (1999). Aggregation and species coexistence of ectoparasites of marine fishes. *International Journal for Parasitology* **29**, 663–672.

PATTERSON, B. D. & ATMAR, W. (1986). Nested subsets and the structure of insular mammalian faunas and archipelagos. *Biological Journal of the Linnean Society* **28**, 65–82.

PERRY, J. N. & TAYLOR, L. R. (1986). Stability of real interacting populations in space and time: implications, alternatives and the negative binomial k_c. *Journal of Animal Ecology* **55**, 1053–1068.

POULIN, R. (1996). Richness, nestedness, and randomness in parasite infracommunity structure. *Oecologia* **105**, 545–551.

POULIN, R. (1997). Species richness of parasite assemblages: evolution and patterns. *Annual Review of Ecology and Systematics* **28**, 341–358.

POULIN, R. & GUÉGAN, J. F. (2000). Nestedness, antinestedness, and the relationship between prevalence and intensity in ectoparasite assemblages of marine fish: a spatial model of species coexistence. *International Journal for Parasitology* **30**, 1147–1152.

ROHDE, K. (1977). A non-competitive mechanism responsible for restricting niches. *Zoologischer Anzeiger* **199**, 164–172.

ROHDE, K. (1979). A critical evaluation of intrinsic and extrinsic factors responsible for niche restriction in parasites. *American Naturalist* **114**, 648–671.

ROHDE, K. (1989). Simple ecological systems, simple solutions to complex problems? *Evolutionary Theory* **8**, 305–350.

ROHDE, K. (1991). Intra- and interspecific interactions in low density populations in resource-rich habitats. *Oikos* **60**, 91–104.

ROHDE, K. (1998). Latitudinal gradients in species diversity. Area matters, but how much? *Oikos* **82**, 184–190.

ROHDE, K. (2001). Spatial scaling laws may not apply to most animal species. *Oikos* **93**, 499–504.

ROHDE, K. (in press). Niche restriction and mate finding in vertebrates. In *Behavioural Ecology of Parasites* (ed. Lewis, E. E., Campbell, J. F. & Sukhdeo, M. V. K.) Wallingford, Oxford, CAB International.

ROHDE, K., HAYWARD, C. & HEAP, M. (1995). Aspects of the ecology of metazoan ectoparasites of marine fishes. *International Journal for Parasitology* **25**, 945–970.

ROHDE, K., HAYWARD, C., HEAP, M. & GOSPER, D. (1994). A tropical assemblage of ectoparasites: gill and head parasites of *Lethrinus miniatus* (Teleostei lethrinidae). *International Journal for Parasitology* **24**, 1031–1053.

ROHDE, K. & HEAP, M. (1998). Latitudinal differences in species and community richness and in community structure of metazoan endo- and ectoparasites of marine teleost fish. *International Journal for Parasitology* **28**, 461–474.

ROHDE, K., WORTHEN, W. B., HEAP, M., HUGUENY, B. & GUÉGAN, J.-F. (1998). Nestedness in assemblages of metazoan ecto- and endoparasites of marine fish. *International Journal for Parasitology* **28**, 543–549.

SIMBERLOFF, D. & BOECKLEN, W. J. (1981). Santa Rosalia reconsidered: size ratios and competition. *Evolution* **35**, 1206–1228.

SIMKOVÁ, A., MORAND, S., MATEJUSOVÁ, I., JURAJDA, P., & GELNAR, M. (2001). Local and regional influences on patterns of parasite species richness of central European fishes. *Biodiversity & Conservation* **10**, 511–525.

TAYLOR, L. R., TAYLOR, R. A. J., WOIWOD, I. P. & PERRY, J. N. (1983). Behavioural dynamics. *Nature* **303**, 801–804.

TAYLOR, L. R., WOIWOD, I. P. & PERRY, J. N. (1978). The density-dependence of spatial behaviour and the rarity of randomness. *Journal of Animal Ecology* **47**, 383–406.

WORTHEN, W. B. & ROHDE, K. (1996). Nested subset analyses of colonization-dominated communities: metazoan ectoparasites of marine fishes. *Oikos* **75**, 471–478.

Cleaning symbioses from the parasites' perspective

A. S. GRUTTER

Department of Zoology and Entomology, School of Life Sciences, University of Queensland, Brisbane, Qld. 4072

SUMMARY

Cleaning behaviour has generally been viewed from the cleaner or client's point of view. Few studies, however, have examined cleaning behaviour from the parasites' perspective, yet they are the equally-important third players in such associations. All three players are likely to have had their evolution affected by the association. As cleaner organisms are important predators of parasites, cleaners are likely to have an important effect on their prey. Little, however, is known of how parasites are affected by cleaning associations and the strategies that parasites use in response to cleaners. I examine here what parasites are involved in cleaning interactions, the effect cleaners have on parasites, the potential counter-adaptations that parasites have evolved against the predatory activities of cleaner organisms, the potential influence of cleaners on the life history traits of parasites, and other factors affected by cleaners. I have found that a wide range of ectoparasites from diverse habitats have been reported to interact with a wide range of cleaner organisms. Some of the life history traits of parasites are consistent with the idea that they are in response to cleaner predation. It is clear, however, that although many cleaning systems exist their ecological role is largely unexplored. This has likely been hindered by our lack of information on the parasites involved in cleaning interactions.

Key words: Cleaning behaviour, cleaner fish, *Labroides dimidiatus*, gnathiidae, evolution.

INTRODUCTION

Cleaning associations involve cleaner organisms that remove ectoparasites and other material, such as mucus, scales and skin, from the body surfaces of other apparently co-operating animals (Feder, 1966). The latter are often referred to as hosts, customers, or clients. Cleaning behaviour is one of the most highly developed inter-specific communication systems known, with clients striking elaborate postures (Feder, 1966) which have generally been assumed to make ectoparasites more accessible to cleaners. A wide range of animals function as cleaners, including crustaceans, ants, birds, fishes, lizards (Nicolette, 1990) and turtles (Krawchuk, Koper & Brooks, 1997).

Although terrestrial examples of cleaning are known, mainly involving cleaner birds (Mooring & Mundy, 1996), most examples are aquatic. Van Tassell, Brito & Bortone (1994) listed 132 fishes and invertebrates, most of which are marine. Cleaning behaviour among fishes occurs in both temperate and tropical waters, with a larger number described for the latter. These are from a wide range of families, with a large proportion belonging to the wrasses (Van Tassell *et al.* 1994). In addition to fish, the clients include octopus (Johnson, 1982), turtles (Vogt, 1979; Losey, Balazs & Privitera, 1994; Krawchuk *et al.* 1997), marine iguanas (Hobson, 1969), and whales (Swartz, 1981).

Cleaner fish are important predators of ectoparasites. They feed on a wide range of ectoparasites in diverse geographic locations. Some cleaners eat large numbers of parasites (Arnal & Morand, 2001; Grutter, 1996a, 1997a) while others, although they may each eat few parasites due to their small size, are found in large numbers (Arnal & Côté, 2000) and thus may, overall, exert a significant effect on their prey.

Interactions between cleaners and parasites can be considered as interactions between enemies and victims, i.e. a predator-prey system (Hastings, 2000; Keeling, Wilson & Pacala, 2000). Predator-prey interactions are a significant component of ecological communities and raise fundamental questions. For example, given that such interactions are inherently unstable, how is it that the species co-exist? Understanding how the responses of the organisms, for example defence against predators, influence the dynamics of the system can provide information on how such systems function. It can also provide insight into the role of evolution in shaping these relationships.

Cleaners, as predators of parasites, are likely to have an effect on their prey. However to what extent they affect parasite populations is only recently being investigated. Although most studies have focused on the effects of cleaners on local parasite size and abundance on fish (Limbaugh, 1961; Youngbluth, 1968; Losey, 1972; Gorlick, Atkins & Losey, 1987; Grutter, 1996b, 1999a) the life-history traits and behaviour of parasites may have also been affected over evolutionary time in response to predation pressure.

Tel: 61-7-3365 7386. Fax: 61-7-3365 1655.
E-mail: a.grutter@mailbox.uq.edu.au

Table 1. Parasitic isopods involved in cleaning interactions. Unless otherwise stated, information was obtained from diet analyses. NA = Not available

Family (order)	Parasite identity	Cleaner species	Family	Location	Comment	Reference
Gnathiidae	juveniles	*Labroides dimidiatus*	Labridae	Lizard Island, Great Barrier Reef (GBR)		Grutter, 1996*a*, 1997*a*
	juveniles	*L. dimidiatus*	Labridae	Heron Island, GBR		Grutter, 1997*a*
	juveniles	*L. dimidiatus*	Labridae	New Caledonia		Grutter, 1999*c*
	juveniles	*L. dimidiatus*	Labridae	Japan		Sano, Shimizu & Nose, 1984
	juveniles	*L. dimidiatus*	Labridae	Society Islands		Randall, 1958
	juveniles	*L. bicolor*	Labridae	Society Islands		Randall, 1958
	juveniles	*L. bicolor*	Labridae	Heron Island, GBR		Randall, 1958
	juveniles	*L. phthirophagus*	Labridae	Hawaii		Randall, 1958
	juveniles	*L. rubrolabiatus*	Labridae	Moorea		Randall, 1958
	juveniles	*Oxyjulis catifornica*	Labridae	California		Hobson, 1971
	juveniles	*Symphodus melanocercus*	Labridae	Mediterranean Sea		Senn, 1979; Arnal & Morand, 2001
	juveniles	*Centrolabrus rupestris*	Labridae	Portugal		Henriques & Almada, 1997
	juveniles	*Ctenolabrus rupestris*	Labridae	Plymouth aquarium, United Kingdom (UK)		Potts, 1973
	juveniles	*Halichoeres cyanocephalus*	Labridae	Brazil		Sazima, Moura & Gasparini, 1998
	juveniles	*Crenilabrus melops*	Labridae	Plymouth aquarium, UK		Potts, 1973
	juveniles	*Centrolabrus exoletus*	Labridae	Spain		Galeote & Otero, 1998
	juveniles	*Thalassoma bifasciatum*	Labridae	Puerto Rico		Losey, 1974

Parasite	Cleaner species	Family	Location	Notes	Reference
juveniles	*Heniochus monoceros*	Chaetodontidae	New Caledonia		Lo, C., unpublished data
juveniles	*Chaetodon citrinellus*	Chaetodontidae	Lizard Island, GBR		Cribb, T., unpublished data
juveniles	*Elacatinus prochilos*	Gobiidae	Bahamas		Arnal & Côté, 2000
juveniles	*Gobiosoma (Elacatinus) illecebrosum*	Gobiidae	Panama		Bohlke & McCosker, 1973
juveniles	*Gobiosoma* spp.	Gobiidae	Puerto Rico		Losey, 1974
juveniles	*Entelurus aequoreus*	Syngnathidae	Plymouth aquarium, UK		Potts, 1973
juveniles	*Syngnathus typhle*	Syngnathidae	Plymouth aquarium, UK		Potts, 1973
juveniles	*S. acus*	Syngnathidae	Plymouth aquarium, UK		Potts, 1973
juveniles	*Brachyistius frenatus*	Embiotocidae	California		Hobson, 1971
juveniles	*Phanerodon atripes*	Embiotocidae	California		Hobson, 1971
Cymothoidae *Anilocra haemuli*	*Periclimines pedersoni*	Palaemonidae	Puerto Rico	Laboratory experiment. Observed eating isopods	Bunkley-Williams & Williams, 1998
'cymothoids'	*Thalassoma bifasciatum*	Labridae	Puerto Rico		Losey, 1974
Codonophilus sp.	*Coris sandageri*	Labridae	New Zealand	2 cm parasite removed from mouth of client by cleaner, disabled by repeatedly striking it against rocks, then eaten	Ayling & Grace, 1971
(Isopoda) 'Larval isopods'	*Echeneis naucrates*	Echenidae	NA		Cressey & Lachner, 1970
'Parasitic isopods'	*Canthidermis maculatus*	Balistidae	Hawaii	Observed biting 'parasitic isopods'	Gooding, 1964

Predation events are characterized by the encounter, detection, identification, approach, subjugation, and consumption of the prey (Endler, 1991). Parasites involved in cleaning interactions therefore should have developed defence mechanisms or counter-adaptations to deal with the above sequence of predation events. Adaptations can be defined as traits directly shaped by selection (Ridley, 1993). Counter-adaptations of prey are numerous and include crypsis, polymorphism, spacing themselves out, mimicry, escape flights, weapons of defense, active defense and toxins (Krebs & Davies, 1993).

How the life history traits of parasites evolved is an important question in parasitology. Life-history strategies mainly evolve in response to pressures of different environments on the survival and fecundity of different age-classes (Partridge & Harvey, 1988; Poulin, 1996). Most of the studies on the evolution of life history traits of parasites have been on how parasites evolved from free-living forms (Poulin, 1995a). Factors, other than those directly related to parasitism, however, are also likely to have played a role in the evolution of parasite life history traits.

Adaptation in defence against predators is one of these. Since it is nearly impossible to observe the development of these processes, their evolution can be extrapolated from existing associations. Although adaptations can only be studied in light of phylogeny (Harvey & Pagel, 1991), there is some circumstantial evidence to support some of the predictions below. Further studies, including phylogenetic approaches and experimental studies are, however, needed.

Much of the work on cleaning is restricted to tropical, semitropical, and warm-temperate regions most likely due, at least in part, to a lack of observations in other areas. Most of the information on the parasites involved in aquatic cleaning interactions is restricted to fishes. For example, although cleaner shrimps are a popular example of cleaner organisms, evidence that they indeed remove parasites is almost nil (Spotte, 1998). This study, therefore, relies heavily on information available from marine fishes including studies involving the use of cleaner fish in commercial fish farms, and to a lesser degree freshwater fishes. When examining what parasites are involved in interactions with cleaners I have also included cleaners other than fish for completeness. Much of the following is speculative. In general, much more evidence is needed to support the proposed hypotheses.

To explore cleaning behaviour from the parasites' perspective, I examined (1) what parasites are involved in cleaning interactions, (2) the effect cleaners have on parasites, (3) the potential counter-adaptations that parasites have made against the predatory activities of cleaner organisms, (4) the influence of cleaners on the life history traits of parasites, and (5) other factors affected by cleaners.

PARASITES INVOLVED IN CLEANING INTERACTIONS

Crustacea

Isopods. Only examples where gnathiids were actually found in the diet were included in this study. Gnathiids, in addition to caligid copepods, are one of the most common parasite groups found in the diet of cleaner fishes (Table 1). Gnathiids are eaten by 20 fish species which include wrasses, gobies, pipefish, butterflyfish and perch from diverse locations including the Indo-Pacific, the Caribbean, Brazil, Europe, and California (Table 1). For the cleaner fish *Labroides dimidiatus* at Lizard Island, gnathiid isopods make up 95% of the items in the diet (Grutter, 1997a). Butterflyfish occasionally have gnathiids in their diet (Table 1), yet this has, as of yet, not been reported in the literature. The above butterflyfish likely obtained gnathiids by cleaning other fish, as *Heniochus monoceros* at the same location that the diet analyses were made have been observed cleaning other fish (T. Cribb, personal communications) and butterflyfish juveniles are known to act as cleaners (Feder, 1966; Youngbluth, 1968; Hobson, 1969; Allen, Steene & Allen, 1998).

Other isopods in the diet of cleaners are cymothoids which are eaten by a Caribbean cleaner shrimp in the laboratory and by wrasse in Puerto Rico and New Zealand. Seabass in the Canary Islands and diskfishes also eat 'isopods'. In addition, triggerfish off drifting objects in Hawaii have been observed biting 'parasitic isopods' on another triggerfish (Table 1).

Copepods. Caligid copepods have been found in the diet of 15 fish species including wrasses, diskfish, jacks, chubs, and sweeps (Table 2). Other copepods also eaten by cleaner fish include pennellids, bomolochids, pandarids, *Hastchekia* sp., and laerneids. Cleaner fish that eat copepods are geographically diverse and include fish from the Indo-Pacific, California, Argentina, Florida, the Caribbean, Europe, and New Zealand (Table 2).

Other parasitic arthropods

Other parasitic arthropods involved in cleaning interactions include barnacles which are removed by fish from grey whales and by crabs from turtles, amphipods removed by fish from grey whales and by crabs from turtles, argulids removed by fish from other fish and ticks removed by crabs from marine iguanas (Table 2).

Parasitic flatworms

Of the parasitic flatworms, only capsalid monogeneans have been found in the diet of cleaner fish (Table 3). There is some evidence that some monogeneans may be affected by cleaners. Cowell *et* al. (1993) showed that the marine cleaner fishes *Gobiosoma oceanops* and *G. genie* significantly affected the abundance of *Neobenedenia melleni* on sea-water cultured tilapia while *Thalassoma bifasciatum* did not. Benedeniine monogeneans were also found in the diet of the cleaner fish *Labroides dimidiatus* (Table 3). This cleaner fish also affected the abundance and size-frequency distribution of *Benedenia lolo* on the wrasse *Hemigymnus melapterus* (Labridae), but this effect varied with the size of the client fish (Grutter, Whittington & Deveney, unpublished observations). Other parasitic flatworms involved in 'circumstantial' cleaning interactions include microbothrids, gyrodactylids, and 'encysting trematodes' (Table 3).

Other parasites

Some leeches may also be affected by cleaners. Wrasses have been observed removing leeches from fish and the scarcity of a leech species has been related to the high abundance of the cleaner *Crenilabrus melops* in some areas of Britain (Table 3). Ciliates may also be involved in cleaning interactions. Although the cleaner fish *L. dimidiatus* ingests the trophont stages of the parasitic ciliate *Cryptocaryon irritans*, the agent of white spot disease, from the cultured stenohaline fish *Lates calcarifer* (Centropomidae), cleaner did not significantly affect the parasite's abundance (Halliday, unpublished observations). This was likely due to the inaccessibility, the small size and the rapid population growth of *C. irritans* and the uncooperative behaviour of the client. The small number of host epithelial cells in the cleaner's gut suggested that cleaners did not remove trophonts beneath the epithelium but rather those trying to burrow into or leaving the host (Halliday, unpublished observations).

Bacteria

While tissue infected with bacteria has been reported as a target of cleaner fishes (Limbaugh, 1961) and there is some evidence that cleaners may assist with wound healing in fish (Foster, 1985), the effect of cleaners on bacterial infections remains unresolved.

THE EFFECT OF CLEANERS ON PARASITES

Earlier studies found no effect of marine fish cleaners on parasites. Youngbluth (1968) and Losey (1972), both found at the same sites in Hawaii, no effect of cleaners on the abundance of parasites or client fish. Similarly, at Lizard Island on the Great Barrier Reef, Grutter (1996b) found no effect of cleaners on parasites of the damselfish *Pomacentrus moluccensis*. The study examined the size, total abundance and number per taxon of parasites. Although Limbaugh (1961) found that the removal of 'all known cleaning organisms' from 2 reefs in the Bahamas resulted in increased infection in the form of 'fuzzy white blotches, swelling, and ulcerated sores and frayed fins' and emigration of clients, the study involved no quantitative data or controls.

Gorlick et al. (1987) found that the abundance of parasites was not affected by cleaners at Enewetak Atoll. They did, however, find that parasitic copepods *Dissonus* sp. were larger in the absence of cleaners. As *Labroides dimidiatus* selectively feed on the larger parasites (Grutter, 1997b), this result would, at first glance, appear to be due to cleaner feeding preferences. The study, however, also suggested density dependent population regulation by the parasites themselves. In the absence of cleaners, most fish had only one large (ca. 1 mm) copepod. In contrast, on reefs with cleaners, fish had several small copepods. This suggested that a large copepod prevented new copepods from recruiting onto fish (Gorlick et al. 1987).

The only study to date to show a quantitative effect of any cleaner on parasite abundance was done using caged fish on reefs with cleaners or with all cleaners removed (Grutter, 1999a). This study found a 3·8 fold increase in gnathiid isopod abundance on reefs without cleaners after 12 days. More interestingly, no differences were found between reefs with and without cleaners when caged fish were sampled at dawn after 12 h. In contrast, when fish were sampled the following sunset after 24 h, there was a 4·5 fold increase in gnathiid abundance on fish from reefs without cleaners. This change in abundance of gnathiids between dawn and sunset is likely due to the fact that cleaners are only active during the day (Grutter, 1996a). It also suggests that cleaner fish predation plays a significant role in the daily decline in gnathiid abundance found on wild fish (Grutter, 1999a, 1999b). Cleaners also affected the abundance of the corallanid isopod *Argathona macronema* on caged fish (Grutter & Lester, in press).

The size of a parasite may influence its likelihood of being eaten by a cleaner. Within the species *L. dimidiatus*, cleaners selectively feed on larger gnathiid isopods thus making larger gnathiids more vulnerable to predation (Grutter, 1997b). However, the maximum size of gnathiids eaten may be limited by the size of the cleaner, or more specifically by its throat width, as more large gnathiids are eaten by large cleaners than by smaller cleaners (Grutter, 2000). Interestingly, cleaners affect parasites (corallanid isopods) that are even larger than gnathiids, but only up to a particular size (< 6 mm). Isopods that are larger, in contrast, are not affected (Grutter, McCallum & Lester, in press). How this occurs is unclear as the smaller corallanids are wider than the throat width of cleaners (Grutter, 2000). The small size of some parasites may explain why they are not affected by cleaners, such as *Cryptocaryon irritans* (350 μm) which are eaten in low numbers while

Table 2. Other parasitic arthropods involved in cleaning interactions

Family (order or class)	Parasite identity	Cleaner species	Family	Location	Comment	Reference
Caligidae	*Caligus hobsoni, Lepeophtheirus* sp.	*Oxyjulus californica*	Labridae	California	Nest-guarding male garibaldi, which attacked all fishes including cleaners, had more *C. hobsoni* during the breeding season than outside the breeding season	Hobson, 1971
	C. hobsoni	*Phanerodon atripes*	Labridae	California		Hobson, 1971
	Caligus sp.	*Centrolabrus exoletus*	Labridae	Spain		Galeote & Otero, 1998
	Caligus elongatus	*Crenilabrus melops*	Labridae	Northern Europe		Costello, 1996; Deady, Varian & Fives, 1995
	'Caligid'	*L. dimidiatus*	Labridae	Japan		Chikasue, M., personal communication
	Caligus spp.	*Symphodus melanocercus*	Labridae	Mediterranean Sea		Arnal & Morand, 2001
	'Calagoid copepods'	*L. bicolor*	Labridae	Society Islands		Randall, 1958
	'Calagoid copepods'	*L. phthirophagus*	Labridae	Hawaii		Randall, 1958
	'Caligoid copepods'	*L. dimidiatus*	Labridae	Japan		Sano, Shimizu & Nose, 1984
	'caligid larvae' & 'Caligidae'	*L. dimidiatus*	Labridae	Lizard Island, Great Barrier Reef		Grutter, 1997a
	'caligid larvae' & 'Caligidae'	*L. dimidiatus*	Labridae	New Caledonia		Grutter, 1999c
	'caligoid' & 'caligid copepods'	*L. dimidiatus*	Labridae	Marshall, Phoenix, & Society Islands		Randall, 1958
	Achtheinus dentatus	*Remora remora*	Echenidae	Argentina		Szidat & Nani, 1951
	Nesippus sp.	'Echeneid fishes'	Echenidae	Central Pacific		Strasburg, 1959
	Gloiopotes sp.	'Echeneid fishes'	Echenidae	Central Pacific		Strasburg, 1959
	'Caligoid copepods'	*Oligoplites saurus*	Carangidae	Florida, USA		Carr & Adams, 1972; Lucas & Benkert, 1983
	'caligid copepods'	*Hermosilla azurea*	Kyphosidae	California, USA		DeMartini & Coyer, 1981
	'caligid copepods'	*Girella nigricans*	Kyphosidae	California, USA		DeMartini & Coyer, 1981
	'caligid copepods'	*Thalassoma bifasciatum*	Labridae	Puerto Rico		Losey, 1974
	'Caligoids'	*Atypichthus strigatus*	Scorpididae	Southern Australia		Glasby & Kingsford, 1994

Family	Parasite	Cleaner	Fish family	Location	Notes	Reference
Bomolochidae	'caligid copepods'	*Elacatinus prochilus*	Labridae	Bahamas		Arnal & Côté, 2000
	'bomolochid'	*L. dimidiatus*	Labridae	Japan		Chikasue, M., personal communication; Arnal & Morand, 2001
Pennellidae	*Peniculus fistula*	*Symphodus melanocercus*	Labridae	Mediterranean Sea		McCutcheon & McCutcheon, 1964
Pennellidae	*Lernaeenicus radiatus*	*Fundulus heteroclitus*	Cyprinodontidae	North Carolina, USA	In laboratory	
Pennellidae	'pennellid'	*L. dimidiatus*	Labridae	Japan		Chikasue, M., personal communication
Pandaridae	*Pandarus armatus* 'pandarid'	*Remora remora* 'Echenied fishes' *Remora* spp.	Echenidae Echenidae Echenidae	Argentina Central Pacific NA		Szidat & Nani, 1951 Strasburg, 1959 Cressey & Lachner, 1970
	Pandarus sp.	*Centrolabrus exoletus*	Labridae	Spain		Galeote & Otero, 1998
Hatschekiidae	*Hatschekia* sp.					
(Copepoda)	'Laerneids'	*Thalassoma bifasciatum*	Labridae	Puerto Rico		Losey, 1974
	'Laerneids'	*L. dimidiatus*	Labridae	Society Islands		Randall, 1958
Coronulidae	'Parasitic copepods' *Cryptolepas rhachianecti*	*Coris sandageri* *Atherinops affinis*	Labridae Atherinidae	New Zealand Baja California, Mexico	Barnacles observed being picked off grey whale and in the diet of cleaner	Ayling & Grace, 1971 Swartz, 1981
Lepadidae	"Goose barnacle cyprids"	*Planes minutus*	Grapsidae	Madeira, Portugal	Removed by crabs from loggerhead turtles *Caretta caretta*	Davenport, 1994
(Cirripedia)	'Small barnacles'	*Thalassoma lunare*	Labridae	Fairfax Island, Great Barrier Reef	Removed from green turtle *Chelonia mydas*	Booth & Peters, 1972
Platylepadidae	*Platylepas hexastylos*	*Thalassoma duperry*	Labridae	Hawaii	Barnacles removed from green turtle *Chelonia mydas*	Losey, Balazs & Privitera, 1994
Amphipoda	*Cyamus* sp.	*Atherinops affinis*	Atherinidae	Baja California, Mexico	Observed being picked off grey whales *Eschrichtius robustus* and in the diet of cleaner fish	Swartz, 1981
Amphipoda	*Podoceros chelophilus*	*Planes minutus*	Grapsidae	Madeira, Portugal	Removed from loggerhead turtles *Caretta caretta*	Davenport, 1994
Arguloidae	*Argulus* sp.	*Oligoplites saurus*	Carangidae	Florida		Carr & Adams, 1972
Ixodidae	*Amblyomma darwini cristatus*	*Amblyrhynchus cristatus*	Grapsidae	Galapagos	Ticks removed by crab from sunbathing marine iguanas	Beebe, 1926

Table 3. Other parasites involved in cleaning interactions

Family (order or class)	Parasite identity	Cleaner species	Family	Location	Comment	Reference
Capsalidae	Neobenedenia melleni	Gobiosoma oceanops	Gobiidae	Bahamas	Laboratory study including diet analysis	Cowell, Watanabe, Head, Grover & Shenker, 1993
Capsalidae	Neobenedenia melleni	G. genie	Gobiidae	Bahamas	Laboratory study including diet analysis	Cowell et al. 1993
Capsalidae	Neobenedenia melleni	Thalassoma bifasciatum	Labridae	Bahamas	Laboratory study including diet analysis	Cowell et al. 1993
Capsalidae	'benedeniine monogeneans'	Labroides dimidiatus	Labridae	Lizard Island, Great Barrier Reef	Accessory sclerites in diet	Deveney, M., unpublished data, see Grutter, 1997a
Capsalidae	Benedenia lolo	L. dimidiatus	Labridae	Australia	Laboratory experiment including diet analysis	Grutter, Whittington & Deveney, unpublished data
Microbothriidae	Psudoleptobothrium aptychotrema	Paramonacanthus oblongus	Monacanthidae	Australia	Parasitic flatworm found severed off a ray was attributed to the leatherjacket fish	Kearn, 1978
Gyrodactylidae	Swingleus	Cyprinodon variegatus	Cyprinodontidae	Virginia, USA	Infected fish seen chafing and posing for cleaner, parasites not found in diet	Able, 1976
Gyrodactilidae	Gyrodactylus	C. variegatus	Cyprinodontidae	Virginia, USA	Infected fish seen chafing and posing for conspecifics	Able, 1976
Gyrodactilidae	Gyrodactylus	Apeltes quadracus	Gasteroidae	Maryland, USA	Infected fish pose for and are cleaned by stickleback	Tyler, 1963
(Trematoda)	'encysting trematodes'	Thalassoma bifasciatum	Labridae	Belize	Bites by fish from skin and fins may have been directed at parasites found on clients which were subsequently collected	Reinthal & Lewis, 1986
Piscicolidae	Calliobdella lophii	Centrolabris exoletus	Labridae	Germany	Fish observed removing leeches in laboratory	Samuelsen, 1981
Piscicolidae	Sanguinothus pinnarum	Crenilabrus melops	Labridae	Britain	Scarcity of leech has been attributed to high incidence of cleaner fish	Hussein & Knight-Jones, 1995
Hymenostomatidae	Cryptocaryon irritans	Labroides dimidiatus	Labridae	Australia	Trophont stages of ciliate on cultured stenohaline fish Lates calcarifer were eaten by cleaner but parasite's abundance not affected by cleaner	Halliday I.A., unpublished data

larger copepod parasites (*Lernaeenicus* sp.) are actively targeted on the same fish (Halliday, unpublished observations).

More information, however, is needed on the long-term effect of cleaners on parasite populations. This is currently being examined for gnathiids using emergence traps on reefs with and without cleaners to sample gnathiids as they emerge from the reef in search of hosts (Murphy and Grutter, unpublished observations).

The above impacts on the local abundance and size of parasites show that some parasites are under pressure to avoid being eaten. Certain strategies, such as counter-adaptations against predators, and altered life-history traits may therefore have evolved to minimize this risk. Surprisingly, this is a relatively little studied area. Some limited information, however, is available to explore these potential strategies.

COUNTER-ADAPTATIONS OF PARASITES AGAINST CLEANERS

Counter-adaptations of prey against predation can be divided into the sequence of events that characterize a predator's behaviour when feeding (see introduction). These are listed below and include the encounter, detection, identification, approach, subjugation and consumption of the prey (Endler, 1991). Prey have a greater advantage in avoiding predation in the early stages of the predation sequence as the probability of getting through all stages is low, the predator is closer in later stages, and defenses used in later stages require more energy expenditure (e.g. toxins, spines) (Endler, 1991).

Encounter

Timing of encounters with predators. During encounters between prey and predators, apparent rarity (Endler, 1991) may be used by prey as an antipredator defence. This may involve differences in predator and prey activity times (Endler, 1991). Thus, parasites may avoid being eaten by cleaners by infecting fish at times when predators are scarce. Most families to which cleaner species belong, such as labrids and chaetodontids, are diurnal (Hobson, 1965, 1972). Of these, the behaviour of *Labroides dimidiatus* is best studied. They are strictly diurnal, returning to a sleeping hole each night (Grutter, 1995 a; Robertson & Choat, 1974). *Gobiosoma evelynae* also only clean during the day (Johnson & Ruben, 1988). Thus the predation risk from many cleaner species should be lower at night.

In California, gnathiids emerge from the reef in search of hosts mainly at night (Hobson & Chess, 1976; Stepien & Brusca, 1985) with fewer emerging during the day (Hobson & Chess, 1976). At Lizard Island, where cleaners are numerous (Green, 1996), some life-stages or species of gnathiids only emerge from the reef at night, indicating they are nocturnal;

diurnal gnathiids are only found during the new and full moon (Grutter, Morgan & Adlard, 2000). More importantly, some gnathiid species only infect fish at night (Grutter, 1999 b; Paperna & Por, 1977; Potts, 1973), although some are diurnal, but to a lesser degree (Grutter, 1999 b). For some of these species, such nocturnal tactics should reduce predation from diurnal cleaners and other diurnal predators such as planktivores. It has been suggested that some shrimp may engage in nocturnal cleaning (Corredor, 1978). Laboratory experiments, however, found that only one cleaner shrimp (*Periclimenes pedersoni*) out of 4 species tested ate cymothoid isopods (Bunkley-Williams & Williams, 1998). Whether it also cleans at night is unknown. Due to the lack of information on the feeding habits of cleaner shrimp the possibility that cleaner shrimp eat gnathiid isopods at night cannot be excluded.

Exposure to predators. One way of reducing the rate of encounter with a predator is to reduce the time prey are exposed to predators. When apparently sampling the host for an appropriate place to feed, gnathiids often land on the host for a few seconds then return to the benthos before re-sampling the fish (Grutter, unpublished observations). Gnathiids also only require up to an hour to feed and then leave the host (Grutter, unpublished observations). Such behaviour reduces the time they are exposed to cleaners. These behaviours are possible because of their high mobility (see below).

Hiding from predators. Another tactic of apparent rarity is the use of hiding or inconspicuous resting places (Endler, 1991). After quickly feeding and leaving the host, gnathiids then return to the benthos and hide in the benthos, most likely in dead coral and sponges (Holdich & Harrison, 1980) where they digest and moult to the next larval stage (Upton, 1987). Again, such behaviour is possible because of their high mobility (see below).

Detection and identification

Immobility. Immobility of prey at certain times is often used as an antipredator defence (Endler, 1991). Observations of gnathiid isopods reveal that when on host fish they often remain immobile and rarely move to other sites by crawling over the body surface (Grutter, unpublished observations).

Cryptic locations (site-specificity). Some of the cryptic lifestyles of some fish parasites may have evolved, in part, as a response to cleaning. Some microhabitats of fish are less likely to be cleaned than others. The copepod *Caligus minimus*, for example, lives inside the opercula and mouth cavity of seabass *Dicentrachus labrax* and thus may be inaccessible to Mediterranean cleaners (Costello *et al.* 1996). Similarly, the copepod *Lepsosphilus labrei*, which dwells within the skin of fish and is thus inaccessible to cleaners, is common on corkwing wrasse that engage in intraspecific cleaning (Costello *et al.* 1996). The parasites eaten by the cleaner fish *Oxyjulis californica* consist mainly of mobile forms, such as gnathiid isopods and caligid copepods, while more specialized forms such as dichelestiid, chondracanthid, and lerneopodid copepods in the branchial and oral cavities escape predation (Hobson, 1971). The former are more similar to the free-living forms requiring few changes in the feeding behaviour of cleaners compared to feeding on the more specialized species (Hobson, 1971).

The potential role of predation in monogenean biology was reviewed by Kearn (1999). Kearn (1994) proposed that predation was likely more significant in the ecology and evolution of monogeneans than previously thought and that predation may have provided the selective pressure which led to the colonization of sites, other than the skin, by ancestral monogeneans. Most likely this would have included the use of sites which are cryptic. For example, benedeniine monogeneans are found in cryptic areas such as the branchial cavity, lip folds, pharyngeal tooth pads, pelvic fins, and branchiostegal membranes of fish (Whittington, 1996; Whittington & Kearn, 1990). On elasmobranchs, the body chambers with external openings, such as the branchial cavity, nasal fossae, cloaca, rectal gland, and coelum of elasmobranch are regularly examined when surveyed for monocotylid monogeneans (Whittington, 1996; Whittington & Kearn, 1990).

The cleaners *Remora* spp. are rarely found on the fins of fish, possibly because of the disturbances of the host's movements (Cressey & Lachner, 1970). Interestingly, the copepods *Pandarus* are abundant on the fins but are not found in the diet of *Remoras* (Cressey & Lachner, 1970). This suggests that the fins may serve as a refuge from predation by diskfishes (Cressey & Lachner, 1970).

Client posing behaviour, such as the opening of mouths and operculae or extension of fins, increases the likelihood that clients are cleaned (Côté, Arnal & Reynolds, 1998) and most likely serves to increase the accessibility of parasites by exposing them to cleaners. For example, sea bass *Centropristes striatus* expose their gills to topminnows *Fundulus heteroclitus* in the laboratory so they can access their parasitic copepods (McCutcheon & McCutcheon, 1964). Whether or not clients engage in posing behaviour and the types of postures they use are likely to affect a parasite's choice of habitat. Finally, cryptic sites may also protect the parasite from the client's own attempts to clean itself, such as by chafing.

Cryptic colour patterns. The colour of prey is cryptic if it resembles a random sample of the visual background as perceived by the predator at the time

and place where the prey is most vulnerable to predation (Endler, 1991). It has been suggested that such crypsis in some parasites may reduce predation from cleaners. Pigmentation has been reported in many monogeneans (Deveney & Whittington, 2001) and may serve as camouflage from predators (Kearn, 1976, 1979, 1994; Roubal & Quartararo, 1992; Whittington, 1996). Some cleaner fish eat monogeneans in aquaria (Cowell *et al.* 1993; Kearn, 1976, 1978) and in the wild (Grutter, 1997*a*). Laboratory experiments show that pigmented monogeneans *Benedenia* spp. suffer decreased predation from *L. dimidiatus* compared with unpigmented species (Deveney, Whittington, & Grutter, unpublished observations). Pigments in the dendritic gut of the monogenean *Dendromonocotyle kuhlii* may serve as a dorsal screen for reproductive organs (Kearn, 1979). An ideal camouflage would be to match pigments with the site of attachment. This appears to occur in *Benedenia lutjani* which can contain yellow pigment that matches the colour of the site of attachment (Whittington, 1996). Pigmentation also varies on the body of some parasites, for example, gnathiids have more pigmentation on their dorsal side, the side more exposed to potential predators (Grutter, unpublished observations). Fishes sometimes alter their colouration while being cleaned (Feder, 1966 and references therein). Possibly, this may make parasites more visible to cleaners, such as in the case of white fungi which were accentuated by a darkening of the fish (Wyman & Ward, 1972). Parasites, in turn, may also have to adjust their camouflage in response to client colour changes.

Many parasites, such as copepods, gnathiid isopods and helminths, also have wholly or partly translucent bodies which may possibly also serve to camouflage the parasite. *Benedenia seriolae*, found on the sides of the fish *Seriola quinqueradiata*, have a highly transparent vitellarium compared to *Entobdella soleae* which are found on the lower side of sole and thus are unlikely to be exposed to predation (Kearn, 1994). Some *Anoplodiscus* monogeneans found on fish fins are also translucent while others are not (Roubal & Quartararo, 1992). It has been suggested that some *Anoplodiscus* may use a combination of pigmentation and translucence to help them blend into the background of the hosts which may reduce predation from cleaners (Roubal & Quartararo, 1992). These authors also pointed out that worms and copepods on the gills or in the buccal cavity of the sparid hosts, areas likely less accessible by predators, tended to be more opaque. Wahlert & Wahlert (1961) proposed that the differences in the colours of copepods of the Mediterranean and North Sea (opaque versus coloured) may be due to fewer potential known cleaners (perciform fishes) in the latter.

The response of the monogenean *Encotyllabe caballeroi* to light may reduce predation from cleaners (Kearn & Whittington, 1992). The rapid contraction of the body in response to brief illumination to light may occur in response to the increase in illumination that occurs when host fishes pose for cleaner fishes by opening their opercula and mouths. If this is the case, then invasion into gill and oral cavities by monogeneans has not necessarily prevented predation by cleaners (Whittington, 1996). Interestingly, benedeniines from the gills of fishes on the Great Barrier Reef do not contain pigments (Whittington, 1996). As benedeniines most likely ingest pigments from pigmented host tissue, gill parasites are unlikely to obtain such pigments, as pigmentation is not common in fish gill chambers (Whittington, 1996). In addition, pigmentation would likely make parasites more conspicuous on gills (Whittington, 1996).

Approach

Mode of fleeing. Speed and high mobility to avoid the approach of predators may be used by prey to avoid predation (Endler, 1991). Such tactics may be adopted by some parasites. Gnathiids readily leave teleost hosts when disturbed (Grutter, 1995*b*) and are rapid swimmers, swimming an estimated 10 to 20 cm.sec^{-1} (Grutter, unpublished observations). Interestingly, gnathiids on sharks and rays (Grutter & Poulin, 1998) are not as mobile as those found on teleosts. They often remain on host gills, even after they have been dissected (Grutter, unpublished observations) or in the case of epaulette sharks, remain on fish in the laboratory for days (Grutter, unpublished observations) and even when disturbed (Heupel & Bennett, 1999). Although there are some reports of cleaning in sharks (Keyes, 1982; Sazima & Moura, 2000), no information on the parasites involved is available.

Consumption

Large size of prey. Some prey may evade or slow down their consumption by predators through large body size. The larger parasitic coralanid isopods escape predation from *Labroides dimidiatus*, most likely because their widths are up to 4 times the throat width (Grutter, 2000) of the cleaner (Grutter & Lester, in press).

Unpalatability. Whether some parasites are unpalatable is yet to be determined. Some parasites, such as gnathiid isopods of the elasmobranchs *Carcharhinus melanopterus*, and *Rhynchobatus djiddensis* from Heron Island, and *Carcharias taurus*, from Umghlanga Rocks, South Africa have bizarre colouration consisting of swirling or banded colour patterns, colourful 'eyes' marked on the dorsal surface and bright yellow pigments (B. Moore, I.

Whittington, N. Smit, respectively, personal communications). Monogeneans also often have brilliant colours (Whittington, 1996). The possibility exists that these colours may be aposematic (I. Whittington, personal communications), whereby the colours are 'warnings' indicating the organism is noxious or at least unpalatable due to distasteful chemicals (Mallet & Joron, 1999).

INFLUENCE OF CLEANERS ON THE LIFE HISTORY TRAITS OF PARASITES

The major factors influencing parasite life history traits, particularly virulence, are transmission rate, transmission mode and host immunity (Clayton & Tompkins, 1994; Ewald, 1995; Frank, 1996; Koella & Agnew, 1999; Koella & Doebeli, 1999). However, little is known about the impact of ecological factors, such as predation risk, on the evolution of ecto-parasite life-traits. There is some evidence that predation risk may affect the body size and age at maturity of some parasites. Whether predation risk affects the fecundity and generation time of parasites eaten by cleaners needs to be explored. In gnathiids, both fecundity and generation time vary (Grutter, unpublished observations), but how this relates to predation risk has not been explored.

Body size of parasites

For isopods, in addition to the effect of habitat characteristics on the evolution of body size (Poulin, 1995*b*), the effect of predators may be important. Isopods in higher latitudes have larger body sizes (Poulin, 1995*b*). For some non-parasitic isopods, this has been explained as due to lower predation intensity by fish at high latitudes (Wallerstein & Brusca, 1982). The cleaner *L. dimidiatus* selectively preys on larger gnathiid isopods (Grutter, 1997*b*). Interestingly, Gnathidae have smaller body sizes than their closest free-living relatives (Poulin, 1995*b*). Selection pressure from cleaners could thus have resulted in gnathiids adopting small size as a refuge from predators.

Conversely, cleaners may also have some difficulty in eating very large parasites. The maximum prey size the cleaner *L. dimidiatus* can exploit appears to be limited by their throat size (Grutter, 2000). This may explain why cleaners only affected the smaller corallanid isopods found on caged fish (Grutter & Lester, in press). Interestingly, parasitic cymothoids and corallanids, in contrast to gnathiids, generally attain larger body sizes than free-living relatives (Poulin, 1995*b*).

Some gnathiids, at first glance, may appear to escape predation through large size. Gnathiids on the benthic epaulette shark *Hemiscyllium ocellatum* are much larger (Heupel & Bennett, 1999) than those found in the diet of *L. dimidiatus* (Grutter, 1997*b*). They are also found mainly on exposed areas of the fish, such as the cloaca and lips, and do not readily leave the fish when disturbed (Heupel & Bennett, 1999). Although common in shallow coral reefs (Last & Stevens, 1994) where cleaners such as *L. dimidiatus* are prevalent (Green, 1996), epaulette sharks are, however, nocturnal (Heupel & Bennett, 1999). They are thus unlikely to encounter cleaner fish that are mainly active during the day (Grutter, 1996*a*). Thus the behaviour of the host may play a more important role than parasite size in the parasite's vulnerability to cleaners.

Age at maturity

Arnal, Charles, Grutter, & Morand (unpublished observations) developed a model that predicted the age of maturity of ectoparasites, a trait that influences the size at maturity and hence the fecundity of ectoparasites, as a function of cleaner fish abundance. There are great local and regional variations in cleaner density (Green, 1996; Losey, 1987; Arnal, Morand & Kulbiki, 1999). Thus, depending on the location, predation pressure on ectoparasite populations should differ: i.e. there should be a high predation risk where cleaner fishes are abundant and a low predation risk where cleaner density is low. Arnal *et al.* (unpublished observations) hypothesised that the evolution of the life history traits of ectoparasites should thus vary as a function of cleaner density. A relationship between age at maturity, body size and fecundity in parasites, has been shown in the case of helminth parasites (Gemmill & Read, 1998; Morand & Sorci, 1998; Read & Allen, 2000) and between body size and fecundity in the case of gnathiids (Tanaka & Aoki, 2000). Arnal *et al.* (unpublished observations) therefore assumed that the fecundity of gnathiid isopods is related to age at maturity. Their model predicts that a high density of cleaner fish selects for an early age in maturity in ectoparasites and thus that fish cleaning behaviour is a selective factor acting on ectoparasites. This is supported by observations showing that gnathiids eaten by *L. dimidiatus* in lower latitude areas where cleaners appear to be more abundant (Côté, 2000) moult earlier and mature earlier (Grutter, 1999*a*; Grutter & Hendrikz, 1999; Paperna & Por, 1977; Tanaka & Aoki, 2000) than do gnathiids in other areas (Klitgaard, 1991; Stoll, 1962; Wägele, 1987). Whether this is just due to geographic and/or temperature variation, however, needs to be tested.

It should be noted, however, that most of the above studies on gnathiids have only dealt with unidentified juvenile stages (gnathiids are generally only identified from males (Cohen & Poore, 1994)). As there is evidence that several gnathiid species are found on a single host (Grutter *et al.* 2000), the

juveniles exhibit a wide range of different colour patterns (Murphy & Grutter, unpublished observations) which has been linked to their species identity (Grutter *et al.* 2000), their size range is large (Grutter, unpublished observations), and their mobility and life-cycle (moulting rates) differ between fish groups (e.g. teleosts and elasmobranchs) (Grutter, unpublished observations) it is highly likely that numerous species are involved in cleaning interactions. This raises the possibility that some cleaners may prefer some species over others which may in turn explain some of the patterns discussed above. Clearly, more precise information is needed on the identity of the parasites involved in cleaning interactions.

OTHER FACTORS AFFECTED BY CLEANERS

Parasite transmission

Cleaners may become infected with the parasites and diseases of their clients. The intimate contact between cleaners and clients may increase their transmission, particularly for parasites with direct transmission between hosts. There is some evidence that cleaners become infected with client parasites. Individual *Oxyjulis californica* engaged in cleaning have similar parasitic copepods to the client species they clean (Hobson, 1971) implying they may get them from their clients. *L. dimidiatus* become infected with gnathiid isopods, their main food source, in the laboratory (Grutter, unpublished observations).

Whether cleaners transmit parasites or diseases is of great concern for cleaners used in fish farms (Costello *et al.* 1996). Corkwing cleaners became infected when held with salmon infected with the bacterium *Vibrio* sp. (Costello *et al.* 1996). Similarly, *Aeromonas salmonicida* has been transmitted from farmed Atlantic salmon (most likely dead) to cleaners *Centrolabrus exoletus* and *C. rupestris* in salmon cages (Treasurer & Cox, 1991; Treasurer & Laidler, 1994). *Centrolabrus rupestris* become infected with infectious pancreatic necrosis virus from salmon (Gibson, Smail & Sommerville, 1998).

However, most cleaners clean conspecifics which may explain the low number of parasites on some cleaners (Costello, 1996; Costello *et al.* 1996). Of 5 species of cleaners, 2 that do not engage in intraspecific cleaning had a greater proportion of external parasites (Costello, 1996; Costello *et al.* 1996).

Less is known of the transmission of disease by cleaners to clients. Parasites on temperate cleaners are common (Costello, 1991; Karlsbakk, Hodneland & Nylund, 1996). The probability of their transmission to farmed salmon is low as most parasites are either specific to labrid fish or require that the host be eaten or be passed on to an invertebrate to complete its life-cycle (Costello *et al.* 1996). One of these cleaners, *Centrolabrus rupestris* may be a source of re-infection of infectious pancreatic necrosis virus in farmed salmon (Gibson *et al.* 1998).

These raise the issue of the host specificity of the parasite infecting the cleaner and whether such parasites actually persist on the cleaner. Interestingly, some cleaners appear to be immune to particular parasites. For example, *L. dimidiatus* avoided infection by the parasitic ciliate *Cryptocaryon irritans*, the agent of white spot disease, even after 3 weeks of exposure in captivity (Halliday, unpublished observations). Whether this was due to host specificity is unclear.

Finally, parasites often have parasites themselves. Cleaners may therefore also play a role in their transmission. For example, gnathiids have hyperparasites such as haemogregarines and larval nematodes (Davies, Eiras & Austin, 1994; Davies & Johnston, 2000; Monod, 1926; Smit, 2000; Smit & Davies, 1999). By eating gnathiids, cleaners may either transmit these or alternatively, reduce their populations.

Predation avoidance in parasites

Some parasites appear to avoid being eaten by cleaners as they make no attempt to conceal themselves. Lernaeid and penellid copepods on the external surface of fish often have trailing egg cases or body parts (Kabata, 1992). Although fish cleaned by the cleaner fish *Oxyjulis californica* have lernaeids, none were found in their diet (Hobson, 1971). Why such highly visible parasites are not eaten is unknown. Losey (1987) proposed that this practice may invite predation of eggs to increase dispersal or infection of an intermediate host. Some lernaeids are, however, eaten by *Labroides phthirophagus* (Randall, 1958; Youngbluth, 1968) and topminnows *Fundulus heteroclitus* (McCutcheon & McCutcheon, 1964). Such behaviours also raise questions about the parasite's palatability. Cymothoid isopods are often also highly exposed; Losey (1987) proposed their armour and size, however, may provide them with some form of protection against predators. Such 'immunity' from predation in parasites may also be linked to the parasite's host specificity.

CONCLUSIONS

A wide range of parasites are involved in cleaning interactions. There is some support for counter-adaptations of parasites against predation from cleaners. The evidence is, however, mainly circumstantial indicating that more studies are needed. It is also clear that in most examples above, the ecological role of parasites in cleaning interactions is little understood. In most of these cases, reports where parasites were involved in cleaning interactions were not studied further. Yet, the diversity of

parasites and wide range of hosts from diverse environments suggest that cleaning may be more common and widespread than previously thought. More observations are needed in temperate and polar seas.

Of these, gnathiid isopods and caligid copepods stand out as common parasites eaten by cleaner fishes. Why these parasites are so commonly eaten deserves attention and would increase our understanding of cleaning behaviour.

When evaluating 'adaptations', it should be noted that adaptations present in one organism are not necessarily adaptations in another organism (Poulin, 1995c). As Poulin (1995c) proposed for the supposed 'adaptive' changes in behaviour of parasitized animals, many of the 'adaptations' of parasites against predators proposed in this study are based only on intuition and not on rigorous criteria. A similar set of conditions used for host behaviour (Poulin, 1995c) could be applied to the 'adaptive' changes in parasites in response to predators: these include complexity, purposive design, convergence among different lineages and fitness benefits to the parasites.

Finally, the role of parasites in cleaning interactions is not one-way, with cleaners only affecting the parasites. Complex interactions between the three main players, the parasites, the cleaners, and the clients occur (Losey, 1987). For example, these include the effect of parasites on the client's cleaning behaviour (Grutter, 2001) and schooling behaviour (Reinthal & Lewis, 1986), the effect of cleaning behaviour on the feeding behaviour of clients (Grutter *et al.* in press), and the effect of cleaners on the parasites of the parasites they feed on (see above). In addition, there are other participants which are affected by cleaning interactions including aggressive mimics (Wickler, 1968) and Batesian mimics of cleaners (Zander & Nieder, 1997).

REFERENCES

ABLE, K. W. (1976). Cleaning behaviour in the cyprinodontid fishes: *Fundulus majalis*, *Cyprinodon variegatus*, and *Lucania parva*. *Chesapeake Science* **17**, 35–39.

ALLEN, G. R., STEENE, R. & ALLEN, M. (1998). *A Guide to Angel Fishes and Butterflyfishes*. Perth, Australia, Van Guard Press.

ARNAL, C. & CÔTÉ, I. M. (2000). Diet of broadstripe cleaning gobies on a Barbadian reef. *Journal of Fish Biology* **57**, 1075–1082.

ARNAL, C. & MORAND, S. (2001). Importance of ectoparasites and mucus in cleaning interactions in the Mediterranean Sea. *Marine Biology* **138**, 777–784.

ARNAL, C., MORAND, S. & KULBICKI, M. (1999). Patterns of cleaner wrasse density among three regions of the Pacific. *Marine Ecology Progress Series* **177**, 213–220.

AYLING, A. M. & GRACE, R. V. (1971). Cleaning symbiosis among New Zealand fishes. *New Zealand Journal of Marine and Freshwater Research* **5**, 205–218.

BEEBE, W. (1926). *Galapagos World End*. New York, G. P. Putnam's Sons.

BOHLKE, J. E. & MCCOSKER, J. E. (1973). Two additional West Atlantic gobies (genus *Gobiosoma*) that remove ectoparasites from other fishes. *Copeia* 1973, 609–610.

BOOTH, J. & PETERS, J. A. (1972). Behavioural studies on the green turtle (*Chelonia mydas*) in the sea. *Animal Behaviour* 20, 808–812.

BUNKLEY-WILLIAMS, L. & WILLIAMS JR, E. H. (1998). Ability of Pederson cleaner shrimp to remove juveniles of the parasitic cymothoid isopod, *Anilocra haemuli*, from the host. *Crustaceana* 71, 862–869.

CARR, W. E. & ADAMS, C. A. (1972). Food habits of juvenile marine fishes: evidence of the cleaning habit in the leatherjacket, *Oligoplites saurus*, and the spottail pinfish, *Diplodus holbrooki*. *Fishery Bulletin* 70, 1111–1120.

CLAYTON, D. H. & TOMPKINS, D. M. (1994). Ectoparasite virulence is linked to mode of transmission. *Proceedings of the Royal Society of London. Series B* 246, 211–217.

COHEN, G. F. & POORE, G. C. B. (1994). Phylogeny and biogeography of the Gnathiidae (Crustacea: Isopoda) with descriptions of new genera and species, most from south-eastern Australia. *Memoirs of the Museum of Victoria* 54, 271–397.

CORREDOR, L. (1978). Notes on the behaviour and ecology of the new fish cleaner shrimp *Brachycarpus biunguiculatus* (Lucas) (Decapoda natantia, Palaemonidae). *Crustaceana* 35, 35–40.

COSTELLO, M. J. (1991). Review of the biology of wrasse (Labridae: Pisces) in Northern Europe. *Progress in Underwater Science* 16, 29–51.

COSTELLO, M. J. (1996). Development and future of cleaner-fish technology and other biological control techniques in fish farming. In *Wrasse: Biology and Use in Aquaculture* (ed. Sayer, M. D. J., Treasurer, J. W. and Costello, M. J.), pp. 171–184. Oxford, Fishing News Books.

COSTELLO, M. J., DEADY, S., PIKE, A. & FIVES, J. M. (1996). Parasites and diseases of wrasse being used as cleaner-fish on salmon farms in Ireland and Scotland. In *Wrasse: Biology and Use in Aquaculture* (ed. Sayer, M. D. J., Treasurer, J. W. and Costello, M. J.), pp. 211–227. Oxford, Fishing News Books.

CÔTÉ, I. (2000). Evolution and ecology of cleaning symbioses in the sea. *Oceanography and Marine Biology: An Annual Review* 38, 311–355.

CÔTÉ, I. M., ARNAL, C. & REYNOLDS, J. D. (1998). Variation in posing behaviour among fish species visiting cleaning stations. *Journal of Fish Biology* 53, 256–266.

COWELL, L. E., WATANABE, W. O., HEAD, W. D., GROVER, J. J. & SHENKER, J. M. (1993). Use of tropical cleaner fish to control the ectoparasite *Neobenedenia melleni* (Monogenea: Capsalidae) on seawater-cultured Florida red tilapia. *Aquaculture* 113, 189–200.

CRESSEY, R. F. & LACHNER, E. A. (1970). The parasitic copepod diet and life history of diskfishes (Echeneidae). *Copeia* 1970, 310–318.

DAVENPORT, J. (1994). A cleaning association between the Oceanic Crab *Planes minutus* and the Loggerhead Sea Turtle *Caretta caretta*. *Journal of the Marine Biological Association of the United Kingdom* 74, 735–737.

DAVIES, A. J., EIRAS, J. C. & AUSTIN, R. T. E. (1994). Investigations into the transmission of *Haemogregarina bigemina* Laveran & Mesnil, 1901 (Apicomplexa: Adeleorina) between intertidal fishes in Portugal. *Journal of Fish Diseases* 17, 283–289.

DAVIES, A. J. & JOHNSTON, M. R. L. (2000). The biology of some intraerythrocytic parasites of fishes, amphibia and reptiles. *Advances in Parasitology* 45, 1–107.

DEADY, S., VARIAN, S. J. A. & FIVES, J. M. (1995). The use of cleaner-fish to control sea lice on two Irish salmon (*Salmo salar*) farms with particular reference to wrasse behaviour in salmon cages. *Aquaculture* 131, 73–90.

DEMARTINI, E. E. & COYER, J. A. (1981). Cleaning and scale-eating in juveniles of the Kyphosid fishes, *Hermosilla azurea* and *Girella nigricans*. *Copeia* 1981, 785–789.

DEVENEY, M. R. & WHITTINGTON, I. D. (2001). A technique for preserving pigmentation in some capsalid monogeneans for taxonomic purposes. *Systematic Parasitology* 48, 31–35.

ENDLER, J. A. (1991). Interactions between predator and prey. In *Behavioural Ecology: An Evolutionary Approach*, 3rd edn., (ed. Krebs, J. R. and Davies, N. B.), pp. 169–196. Oxford, Blackwell Science Publications.

EWALD, P. W. (1995). The evolution of virulence: a unifying link between parasitology and ecology. *Journal of Parasitology* 81, 659–669.

FEDER, H. M. (1966). Cleaning symbiosis in the marine environment. In *Symbiosis* (ed. Henry, S. M.), pp. 327–380. New York, Academic Press.

FOSTER, S. A. (1985). Wound healing: a possible role of cleaning stations. *Copeia* 1985, 875–880.

FRANK, S. A. (1996). Models of parasite virulence. *Quarterly Review of Biology* 71, 37–78.

GALEOTE, M. D. & OTERO, J. G. (1998). Cleaning behaviour of rock cook, *Centrolabrus exoletus* (Labridae), in Tarifa (Gibraltar Strait Area). *Cybium* 22, 57–68.

GIBSON, D. I., SMAIL, D. A. & SOMMERVILLE, C. (1998). Infectious pancreatic necrosis virus: experimental infection of goldsinny wrasse, *Ctenolabrus rupestris* L. (Labridae). *Journal of Fish Diseases* 21, 399–406.

GEMMILL, A. W. & READ, A. F. (1998). Counting the cost of disease resistance. *Trends in Ecology and Evolution* 13, 8–9.

GLASBY, T. M. & KINGSFORD, M. J. (1994). *Atypichthys strigatus* (Pisces: Scorpididae): An opportunistic planktivore that responds to benthic disturbances and cleans other fishes. *Australian Journal of Ecology* 19, 385–394.

GOODING, R. M. (1964). Observations of fish from a floating observation raft at sea. *Proceedings of the Hawaiian Academy of Science* 39, 27.

GORLICK, D. L., ATKINS, P. D. & LOSEY, G. S. (1987). Effect of cleaning by *Labroides dimidiatus* (Labridae) on an ectoparasite population infecting *Pomacentrus vaiuli* (Pomacentridae) at Enewetak Atoll. *Copeia* 1987, 41–45.

GREEN, A. L. (1996). Spatial, temporal and ontogenetic patterns of habitat use by coral reef fishes (Family Labridae). *Marine Ecology Progress Series* 133, 1–11.

GRUTTER, A. S. (1995*a*). Relationship between cleaning rates and ectoparasite loads in coral reef fishes. *Marine Ecology Progress Series* 118, 51–58.

GRUTTER, A. S. (1995*b*). Comparison of methods for sampling ectoparasites from coral reef fishes. *Marine and Freshwater Research* **46**, 897–903.

GRUTTER, A. S. (1996*a*). Parasite removal rates by the cleaner wrasse *Labroides dimidiatus*. *Marine Ecology Progress Series* **130**, 61–70.

GRUTTER, A. S. (1996*b*). Experimental demonstration of no effect by the cleaner wrasse *Labroides dimidiatus* (Cuvier and Valenciennes) on the host fish *Pomacentrus moluccensis* (Bleeker). *Journal of Experimental Marine Biology and Ecology* **196**, 285–298.

GRUTTER, A. S. (1997*a*). Spatio-temporal variation and feeding selectivity in the diet of the cleaner fish *Labroides dimidiatus*. *Copeia* **1997**, 346–355.

GRUTTER, A. S. (1997*b*). Size-selective predation by the cleaner fish *Labroides dimidiatus*. *Journal of Fish Biology* **50**, 1303–1308.

GRUTTER, A. S. (1999*a*). Cleaner fish really do clean. *Nature* **398**, 672–673.

GRUTTER, A. S. (1999*b*). Infestation dynamics of parasitic gnathiid isopod juveniles on a coral reef fish *Hemigymnus melapterus*. *Marine Biology* **135**, 545–552.

GRUTTER, A. S. (1999*c*). Fish cleaning behaviour in Noumea, New Caledonia. *Marine and freshwater Research* **50**, 209–212.

GRUTTER, A. S. (2000). Ontogenetic variation in the diet of the cleaner fish *Labroides dimidiatus* and its ecological consequences. *Marine Ecology Progress Series* **197**, 241–246.

GRUTTER, A. S. (2001). Parasite infection rather than tactile stimulation is the proximate cause of cleaning behaviour in reef fish. *Proceedings of the Royal Society of London. Series B* **268**, 1361–1365.

GRUTTER, A. S. & HENDRIKZ, J. (1999). Diurnal variation in the abundance of parasitic gnathiid isopod larvae on coral reef fish: its implications in cleaning interactions. *Coral Reefs* **18**, 187–191.

GRUTTER, A. S. & LESTER, R. J. G. (in press). Cleaner fish *Labroides dimidiatus* reduce *Argathona macronema* (Corallanidae) isopod infection on the coral reef fish *Hemigymnus melapterus*. *Marine Ecology Progress Series*.

GRUTTER, A. S., McCALLUM, H. I. & LESTER, R. J. G. (in press). Optimising cleaning behaviour: minimising the costs and maximising ectoparasite removal. *Marine Ecology Progress Series*.

GRUTTER, A. S., MORGAN, J. A. T. & ADLARD, R. D. (2000). Characterising parasitic gnathiid isopod species and matching life stages using ribosomal DNA ITS2 sequences. *Marine Biology* **136**, 201–205.

GRUTTER, A. S. & POULIN, R. (1998). Intraspecific and interspecific relationships between host size and the abundance of parasitic larval gnathiid isopods on coral reef fishes. *Marine Ecology Progress Series* **164**, 263–271.

HARVEY, P. H. & PAGEL, M. D. (1991). *The Comparative Method in Evolutionary Biology*. Oxford, Oxford University Press.

HASTINGS, A. (2000). The lion and the lamb find closure. *Science* **290**, 712–713.

HENRIQUES, M. & ALMADA, V. C. (1997). Relative importance of cleaning behaviour in *Centrolabrus*

exoletus and other wrasse at Arrabida, Portugal. *Journal of the Marine Biological Association of the United Kingdom* **77**, 891–898.

HEUPEL, M. R. & BENNETT, M. B. (1999). The occurrence, distribution and pathology associated with gnathiid isopod larvae infecting the epaulette shark, *Hemiscyllium ocellatum*. *International Journal for Parasitology* **29**, 321–330.

HOBSON, E. S. (1965). Diurnal-nocturnal activity of some inshore fishes in the Gulf of California. *Copeia* **1965**, 291–302.

HOBSON, E. S. (1969). Remarks on aquatic habits of the Galapagos marine iguana, including submergence times, cleaning symbiosis, and the shark threat. *Copeia* **1969**, 401–402.

HOBSON, E. S. (1971). Cleaning symbiosis among California inshore fishes. *Fishery Bulletin* **69**, 491–523.

HOBSON, E. S. (1972). Activity of Hawaiian reef fishes during the evening and morning transitions between daylight and darkness. *Fishery Bulletin* **70**, 715–740.

HOBSON, E. S. & CHESS, J. R. (1976). Trophic interactions among fishes and zooplankters near shore at Santa Catalina Island, California. *Fishery Bulletin* **74**, 567–598.

HOLDICH, D. M. & HARRISON, K. (1980). The crustacean isopod genus *Gnathia* Leach from Queensland waters with descriptions of nine new species. *Australian Journal of Marine and Freshwater Research* **31**, 215–240.

HUSSAIN, N. A. & KNIGHT-JONES, E. W. (1995). Fish and fish-leeches on rocky shores around Britain. *Journal of the Marine Biological Association of the United Kingdom* **75**, 311–322.

JOHNSON, W. S. (1982). A record of cleaning symbiosis involving *Gobiosoma* sp. and a large Caribbean octopus. *Copeia* **1982**, 712–714.

JOHNSON, W. S. & RUBEN, P. (1988). Cleaning behavior of *Bodianus rufus*, *Thalassoma bifasciatum*, *Gobiosoma evelynae*, and *Periclimenes pedersoni* along a depth gradient at Salt River Submarine Canyon, St Croix. *Environmental Biology of Fishes* **23**, 225–232.

KABATA, Z. (1992). *Copepods Parasitic on Fishes*, Oegstgeest, The Netherlands, Universal Book Services/Dr W. Backhuys.

KARLSBAKK, E., HODNELAND, K. & NYLUND, A. (1996). Health status of goldsinny wrasse, including a detailed examination of the parasite community at Flodevigen, Southern Norway. In *Wrasse: Biology and Use in Aquaculture* (ed. Sayer, M. D. J., Treasurer, J. W. and Costello, M. J.), pp. 228–239. Oxford, Fishing News Books.

KEARN, G. C. (1976). *Body Surface of Fishes*. Amsterdam, North-Holland Publishing Company.

KEARN, G. C. (1978). Predation on a skin-parasitic monogenean by a fish. *Journal of Parasitology* **64**, 1129–1130.

KEARN, G. C. (1979). Studies on gut pigmentation in skin-parasitic monogeneans, with special reference to the monocotylid *Dendromonocotyle kuhlii*. *International Journal for Parasitology* **9**, 545–552.

KEARN, G. C. (1994). Evolutionary expansion of the Monogenea. *International Journal for Parasitology* **24**, 1227–1271.

KEARN, G. C. (1999). The survival of monogenean (platyhelminth) parasites on fish skin. *Parasitology* **119** (Suppl.), S57–S88.

KEARN, G. C. & WHITTINGTON, I. D. (1992). A response to light in an adult encotyllabine (Capsalid) monogenean from the pharyngeal tooth pads of some marine teleost fishes. *International Journal for Parasitology* **22**, 119–121.

KEELING, M. J., WILSON, H. B. & PACALA, S. W. (2000). Reinterpreting space, time lags, and functional responses in ecological models. *Science* **290**, 758–1761.

KEYES, R. S. (1982). Sharks: an unusual example of cleaning symbiosis. *Copeia* **1982**, 225–227.

KLITGAARD, A. B. (1991). *Gnathia abyssorum* (G. O. Sars, 1872) (Crustacea, Isopoda) associated with sponges. *SARSIA* **76**, 33–40.

KOELLA, J. C. & AGNEW, P. (1999). A correlated response of a parasite's virulence and life cycle to selection on its host's life history. *Journal of Evolutionary Biology* **21**, 70–79.

KOELLA, J. C. & DOEBELI, M. (1999). Population dynamics and the evolution of virulence in epidemiological models with discrete host generations. *Journal of Theoretical Biology* **198**, 461–475.

KRAWCHUK, M. A., KOPER, N. & BROOKS, R. J. (1997). Observations of a possible cleaning symbiosis between painted turtles, *Chrysemys picta*, and snapping turtles, *Chelydra serpentina*, in Central Ontario. *The Canadian Field-Naturalist* **111**, 315–317.

KREBS, J. R. & DAVIES, N. B. (1993). *An Introduction to Behavioural Ecology*. Third edition. Oxford, UK, Blackwell Science.

LAST, P. R. & STEVENS, J. D. (1994). *Sharks and Rays of Australia*. Melbourne, CSIRO Australia.

LIMBAUGH, C. (1961). Cleaning Symbiosis. *Scientific American* **205**, 42–49.

LOSEY, G. S. (1972). The ecological importance of cleaning symbiosis. *Copeia* **1972**, 820–833.

LOSEY, G. S. (1974). Cleaning symbiosis in Puerto Rico with comparison to the tropical pacific. *Copeia* **1974**, 960–970.

LOSEY, G. S. (1987). Cleaning Symbiosis. *Symbiosis* **4**, 229–258.

LOSEY, G. S., BALAZS, G. H. & PRIVITERA, L. A. (1994). Cleaning symbiosis between the wrasse, *Thalassoma duperry*, and the green turtle, *Chelonia mydas*. *Copeia* **1994**, 684–690.

LUCAS, J. R. & BENKERT, K. A. (1983). Variable foraging and cleaning behavior by juvenile leatherjackets, *Oligoplites saurus* (Carangidae). *Estuaries* **6**, 247–250.

MALLET, J. & JORON, M. (1999). Evolution of diversity in warning color and mimicry: polymorphisms, shifting balance, and speciation. *Annual Review of Ecology and Systematics* **30**, 201–233.

McCUTCHEON, R. H. & McCUTCHEON, A. E. (1964). Symbiotic behavior among fishes from temperate ocean waters. *Science* **145**, 948–949.

MONOD, T. (1926). Les gnathiidae. Essai monographique (morphologie, biologie, systématique). *Mémoires de la Société des Sciences Naturelles du Maroc* **13**, 1–661.

MOORING, M. S. & MUNDY, P. J. (1996). Interactions between impala and oxpeckers at Matobo National Park, Zimbabwe. *African Journal of Ecology* **34**, 54–65.

MORAND, S. & SORCI, G. (1998). Determinants of life-history evolution in nematodes. *Parasitology Today* **14**, 193–196.

NICOLETTE, P. (1990). *Symbiosis: Nature in Partnership*. London, Blandford.

PAPERNA, I. & POR, F. D. (1977). Preliminary data on the Gnathiidae (Isopoda) of the Northern Red Sea, the Bitter Lakes and the Eastern Mediterranean and the Biology of *Gnathia piscivora* n. sp. *Rapports de la Commission Internationale pour la Mer Méditerranée* **24**, 195–197.

PARTRIDGE, L. & HARVEY, P. H. (1988). The ecological context of life history evolution. *Science* **241**, 1449–1454.

POTTS, G. W. (1973). Cleaning symbiosis among British fish with special reference to *Crenilabrus melops* (Labridae). *Journal of the Marine Biological Association of the United Kingdom* **53**, 1–10.

POULIN, R. (1995*a*). Evolution of parasite life history traits: Myths and reality. *Parasitology Today* **11**, 342–345.

POULIN, R. (1995*b*). Evolutionary influences on body size in free-living and parasitic isopods. *Biological Journal of the Linnean Society* **54**, 231–244.

POULIN, R. (1995*c*). "Adaptive" changes in the behaviour of parasitized animals: A critical review. *International Journal for Parasitology* **25**, 1371–1383.

POULIN, R. (1996). The evolution of life history strategies in parasitic animals. *Advances in Parasitology* **37**, 107–134.

RANDALL, J. E. (1958). A review of the labrid fish genus *Labroides*, with description of two new species and notes on ecology. *Pacific Science* **12**, 327–347.

READ, A. F. & ALLEN, J. E. (2000). The economics of immunity. *Science* **290**, 1104–1105.

REINTHAL, P. N. & LEWIS, S. M. (1986). Social behaviour, foraging efficiency and habitat utilization in a group of tropical herbivorous fish. *Animal Behaviour* **34**, 1687–1693.

RIDLEY, M. (1993). *Evolution*. Oxford, Blackwell Scientific Publications.

ROBERTSON, D. R. & CHOAT, J. H. (1974). Protogynous hermaphroditism and social systems in labrid fish. *Proceedings of the Second International Coral Reef Symposium 1. Great Barrier Reef Committee*, 217–225.

ROUBAL, F. R. & QUARTARARO, N. (1992). Observations on the pigmentation of the monogeneans, *Anoplodiscus* spp. (Family Anoplodiscidae) in different microhabitats on their sparid teleost hosts. *International Journal for Parasitology* **22**, 459–464.

SAMEULSEN, T. J. (1981). Der seeteufel (*Lophius piscatorius* L.) in Gefangenschaft. *Zeitschrift Kolner Zoo* **24**, 17–19.

SANO, M., SHIMIZU, M. & NOSE, Y. (1984). Food habits of teleostean reef fishes in Okinawa Island, southern Japan. *The University Museum, The University of Tokyo, Bulletin* **25**, 1–128.

SAZIMA, I. & MOURA, R. L. (2000). Shark (*Carcharhinus perezi*), cleaned by the goby (*Elacatinus randalli*), at Fernando de Noronha Archipelago, Western South Atlantic. *Copeia* **2000**, 297–299.

SAZIMA, I., MOURA, R. L. & GASPARINI, J. L. (1998). The wrasse *Halichoeres cyanocephalus* (Labridae) as a specialized cleaner fish. *Bulletin of Marine Science* **63**, 605–610.

SENN, D. G. (1979). Zur Biologie des Putzerfisches *Crenilabrus melanocercus* (Risso). *Senckenbergiana maritima* **11**, 23–38.

SMIT, N. J. (2000). A trypanosome from the silver catfish (*Schilbe intermedius*) in the Okavango Delta, Botswana. *Bulletin of the European Association of Fish Pathologists* **20**, 116–119.

SMIT, N. J. & DAVIES, A. J. (1999). New host records for *Haemogregarina bigemina* from the coast of southern Africa. *Journal of the Marine Biological Association of the United Kingdom* **79**, 933–935.

SPOTTE, S. (1998). "Cleaner" shrimps? *Helgolander Meeresuntersuchungen* **52**, 59–64.

STEPIEN, C. A. & BRUSCA, R. C. (1985). Nocturnal attacks on nearshore fishes in southern California by crustacean zooplankton. *Marine Ecology Progress Series* **25**, 91–105.

STOLL, C. (1962). Cycle évolutif de *Paragnathia formica* (Hesse) (Isopode – Gnathiidae). *Cahiers de Biologie Marine* **3**, 401–416.

STRASBURG, D. W. (1959). Notes on the diet and correlating structures of some central Pacific echeneid fishes. *Copeia* **1959**, 244–248.

SWARTZ, S. L. (1981). Cleaning symbiosis between topsmelt, *Atherinops affinis*, and Gray Whale, *Eschrichtius robustus*, in Laguana San Ignacio, Baja California Sur, Mexico. *Fishery Bulletin* **79**, 360.

SZIDAT, L. & NANI, A. (1951). Las remoras del Atlantico Austral con un estudio de su nutricion natural y de parasitos (Pisc. Echeneidae). *Revista del Museo Argentino de Ciencias Naturales* **2**, 385–417.

TANAKA, K. & AOKI, M. (2000). Seasonal traits of reproduction in a gnathiid isopod *Elaphognathia cornigera* (Nunomura, 1992). *Zoological Science* **17**, 467–475.

TREASURER, J. W. & COX, D. (1991). The occurrence of *Aeromonas salmonicida* in wrasse (Labridae) and implications for Atlantic salmon farming. *Bulletin of the European Association for Fish Pathology* **11**, 208–210.

TREASURER, J. W. & LAIDLER, L. A. (1994). *Aeromonas salmonicida* infection in wrasse (Labridae), used as cleaner fish, on an Atlantic salmon, *Salmo salar* L., farm. *Journal of Fish Diseases* **17**, 155–161.

TYLER, A. V. (1963). A cleaning symbiosis between the rainwater fish, *Lucania parva* and the stickleback, *Apeltes quadracus*. *Chesapeake Science* **4**, 105–106.

UPTON, N. P. D. (1987). Asynchronous male and female life cycles in the sexually dimorphic, harem-forming isopod *Paragnathia formica* (Crustacea: Isopoda). *Journal of Zoology* **212**, 677–690.

VAN TASSELL, J. L., BRITO, A. & BORTONE, S. A. (1994). Cleaning behavior among marine fishes and invertebrates in the Canary Islands. *Cybium* **18**, 117–127.

VOGT, R. C. (1979). Cleaning/feeding symbiosis between grackles (*Quiscalus*: Icteridae) and map turtles (*Graptemys*: Emydidae). *Auk* **96**, 608–609.

WÄGELE, J. W. (1987). Description of the postembryonal stages of the Antarctic Fish Parasite *Gnathia calva* Vanhoffen (Crustacea: Isopoda) and synonymy with *Heterognathia* Amar & Roman. *Polar Biology* **7**, 77–92.

WAHLERT, G. V. & WAHLERT, H. V. (1961). Le comportement de nettoyage de *Crenilabrus melanocercus* (Labridae, Pisces) en Mediterranee. *Vie et Milieu* **12**, 1–10.

WALLERSTEIN, B. R. & BRUSCA, R. C. (1982). Fish predation: a preliminary study of its role in the zoogeography and evolution in shallow water idoteid isopods (Crustacea: Isopoda: Idoteidae). *Journal of Biogeography* **9**, 135–150.

WHITTINGTON, I. D. (1996). Benedeniine capsalid monogeneans from Australian fishes: pathogenic species, site-specificity and camouflage. *Journal of Helminthology* **70**, 177–184.

WHITTINGTON, I. D. & KEARN, G. C. (1990). Effects of urea analogs on egg hatching and movement of unhatched larvae of monogenean parasite *Acanthocotyle lobianchi* from skin of *Raja montagui*. *Journal of Chemical Ecology* **16**, 3523–3529.

WICKLER, W. (1968). The origin of the cleaner mimic. In *Mimicry in Plants and Animals* (ed. Wickler, W.), pp. 157–176. London, Weidenfeld and Nicolson.

WYMAN, R. L. & WARD, J. A. (1972). A cleaning symbiosis between the cichlid fishes *Etroplus maculatus* and *Etroplus suratensis*. I. Description and possible evolution. *Copeia* **1972**, 834–838.

YOUNGBLUTH, M. J. (1968). Aspects of the ecology and ethology of the cleaning fish, *Labroides phthirophagus* Randall. *Zeitschrift für Tierpsychologie* **25**, 915–932.

ZANDER, C. D. & NIEDER, J. (1997). Interspecific associations in Mediterranean fishes: feeding communities, cleaning symbioses and cleaner mimics. *Vie et Milieu* **47**, 203–212.

Food webs and the transmission of parasites to marine fish

D. J. MARCOGLIESE

St. Lawrence Centre, Environment Canada, 105 McGill, 7th Floor, Montreal, Quebec, Canada H2Y 2E7

SUMMARY

Helminth parasites of fish in marine systems are often considered to be generalists, lacking host specificity for both intermediate and definitive hosts. In addition, many parasites in marine waters possess life cycles consisting of long-lived larval stages residing in intermediate and paratenic hosts. These properties are believed to be adaptations to the long food chains and the low densities of organisms distributed over broad spatial scales that are characteristic of open marine systems. Moreover, such properties are predicted to lead to the homogenization of parasite communities among fish species. Yet, these communities can be relatively distinct among marine fishes. For benthos, the heterogeneous horizontal distribution of invertebrates and fish with respect to sediment quality and water depth contributes to the formation of distinct parasite communities. Similarly, for the pelagic realm, vertical partitioning of animals with depth will lead to the segregation of parasites among fish hosts. Within each habitat, resource partitioning in terms of dietary preferences of fish further contributes to the establishment of distinct parasite assemblages. Parasite distributions are predicted to be superimposed on distributional patterns of free-living animals that participate as hosts in parasite life cycles. The purpose of this review is first, to summarize distribution patterns of invertebrates and fish in the marine environment and relate these patterns to helminth transmission. Second, patterns of transmission in marine systems are interpreted in the context of food web structure. Consideration of the structure and dynamics of food webs permits predictions about the distribution and abundance of parasites. Lastly, parasites that influence food web structure by regulating the abundance of dominant host species are briefly considered in addition to the effects of pollution and exploitation on food webs and parasite transmission.

Key words: Marine, transmission, food webs, parasites, fish, zooplankton, benthos.

INTRODUCTION

The marine environment has more higher taxa and twice the number of phyla than tropical rainforests (Suchanek, 1994). Of the 32 phyla occurring in the oceans, 21 are exclusively marine, of which 10 are endemic to the benthos and 1 to the pelagic realm (May, 1994). Generally, species diversity increases along a gradient from boreal estuaries, through boreal shallow marine, with tropical shallow marine and deep-sea benthos being the most diverse (Hessler & Sanders, 1967; Sanders, 1968). Temperate coastal marine communities are among the most productive and diverse ecosystems on earth (Suchanek, 1994) and inshore and shelf habitats possess the most species on a large geographic scale (May, 1994). Species diversity in the deep sea alone may resemble that of tropical rainforests, though most species are rare (Snelgrove & Grassle, 1995). These organisms are integrated together into complex food webs of long average chain lengths (Schoener, 1989). Helminth parasites track these food web interactions in order to propagate and maintain themselves in the marine milieu (Marcogliese & Cone, 1997). The unique fauna found in the oceans, conceivably, should provide new pathways on which parasites may capitalise within food webs, ultimately facilitating parasite diversification.

Tel: 514-283-6499 Fax: 514-496-7398.
E-mail: david.marcogliese@ec.gc.ca

Numerous characteristics of parasites in marine systems appear at first glance to homogenize their distributions among hosts, leading to undifferentiated communities within regions. Parasites of marine fish tend to be generalists for both intermediate and definitive hosts (Polyanski, 1961; Holmes, 1990); this may be an adaptation to spread the risk among hosts to ensure transmission in a dilute environment (Bush, 1990). Marine parasites tend to be long lived (Campbell, 1983) and many marine parasites indiscriminantly infect paratenic and transport hosts, also possibly as an adaptation to the longer food chains and dilute oceanic environment (Marcogliese, 1995). Among hosts, marine vertebrates tend to be large, extremely vagile and gregarious, with generalized broad diets. This generalized feeding mechanism allows for prey switching and dietary overlap, creating a highly diversified diet in comparison with terrestrial environments. Fish spawning migrations can further obliterate the local character of parasite faunas (Polyanski, 1961). All these factors together would theoretically lead to the homogenization of parasite communities among host species.

Yet, parasite communities differ substantially among fish hosts, even in the same geographic area. The abundant evidence that parasites can be employed as indicators of fish stocks or populations (Williams, MacKenzie & McCarthy, 1992; Arthur, 1997) demonstrates the variability of the marine parasite fauna among host species. Fish and inverte-

Table 1. Some examples of intermediate hosts of benthically-transmitted parasites in the marine environment. Examples were chosen to represent different trophically-transmitted pathways used by parasites and a diversity of invertebrate intermediate hosts. Where possible, general sources were used as references. Only parasites that require ingestion of intermediate hosts by definitive hosts are included

Parasite	Definitive host	1st intermediate host	2nd host	Reference
Digenea				
Aporocotyle simplex	Flatfishes	Gastropods	Polychaetes	Williams & Jones (1994)
Cryptocotyle lingua	Gulls	Gastropods (*Littorina littorea*)	Teleosts	Køie (1983)
Curtuteria australis	Oystercatchers	Gastropods	Bivalves (*Austrovenus stutchburyi*)	Poulin, Hecker & Thomas (1998)
Derogenes varicus	Teleosts	Gastropods (*Natica* spp.)	Harpacticoid copepods, decapods	Williams & Jones (1994)
Fellodistomum fellis	Wolffish (*Anarhichas lupus*)	Bivalves (*Nucula tenuis*), brittle stars (*Ophiura* spp.)	—	Williams & Jones (1994)
Lecithochirium furcolabiatum	Blennies, gobies	Gastropods (*Gibbula ambilicalis*)	Harpacticoids (*Tigriopus brevicornis*)	Williams & Jones (1994)
Lepidapedon elongatum	Gadids (*Gadus morhua*)	Gastropods (*Onoba aculeus*)	Polychaetes, molluscs	Køie (1985)
Maritrema subdolum	Shorebirds	Gastropods (*Hydrobia* spp.)	Amphipods, isopods, crabs	Kostadinova & Gibson (1994), Gollasch & Zander (1995)
Microphallus pygmaeus	Gulls, eider ducks	Gastropods (*Littorina littorea*)		Granovitch & Johannesson (2000)
Microphallus similis	Gulls	Gastropods (*Littorina* spp.)	Shore crab (*Carcinus maenas*)	Granovitch & Johannesson (2000)
Podocotyle reflexa	Teleosts (gadids)	Gastropods (*Buccinum undatum*)	Shrimps (*Crangon crangon*, *Pandalus* spp.), mysids	Williams & Jones (1994)
Proctoeces maculatus	Labrid and sparid fishes	Bivalves (*Mytilus* spp.)	Polychaetes, sea urchins, gastropods, octopods	Williams & Jones (1994)
Ptychogonimus megastomus	Sharks	Scaphopod molluscs (*Antalis* spp.)	Crabs	Williams & Jones (1994)
Renicola roscovita	Gulls	Gastropods (*Littorina* spp.)	Bivalves (*Mytilus* spp.)	Granovitch & Johannesson (2000)
Stephanostomum baccatum	Cottids	Gastropods (*Buccinum undatum*, *Neptunea* spp.)	Flatfishes	Køie (1983)
Zoogonoides viviparus	Flatfishes	Gastropods (*Buccinum undatum*)	Brittle stars (*Ophiura* spp.), plus polychaetes, bivalves, gastropods	Williams & Jones (1994)

Parasite	Definitive host	Intermediate host	Transport/paratenic host	Reference
Cestoda				
Acanthobothrium hispidum	Elasmobranchs	Harpacticoids (*Tigriopus fulvus*)	Teleosts, cephalopods	Williams & Jones (1994)
Bothrimonus sturionis	Teleosts	Gammarid amphipods	—	Williams & Jones (1994)
Grillotia erinaceus	Leopard shark (*Triakis semifasciata*)	Harpacticoids (*Tigriopus californicus*)	—	Williams & Jones (1994)
Lacistorhynchus tenuis	Leopard shark (*Triakis semifasciata*)	Harpacticoids (*Tigriopus californicus*)	—	Williams & Jones (1994)
Nematoda				
Ascarophis morhua	Gadids	Crabs (*Carcinus maenas*), hermit crabs (*Pagurus* spp.)	—	Williams & Jones (1994)
Contracaecum spiculigerum	Cormorants, pelicans	Harpacticoids (*Tigriopus californicus*)	Teleosts	Anderson (1992)
Cucullanus cirratus	Gadids	Gobies, cod fry	—	Koie (2000*a*)
Cucullamus heterochrous	Flatfishes	Polychaetes (*Nereis diversicolor*)	—	Koie (2000*b*)
Hysterothylacium aduncum	Teleosts	Various crustaceans, polychaetes	Many invertebrates and fish	Williams & Jones (1994)
Paracuaria adunca	Piscivorous birds	Amphipods, mysids	Teleosts	Jackson *et al.* (1997)
Pseudanisakis rotunda	Elasmobranchs (*Raja radiata*)	Decapods (*Lithodes* sp.)	Flatfishes, gadids	Williams & Jones (1994)
Pseudoterranova decipiens	Pinnipeds	Harpacticoids, amphipods	Crustaceans, polychaetes, teleosts	Anderson (1992)
Sulcascaris sulcata	Marine turtles	Scallops, gastropods	—	Anderson (1992)
Acanthocephalans				
Corynosoma spp.	Pinnipeds	Amphipods	Teleosts	Valtonen & Niinimaa (1983), Zdzitowiecki (2001)
Echinorhynchus gadi	Teleosts	Gammarid and caprellid amphipods	—	Schmidt (1985)
Echinorhynchus lagenformis	Flatfish (*Platichthyes stellatus*)	Amphipods (*Corophium spinicorne*)	—	Schmidt (1985)
Profilicollis botulus	Eider ducks	Crabs (*Hyas araneus*) hermit crabs (*Pagurus pubescens*)	—	Uspenskaja (1960)

**Depth
zone**

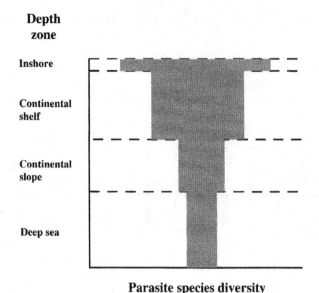

Parasite species diversity

Fig. 1. Theoretical schematic representing relative species diversity of parasites of demersal fish inhabiting different depth zones ranging from inshore to the deep sea. Species diversity decreases with increasing depth.

brates partition the physical habitat and prey resources in space and time. Resource partitioning contributes to the formation of distinct parasite assemblages among host species. Herein, I evaluate patterns of transmission of trophically-transmitted helminth parasites and discuss the relationship with spatial variations in oceanographic conditions that lead to the formation of distinct animal communities. The various roles of community members in the local food web contribute to the development of distinct parasite faunas in marine organisms. Emphasis is placed on research completed in waters off northeastern North America, although other sources are used when pertinent. Lastly, I briefly consider effects of pollution, exploitation and the potential role of keystone parasites on food webs and parasite transmission.

BENTHIC PATTERNS OF DISTRIBUTION

Free-living animals

Marine organisms display horizontal gradients in their patterns of distribution. Proceeding offshore, the composition of the benthic macroinvertebrate fauna (organisms retained by a 500 μm sieve) changes with salinity, substrate, temperature and depth (McLusky & McIntyre, 1988). Particle size diversity of the sediments is a measure of habitat complexity and generally reflects species diversity, with coarse heterogenous sands being more species rich than well-sorted homogenous substrates (Gray, 1974). Continental shelves tend to be impoverished, but slope waters possess high diversity with marked

faunal changes at 100–300 m (Sanders, 1968). While community composition is related somewhat to sediment composition, it is also mediated by interactions between organisms and sediments (Gray, 1974).

Numerous studies document the relationship between substrate texture and quality, depth and the type of benthic community that occurs in Atlantic waters off North America (Wigley & McIntyre, 1964; Day, Field & Montgomery, 1971; Maurer & Wigley, 1984; Sherman *et al.* 1988; Weston, 1988), Europe (Basford, Eleftheriou & Raffaeilli, 1989; Eleftheriou & Basford, 1989; Flach & Thomsen, 1998) and elsewhere (Field, 1970). The dominant taxa are often polychaetes or, occasionally, crustaceans. Bivalves and echinoderms are also common on continental shelves and slope waters. Different taxa occur on particular sediment types and their distribution and relative dominance with depth varies accordingly. Species richness of epifaunal organisms, or those living on the sediment surface, tends to be highest on gravel, while infaunal organisms (or those living in the sediments) are most numerous on muddy or silty sand (Rhoads & Young, 1970; Gray, 1974; Eleftheriou & Basford, 1989).

Within the meiofauna (organisms passing through a 500 μm sieve but retained on a 45 μm sieve), nematodes are the most common organisms, with harpacticoid copepods ranking second in most systems (Coull, 1970). Like benthic macrofauna, the composition of copepods and other meiofauna varies with substrate texture and grain size (Coull, 1970). Off northeastern North America, density and diversity of both macro- and meiobenthic organisms decrease with depth (Wigley & McIntyre, 1964; Sherman *et al.* 1988).

Thus, based on the spatial distribution of both meiofauna and macrofauna, we can expect the parasites transmitted in a habitat to vary with substrate texture and depth according to their life histories and benthic invertebrate species composition.

Parasites

Benthic invertebrates act as intermediate hosts for digeneans, nematodes, acanthocephalans and, to a lesser extent, cestodes. Crustaceans are the most common intermediate hosts in parasite life cycles, but it must be noted that digeneans require molluscs as obligate first intermediate hosts. Polychaetes and echinoderms participate in a limited number of helminth life cycles. A summary of life cycle patterns and types of intermediate hosts for parasites infecting benthic marine invertebrates is found in Table 1.

Horizontal variations in abiotic parameters and the biota are important in determining the distribution of parasites. A wider range of intermediate hosts occurs in shallow waters. Flatfish, for example, inhabiting deeper waters have fewer parasites than those species in shallow waters (Scott & Bray, 1989; Lile, 1998). Variation among local habitats also affects parasite species composition (Polyanski, 1961; Thoney, 1991). Thus, it can be expected that the variation in spatial distribution of the different benthic invertebrate taxa will reflect the distribution of parasites that use them as intermediate hosts.

For instance, Campbell, Haedrich & Munroe (1980) examined 52 deep-sea benthic fishes at depths from 50–5000 m off the New York Bight. The species composition of their parasite fauna reflect dietary differences. The abundance and prevalence of parasites with complex life cycles depend directly on the abundance of free-living fauna, and parasite species richness declines with depth (Fig. 1) (Campbell *et al.* 1980; Campbell, 1983). The parasite fauna in certain benthic fish suggest that prey fish migrate from midwater to the bottom, implying that some parasites pass between habitats or communities as they complete their life cycles (Fig. 2).

Perhaps the most extensive examination of marine parasitic fauna ever undertaken is the benthic survey of Uspenskaja (1960) in the Barents Sea from 1949–54. A total of 10 species of digenean, 6 cestodes, 6 nematodes and 2 acanthocephalans, all larval forms, were found in 31 species of crustaceans. Similarly, Gollasch & Zander (1995) examined for helminths over 36000 crustaceans from the Schlei fjord in the Baltic Sea, in which were found 4 digenean, 1 cestode, 1 nematode and 1 acanthocephalan species. These studies illustrate the complex web of parasite transmission that occurs through the crustaceans alone in benthic communities.

PELAGIC PATTERNS OF DISTRIBUTION

Free-living animals

In the open ocean, the water column is divided into the epipelagic zone, the mesopelagic zone below the thermocline between 200–1000 m, and the bathypelagic zone below 1000 m (Madin & Madin, 1995). The bathypelagic zone comprises 88% of the total global oceanic area, and is characterized by the absence of light and a constant temperature of 4 °C. Its inhabitants tend to display different, sluggish lifestyles compared to other pelagic animals. While a few species are common, over 80% are considered rare (Madin & Madin, 1995).

Within the pelagic zone, zooplankton consist of primarily calanoid copepods, but also other crustaceans including euphausiids, cyclopoid copepods and hyperiid amphipods. Soft-bodied zooplankters are typically predators on other zooplankton and fish larvae, and include chaetognaths, coelenterates and ctenophores (Marcogliese, 1995).

The bottom waters 1 m above the sediments are referred to as the hyperbenthos (or suprabenthos or nektobenthos) (Hamerlynck & Mees, 1991). This fauna has often been neglected because of the difficulty of quantitative sampling (Hamerlynck & Mees, 1991). Its main constituents are decapod larvae, mysids and other crustaceans, fish eggs and larvae, but also include amphipods, isopods and chaetognaths (Hamerlynck & Mees, 1991; Mees, DeWicke & Hamerlynck, 1993). Bottom waters and the sediment surface also are termed the benthic boundary layer (Wildish, Wilson & Frost, 1992). Within this layer, organisms segregate vertically by depth, with densities increasing towards the bottom (Oug, 1977; Chevrier, Brunel & Wildish, 1991). Mysids, amphipods, isopods and chaetognaths are found year-round in the hyperbenthos, with mysids the most abundant (Mees *et al.* 1993). These organisms are an important source of nutrition for fish and this has important ramifications for parasite transmission.

Certain benthic organisms such as harpacticoid copepods, amphipods and cumaceans as well as planktonic copepods migrate in abundance into the hyperbenthos at night (Oug, 1977; Sibert, 1981; Kaartvedt, 1986; Chevrier *et al.* 1991). In addition, mysids can migrate into the zooplankton. These vertical movements create trophic linkages between zones and opportunities for parasites to traverse habitats (Fig. 2).

In addition, surface and subsurface habitats are linked. Epifaunal predation on infaunal organisms is more important than infaunal predation (Virnstein, 1977; Ambrose, 1991), though polychaetes are common infaunal predators in many systems (Wilson, 1991). Benthic copepods dominate the diets of predators, especially juvenile fishes, even though they comprise only 2–20% of meiofauna (Virnstein, 1977; Coull, 1990). Mysids such as *Neomysis integer* also consume meiobenthos, in particular harpacticoid copepods (Johnston & Lasenby, 1982). Thus, the benthic boundary layer provides ample opportunities for pelagic-benthic coupling and cycling of parasites (Fig. 2).

Parasites

Transmission of parasites from zooplankton to fish was reviewed by Marcogliese (1995). Diversity of helminths using zooplankton as intermediate hosts is relatively high in marine systems, mainly due to the presence of hemiuroid, lepocreadoid and didymozooid trematodes, in addition to trypanorhynch and tetraphyllidean cestodes (Marcogliese, 1995). However, infection rates are extremely low owing to the dilute nature of the pelagic realm. The addition of an extra trophic level of planktonic predators and the capacity of many marine helminths to use organisms as paratenic hosts serves to promote transmission and maintain parasites in the environment even at low densities (Marcogliese, 1995). For

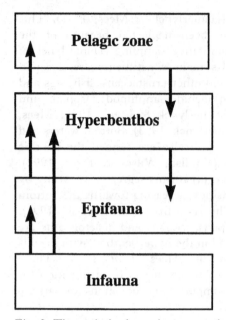

Fig. 2. Theoretical schematic representing vertical flow of parasites among different depth strata based on presumed trophic interactions. Vertical exchange may also occur within the layers that constitute the pelagic zone.

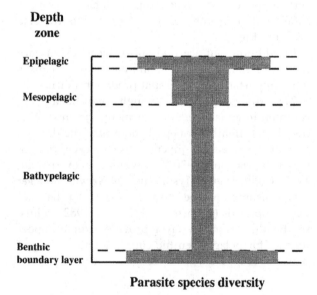

Parasite species diversity

Fig. 3. Theoretical schematic representing relative species diversity of parasites of pelagic fish inhabiting different depth zones in the water column. Species diversity decreases with depth to a minimum in the bathypelagic zone, and then increases in the benthic boundary layer.

examples of zooplankton intermediate hosts used by marine helminths, consult Table 1 in Marcogliese (1995).

Zonation with depth into littoral, sublittoral, pelagic and bathypelagic habitats is an important determinant of the parasite fauna of fish (Polyanski, 1961). With depth there occurs a corresponding change in the intermediate host fauna. Plankton-

feeding fishes do not usually acquire parasites from benthic invertebrates. Their parasite fauna, acquired mainly from copepods and chaetognaths, is distinct and impoverished (Polyanski, 1961), in accordance with the low infection rates in intermediate hosts in dilute waters. Similarly, mesopelagic and bathy-pelagic fish possess poor parasite faunas consisting of few adult forms, few digeneans, and mainly juvenile nematodes and cestodes (Campbell, 1983; Gartner & Zwerner, 1989). Away from the continental shelf slope waters, parasite diversity generally decreases with depth (Campbell *et al.* 1980; Campbell, 1983; Gartner & Zwerner, 1989). Biomass also decreases with depth in the pelagic zone; thus, there are fewer prey to serve as intermediate hosts for parasites of bathypelagic fish, accounting for the relatively impoverished nature of their parasite fauna (Camp-bell *et al.* 1980; Campbell, 1983; Gartner & Zwerner, 1989). Species richness and intensity of infection are highest in epipelagic and benthic zones, decrease in vertically-migrating mesopelagics and are lowest in deep nonmigratory mesopelagic and bathypelagic fishes (Fig. 3). Mesopelagic and bathypelagic fishes possess impoverished parasite communities com-pared to those of benthic fishes, which possess more diverse adult and larval helminths (Campbell *et al.* 1980; Campbell, 1983). The high diversity, density and longevity of benthic invertebrates compared to pelagic ones promote parasite transmission (Camp-bell *et al.* 1980; Campbell, 1983). These factors account for the relative high diversity of parasites in benthic fishes compared to those found higher in the water column (Fig. 3).

Though cycling of helminths is primarily within the benthic boundary layer in deep-sea fishes (Campbell *et al.* 1980), the temporary spatial overlap of vertically-migrating organisms and trophic inter-actions between predators and prey in the water column, hyperbenthos, epifauna and infauna would be expected to promote some exchange of parasites between zones and permit their movement vertically within marine systems (Fig. 2).

DISTRIBUTION AND DIETS OF FISHES

Like benthic communities (Sanders, 1968; Gray, 1974), fish communities vary along a gradient with depth (Markle, Dadswell & Halliday, 1988). How-ever, seasonal inshore-offshore migrations link food webs across depths (Ojeda & Dearborn, 1990). This has implications for parasite transmission, poten-tially linking parasite faunas from different areas, but tends to blur the distinctiveness of parasite com-munities.

On the Scotian Shelf off Nova Scotia there are five different bottom types based on grain size that are correlated with water depth and fish distributions (Scott, 1982a). Coarse sands are shallow and fine sands deeper, with the different bottom types having characteristic invertebrate communities that support a distinct fish fauna. The highest fish diversity occurs on mixed sediment types. Generally, the most widespread predators possess the highest dietary diversity (Richards, 1963; Hacunda, 1981).

Much diet overlap occurs among fishes off northeastern North America and elsewhere (Langton & Bowman, 1980; Langton, 1982; Martell & McClelland, 1994). Dietary overlap is frequently higher between species than between size classes of the same species (Garrison & Link, 2000), as diet shifts are common with age and are virtually ubiquitous (Tyler, 1972; Braber & de Groot, 1973; Langton, 1982; Gibson & Ezzi, 1987). Much of the overlap results from similarities in mouth morphology and gape size, even among unrelated species (Hacunda, 1981; Gibson & Ezzi, 1987). A large portion of this overlap is due to the prominence of crustaceans in fish diets. Crustaceans and epifauna are often over-represented in fish diets, while taxa such as polychaetes and members of the infauna are under-represented (Richards, 1963; Hacunda, 1981; Langton, 1982; Macdonald & Green, 1986).

Despite this overlap in diet, resource partitioning has been repeatedly demonstrated in marine fish off North America and Europe (Tyler, 1972; Langton & Bowman, 1980; Hacunda, 1981; Gibson & Ezzi, 1987; Langton & Watling, 1990; Martell & McClelland, 1994). Diets may vary seasonally, geographically and with depth (Richards, 1963; Langton & Bowman, 1980; Hacunda, 1981; Langton & Watling, 1990). Feeding guilds may be derived from fish diets and dividing prey into functional groups according to lifestyles, sediment preferences and depth off bottom. Guilds may include piscivores, mixed feeders on crustaceans and fish, mesoplankton and macroplankton feeders, nekton feeders and those feeding on benthic and/or hyperbenthic prey (Langton & Bowman, 1980; Mattson, 1981; Vinogradov, 1984; Gibson & Ezzi, 1987; Langton & Watling, 1990; Garrison & Link, 2000). The partitioning of prey should promote segregation of parasites among guilds of hosts, countering the homogenizing influences that were discussed earlier. As a consequence, guilds of parasites may result that are transmitted by suites of invertebrates characteristic of particular habitats to suites of fish in those habitats (Zander *et al.* 2000; Zander, 2001). For example, local differences in ecology are reflected in parasite species composition in the Barents Sea, where parasites are divided into broad host categories (littoral and coastal species, planktivores, benthophagous fish and migratory species) and the Bering Sea (planktivores, piscivores-planktivores and piscivore-benthivores) (Polyanski, 1961). Most parasites are associated with a specific ecological niche and diet.

In a number of demersal fish off Nova Scotia, many species displayed modifications in diet with growth that were accompanied by a change in the parasite fauna (Scott, 1975, 1981, 1982*b*, 1985; Scott & Bray, 1989; see also Thoney, 1991, 1993). For example, American plaice shifted from small crustaceans to echinoderms (Scott, 1975), haddock from plankton to benthos (Scott, 1981), pollock from crustaceans to fish (Scott, 1985), and Atlantic and Greenland halibuts to increasing amounts of fish (Scott & Bray, 1989).

Among related species such as flatfish, there exists some host specificity, but evidence suggests that factors controlling species composition of parasites are ecological, not physiological or phylogenetic (Polyanski, 1961; Scott, 1982*b*; Lile, 1998). However, Pacific halibut (*Hippoglossus stenolepis*) are unusual for temperate marine fish in that their parasite fauna is extremely speciose (Blaylock, Holmes & Margolis, 1998). Most of their parasites are generalists, but these authors were able to detect a phylogenetic component to parasite species composition. As in other species and studies, the parasite fauna changed with age corresponding to a shift from shallow inshore habitats and invertebrate diet to deeper offshore habitats and piscivorous diet. Clearly, the invertebrate and fish fauna that in turn reflect habitat characteristics are important factors in dictating the local composition of the parasite fauna.

PATTERNS IN MARINE FOOD WEBS

Food web structure varies among ecosystems in different parts of the world. Thus, it is worthwhile to briefly consider some of the variations in structure as these differences will impact on the flow of parasites along food chains in these ecosystems.

In marine waters, there are basically three types of production: temperate seas, upwelling areas and oligotrophic oceans (Cushing, 1988). The pelagic realm is divided on the basis of topography, hydrography and latitude, such that distinct ecosystems are formed (Smetacek, 1988). Continental shelves are the most productive of the extensive areas in the oceans. Diversity is highest on the boreal continental shelf in summers, with high copepod production, and high abundance of invertebrate planktonic predators like chaetognaths, ctenophores and coelenterates (Smetacek, 1988). In shelf waters, Atlantic cod and herring are at the top of the demersal and pelagic food chains respectively, but cod also eat herring (Cushing, 1988), thus integrating pelagic and benthic food webs. Copepods are the most important taxa in terms of biomass in the water column (Smetacek, 1988).

Tropical waters have continuously high diversity and possess complex ecosystems. Food chains differ in strong versus weak upwellings. In weak ones, anchoveta and zooplankton both consume phytoplankton and are in turn ingested by carnivores. In strong upwellings, anchoveta eat zooplankton and are then eaten by carnivores (Cushing, 1988).

Productivity in polar systems can rival the tropics, but it is limited seasonally. In the Antarctic, food chains are short, euphausiids being primary consumers and whales, secondary consumers. Seabirds dominate northern food webs such as at Svalbard, Spitzbergen (Weslawski & Kwasniewski, 1990). Small fish are common and abundant, as are seals. Predaceous hyperiid amphipods are potentially important links in parasite transmission, especially in the Arctic where they are among the most common prey of fish and birds (Weslawski & Kwasniewski, 1990).

Interestingly, and perhaps counter-intuitively, benthic production is highest at high latitudes. The productivity ratio of plankton:benthos is 6:1 in the Indian Ocean, 3:1 in the North Sea, and 1:1 off Greenland (McLuskey & McIntyre, 1988). Pelagic organisms are more important in tropical fisheries and demersal ones in the subarctic (Petersen & Curtis, 1980). In tropical waters, zooplankton are a dominant trophic link to pelagic fish. In contrast, on northern continental shelves, 80% of animal species are benthic, with the remainder being pelagic zooplankton (Curtis, 1975). Marine mammals and birds that prey almost exclusively on benthos occur in subarctic and Arctic waters, but not in the tropics (Petersen & Curtis, 1980). Thus, the basic structure of oceanic food webs varies with latitude. Pelagic parasites are probably more common in subtropical and tropical waters, whereas benthic parasites are more important at higher latitudes.

In northern Norway, there are two fundamentally different pathways within the food web. The first is short. In pelagic waters, euphausiids are the primary consumers and are eaten by Atlantic cod. A similar near-bottom food chain exists where prawns, mysids and other crustaceans are the primary consumers, and demersal Atlantic cod their main predators. However, the most important pathway consists of copepods and euphausiids as primary consumers, capelin (*Mallotus villosus*) as secondary consumers and Atlantic cod as the apex predator (Falk-Petersen, Hopkins & Sargent, 1990). In contrast, in the Barents Sea, ctenophores are important copepod predators and they almost act as a trophic dead end, although they can be ingested by *Beroe* sp., another cteno-phore that is then preyed upon by Atlantic cod. Ctenophores actually consume much of the copepod production on the continental shelf. This may limit parasite transfer from copepods to fish compared to Spitzbergen, where capelin is the most important copepod predator, although some are also eaten by chaetognaths. These in turn may act as paratenic hosts for numerous parasites, thus facilitiating transmission.

Primary production on the shallow and cold continental shelf off Nova Scotia is less than that on the slope waters, which are deeper, warmer and more nutrient-rich. However, demersal fish production is highest on the shelf but declines further offshore, whereas pelagic fish production is highest on the slope waters and declines inshore (Mills & Fournier, 1979). The fish community on the Scotian Shelf is composed largely of cod, haddock, pollock and flatfish, whereas hake, redfish, grenadiers and argen-tines are more common on the slope (Mills & Fournier, 1979). Thus, not only is there a transition in the fish fauna at the shelf-slope interface, but there should be a transition in the parasite fauna as well. Pelagic parasites are predicted to be more common in slope waters, and benthic parasites on the shelf (Mills & Fournier, 1979). These authors further suggest that there are fundamental differences in the food chains between Nova Scotian waters and those of the North Sea, with the pelagic web being more important in the latter. This implies that pelagically-transmitted parasites that use zooplankton as in-termediate hosts (such as cestodes) are predicted to be more prevalent in the North Sea, whereas abundance of parasites transmitted via the benthos should be similar as demersal catches are similar between the two areas.

More generally, Hairston & Hairston (1993) suggest that trophic structure controls energy flow within food webs. Using freshwater systems as an example, they note that pelagic food webs possess 1–2 extra trophic levels compared to terrestrial webs. In freshwater pelagic systems, there are 2 distinct levels of predation, those being zooplanktivorous fish and piscivorous fish, both gape-limited, whereas in terrestrial systems there is only 1 trophic level of carnivores that is functionally significant (Hairston & Hairston, 1993). In some freshwater systems and most marine systems, there exists an additional level of predatory zooplankton (Marcogliese, 1995). Coas-tal systems differ from open ocean systems in that they possess shorter food chains (Hairston & Hair-ston, 1993). The least restricted webs are those of the marine pelagic. They are expanded 3-dimensionally in space and contain the longest number (5) of food chain lengths. Marine benthic webs possess slightly shorter chain lengths (4) on average, and are more spatially restricted. Estuarine webs are the most constrained and possess short food chain lengths (3) (Schoener, 1989). These generalizations must be interpreted with caution. For example, well-studied speciose estuarine webs such as the Ythan possess distinctly longer webs (Hall & Raffaelli, 1991; Huxham, Raffaelli & Pike, 1995). A major problem in food-web construction is whether to draw a link between 2 components (Schoener, 1989). Parasites may help verify links, and possibly could help determine the strength of the linkage (Marcogliese & Cone, 1997). Indeed, one model of the Ythan food web postulates links between predators and prey based on parasites present in the system and their life cycles, in the absence of diet data (Huxham *et al.* 1995).

PARASITES IN THE FOOD WEB

The different organisms that occur in or on a particular habitat are integrated together into food webs through trophic interactions. The webs found across different habitats blend together gradually much like the distribution of their component organisms in the seas, with gradients in diversity and interactions. Numerous studies suggest that parasite diversity is related to the diversity of free-living organisms or diversity of the host's diet (Campbell *et al.* 1980; Scott, 1981; Campbell, 1983; Lile, 1998). For example, the White Sea is oceanographically similar to the Barents Sea, but its fish possess fewer parasite due to a reduction in its free-living fauna that act as intermediate hosts (Polyanski, 1961).

An examination of the parasite fauna of Atlantic cod illustrates the diversity of parasitism and the role of cod within the food web (Hemmingsen & MacKenzie, 2001). A total of 107 species of parasites have been reported in cod, of which only 7 are species specific, and another 17 specific to gadids. The remaining 83 are generalists. Atlantic cod acts as intermediate, paratenic or definitive host to these parasites. Eight species mature in marine mammals and another 12 in large piscivorous fish, showing that cod can be an important link to top predators in the food chain. In contrast, only 5, of which 2 are rare, mature in birds, effectively demonstrating that avian predation on cod is limited on a global scale. The fact that Atlantic cod is definitive host to many species of parasites indicates that cod is a major predator on many taxa including fish, crustaceans, polychaetes, coelenterates and chaetognaths.

Within a fish species, parasites will vary between inshore and offshore stocks (Polyanski, 1961; Thoney, 1993; Hemmingsen & MacKenzie, 2001), reflecting differences in the respective food webs. Parasites transmitted by intertidal snails, such as *Cryptocotyle lingua*, and those infecting pinnipeds, such as *Pseudoterranova decipiens*, will be more common in inshore fish. Those using whales as definitive hosts, such as *Anisakis simplex*, will be more abundant in offshore fish. Inshore-offshore differences also occur in parasites of seabirds. Those feeding inshore acquire the acanthocephalan *Corynosoma* spp. from feeding littorally on small to medium nototheniid fishes or amphipods. This parasite does not occur in zooplanktivorous seabirds offshore that prey primarily on euphausiids (Hoberg, 1985, 1986).

Long-term changes in the prevalence and abundance of parasites indicate the variations in density of animals that serve as intermediate or definitive hosts, and/or changes in environmental conditions that subsequently affect food webs (Hemmingsen & MacKenzie, 2001). For example, an increase in *Anisakis simplex* may reflect more predation on capelin, while a decrease in *Contracaecum osculatum* may indicate reduced seal predation on cod. Long-

term changes in the food chain have lead to shifts in the helminth fauna of seabirds in the Seven Islands archipelago of the Barents Sea. Between 1940–41 and 1991–93 the species richness of digeneans declined drastically in common gulls, herring gulls and great black-backed gulls. This decline is attributed to a decrease of molluscs in the birds' diets (Galaktionov, 1995). The species composition of cestodes also changed, reflecting a diet shift from fish to crustaceans as a result of the collapse of the herring and capelin populations (Galaktionov, 1995).

Few studies have incorporated parasites into food webs (Marcogliese & Cone, 1997). In an open water pelagic example, two species of short-finned squid (*Illex coindetii* and *Todaropsis eblanae*) are sympatric off the coast of Spain. Their parasites suggest that they serve as prey for large predators. The squid *I. coindetii* is infected by tetraphyllideans and trypanorhynchs, obtained from feeding on planktonic invertebrates, suggestive that sharks are one of its important predators. In contrast, *T. eblanae* is infected with anisakid nematodes, implying that it feeds on micronekton (small fish and squid) and that it in turn is preyed upon by marine mammals (Pascual *et al.* 1996). Curiously, these squid are not infected with didymozoid digeneans. Normally, squid are important in the diet of top carnivorous teleosts such as swordfish, but the absence of these parasites indicate that tuna, sailfish and swordfish are absent from these waters (Pascual *et al.* 1996). Thus, in this case parasites provide information about the role of squid in the food web, and about the structure of the web itself.

Two species of *Contracaecum* follow distinct pathways to the same seal hosts in Antarctic food webs. Both *C. radiatum* and *C. osculatum* infect Weddell seals (*Leptonychotes weddellii*) as definitive host, but *C. radiatum* is transmitted via a pelagic pathway while *C. osculatum* uses a benthic food chain. The two pathways are integrated through the diet of piscivorous bentho-pelagic channichthyids, which are the seal's main prey (Klöser *et al.* 1992).

Extensive studies by Zander and colleagues have documented patterns of transmission in the inshore brackish waters of the Baltic Sea, and elaborated parasite transmission pathways in food webs (Zander, 1992/93, 1998). Within the parasite communities of small fish (gobies, sprat, sand eels, sticklebacks) in the middle of the food web, over 50% of parasite specimens were larvae, small fish being important in the transfer of parasites from the first intermediate host to the definitive host (Zander, 1998). High prevalences of microphallid digeneans in snails and benthic crustaceans indicates that seabirds are common in the food web (Zander *et al.* 2000). Shallow inshore areas provide more opportunities for birds to feed, thus promoting allogenic life cycles (Campbell, 1983; Zander, Reimer & Barz, 1999). The different species of fish partition planktonic and benthic prey resources, as indicated by their parasite fauna (Zander, Strohbach & Groenewold, 1993; Zander, Groenewold & Strohbach, 1994; Zander *et al.* 2000). Conditions in the Wadden Sea are optimal for 2–3 host life cycles: there is a high density of intermediate hosts, short infection pathways and infection times, and high encounter rates between intermediate and definitive hosts (Groenewold, Berghahn & Zander, 1996). As a result, paratenic hosts are less important and less common in this type of habitat and parasite life cycles are thus shorter than in open marine waters. This host–parasite system reflects the conjecture that coastal systems have short food chains (Schoener, 1989; Hairston & Hairston, 1993).

Information from littoral communities at higher latitudes suggests that shorter life cycles are favoured in harsh environments. Fourteen species of digeneans are found in *Littorina* spp. and other snails on the Norwegian and Russian coasts, all but one of which use birds as definitive hosts. The parasites can be divided into 2 groups, those that employ a single intermediate host without any free-living stages (microphallids), and those that have free-living infective stages and more than 1 intermediate host (Galaktionov & Bustnes, 1999). The former tend to occur along the Russian coast, which has a harsher climate, while the latter are more common in Norway. The authors postulate that parasites with free-living stages might not be favoured in harsher climates, whereas longer life cycles may proceed to completion more successfully in milder climates. The dichotomy might also represent differences in food chain lengths between the two environments, with shorter food chains in harsher littoral habitats. There is a tendency to shorten life cycles in brackish waters as well (Kesting, Gollasch & Zander, 1996).

Among the most detailed studies of parasitism in relation to the ecosystem and food web structure are those of deep-sea mesopelagic and bathypelagic fishes off North America by Campbell and colleagues. They found that parasites are appropriate indicators of community interactions and host biology in that parasitic helminths use the food chain and reflect the diversity of the host's diet (Campbell *et al.* 1980; Campbell, 1983). Effectively, the composition of the helminth fauna in a host indicates that host's role in the food web. This information was corroborated with diet data and stomach content analysis. Parasites, for example, clearly demonstrate a shift from benthic invertebrates to more pelagic cephalopods and fish with depth (Campbell *et al.* 1980). Prevalence of infection and species richness are higher in those deep-sea fishes that feed on benthos compared to those foraging on pelagic or planktonic prey (Zubchenko, 1981; Houston & Haedrich, 1986). In a comparable study, Gollasch & Zander (1995) and Kesting *et al.* (1996) demonstrate the flow of parasites through benthic, hyperbenthic and planktonic intermediate hosts into small brackish fish and other vertebrates in the Schlei fjord of the Baltic Sea.

Given that cestodes, nematodes and acanthocephalans rely exclusively on trophic transmission (George-Nascimento, 1987; Zander, 1992/93), George-Nascimento (1987) suggests that parasites may be useful indicators of persistent food web interactions. He further notes that in fish, larvae typically have a greater host range than adult parasites in the marine environment. The highest species richness should be found at intermediate trophic levels where fish can serve as intermediate, paratenic or definitive hosts, with the proportion of adult digeneans increasing with trophic level.

Marine fish parasites tend to be generalists, as an adaptation for completing their life cycles in a dilute open system (Bush, 1990). Often the most common parasites in fish are juveniles which often are generalists and are commonly transferred trophically from one fish to another via predation. Thus, fish are important and frequently used as intermediate or paratenic hosts (Marcogliese, 1995; Blaylock *et al.* 1998). Many marine parasites possess the ability to maintain themselves without development in paratenic hosts be they invertebrates or fish (Marcogliese, 1995). This too is considered an adaptation to survival in a dilute environment where intermediate hosts such as copepods are relatively short-lived and encounter rates between definitive host predators and intermediate host prey are limited. Colonization success is enhanced and local extinction reduced if a resting stage is present in a parasite's life cycle (Kennedy, 1994).

The large invertebrate predators that commonly occur in marine systems offer unique opportunities for parasites to be transmitted in packets (Bush, Heard & Overstreet, 1993; Lotz, Bush & Font, 1995). Intertidal and salt marsh crabs may be infected with up to 6 species of microphallid digeneans that can be transmitted together as 'source' communities to birds and mammals (Bush *et al.* 1993; Lotz *et al.* 1995). Similarly, co-occurring anisakid nematodes may be transferred simultaneously from fish to fish or from fish to pinnipeds (Marcogliese, 2001*a*). While not as species rich, numerous co-occurring parasites can be transmitted together from infected planktonic invertebrates such as ctenophores infected with *Scolex pleuronectis* and *Opechoena bacillaris* to pelagic fish (Yip, 1984).

For marine mammals, which are apex predators in food webs, parasite species richness is less than that of their piscine prey. Their parasite faunas are considered impoverished compared to those of their terrestrial ancestors, probably because many ancestral helminth species failed to adapt to the marine habitat during the evolution of marine mammals, and to their isolation from existing mammal parasites (George-Nascimento, 1987; Balbuena & Raga, 1993;

Aznar, Balbuena & Raga, 1994). Thus, they are poor integrators of parasites and food web processes. Marine mammal parasite communities are impoverished despite the fact that they possess many of the criteria hypothesized to lead to species rich assemblages (endothermy, large size, longevity, gut complexity, vagility, catholic diet) (Kennedy, Bush & Aho, 1986).

Observations in fresh waters may also be applied to marine systems. For example, planktivores are typically dominated by cestodes and benthivores by digeneans (Dogiel, 1961). Among tropical freshwater fish, herbivores, algal feeders and zooplanktivores have more impoverished enteric helminth communities than do piscivorous and benthophagous fish (Choudhury & Dick, 2000). The richest communities are associated with fish possessing a mixed carnivorous diet of invertebrates and fish. This is similar to the conclusion of George-Nascimento (1987) that small fish in the middle of the food web will have the most speciose parasite communities.

POLLUTION, FOOD WEBS AND PARASITISM

Contamination in the marine environment has an impact on the species composition of benthic macroinvertebrates (Warwick & Clarke, 1995). Annelids tend to be more tolerant of polluted or stressed conditions than are echinoderms, molluscs or crustaceans (Warwick, 1988; Warwick & Clarke, 1993). In the Gulf of Mexico estuaries, sediment contamination with trace metals or organic chemicals decreases taxonomic and trophic diversity, affecting macrobenthic community structure and function (Rakocinski *et al.* 1997). Shrimps and amphipods appear sensitive to contaminants, while certain ophiuroids and crustaceans are more tolerant. Such overall changes in species composition and abundance will, no doubt, affect parasite transmission dynamics.

Indeed, parasites have been proposed as effective indicators of marine pollution (MacKenzie *et al.* 1995; MacKenzie, 1999). Parasites with complex life cycles can be affected at any stage in their life history. As a general rule of thumb, infections with endoparasites decrease and those of ectoparasites increase with pollution (MacKenzie, 1999).

The Baltic Sea has become increasingly eutrophic over the last 40 years, resulting in more parasites with short life cycles involving no or a single intermediate host (Reimer, 1995). Among 4 species of goby, there is a balance between benthic-transmitted and pelagic-transmitted parasites under normal conditions (Zander & Kesting, 1996). In stressed areas, the marine fauna becomes impoverished, with subsequent effects on the parasite fauna. Among parasites, stress leads to fewer specialists, more autogenic parasites and a dominance of plankton-transmitted species (Zander & Kesting, 1996).

Parasites benefit from eutrophication at first, as it promotes plant growth and herbivores and detritivores, such as snails and crustaceans that act as intermediate hosts for parasites (Zander *et al.* 1999; Zander *et al.* 2000; Marcogliese, 2001*b*). These areas are also attractive to fish and birds, and a balance exists between parasites with pelagic and benthic life cycles (Zander *et al.* 2000). As it proceeds further, eutrophication affects invertebrate intermediate hosts through oxygen depletion (Zander, 1998; Marcogliese, 2001*b*). Eutrophication favours generalist parasites (Zander & Kesting, 1998), and promotes 1–2 host life cycles compared to longer ones. A comparison of parasite faunas over 18 years in a region subject to eutrophication indicates a reduction in parasite species richness, a preponderance of simple parasite life cycles, and an increase of planktonic and hyperbenthic intermediate hosts over benthic hosts (Kesting & Zander, 2000). Comparable results were obtained for flounder (*Platichthys flesus*) and eelpout (*Zoarces viviparus*) in areas of the southeastern Baltic polluted with industrial sewage and experiencing eutrophication (Sulgostowska, Banaczyk & Grabda-Kazuska, 1987; Sulgostowska, Jerzewska & Wicikowski, 1990).

EFFECTS OF EXPLOITATION ON FOOD WEBS AND PARASITES

Commercial fishing may have profound impacts on community structure and food web organization. A meta-analysis comparing fishing methods from 56 international studies suggests that intertidal dredging and scallop dredging have the largest impact (Collie *et al.* 2000). Overall conclusions demonstrate that reductions of 27% in species number have occurred, with anthozoans and malacostracans the most affected. Other results indicate that holothurians and ophiuroids are the most sensitive of the echinoderms, gastropods are more sensitive than bivalves and oligochaetes are more vulnerable than polychaetes, the least impacted of all taxa. It is noteworthy to point out that no taxa increased in abundance (Collie *et al.* 2000).

In heavily-trawled areas such as the Gulf of Maine, Georges and the Grand Banks, benthic food webs are altered due to excessive removal of Atlantic cod and flatfish, resulting in increased numbers of skates, rays and longhorn sculpin (Smith *et al.* 2000). Changes have propagated through benthic invertebrates, with the macrofauna changing from echinoids and large clams to opportunistic brittle stars and polychaetes. Biomass of large epibenthic organisms including decapods and echinoderms was reduced by otter trawling on the Grand Banks, Newfoundland (Prena *et al.* 1999). Presumably, alterations of this magnitude on the food web will affect parasite transmission, abundance and diversity. Reduction of benthos should impede transmission of parasites that use benthic intermediate hosts, in particular those found in echinoderms. A long-term study on Georges Bank documents drastic changes in fish abundance by species, with a shift in dominance from demersal to pelagic species (Garrison & Link, 2000). Weakly exploited species such as spiny dogfish (*Squalus acanthus*) and skates increased in range and spatial and dietary overlap. Thus, we might expect significant changes to the parasite fauna as a result of changes in species composition, including an increase in parasites of elasmobranchs.

On a global scale, commercial fishing has affected the structure of entire food webs. As a result of depletion of the top piscivores, the preferred catch in many fisheries, the species now targeted are smaller and the mean trophic level fished has shifted downwards (Pauly *et al.* 2000). Fishing down food webs conceivably will reduce or eliminate top piscivorous fish that serve as definitive hosts for numerous parasites, and their incidence is predicted to decrease. Increasing the relative abundance of pelagic fish, as observed in the northwestern Atlantic, may also shift the parasite fauna from a speciose assemblage of benthic-transmitted forms to a more impoverished pelagic parasite fauna. These ideas can be tested in areas where historic data sets on parasites exist, as in parts of North America and Europe.

KEYSTONE PARASITES IN THE MARINE ENVIRONMENT

A keystone parasite basically has an impact on an entire community through regulation of an important predator or prey species (Minchella & Scott, 1991). Most keystone parasites have been documented for terrestrial systems (Dobson & Hudson, 1986; Marcogliese & Cone, 1997). However, populations of the green sea urchin (*Strongylocentrotus droebachiensis*) are controlled locally by outbreaks of parasites on both sides of the Atlantic Ocean by, curiously, two completely different parasitic organisms. This sea urchin decimates kelp beds and creates barrens, thus completely altering coastal ecosystems. In Norwegian waters, populations of sea urchins are reduced by a nematode (*Echinomermella matsi*), while in Nova Scotia they are limited by a protozoan parasite (*Paramoeba invadens*) (Hagen, 1992, 1996; Scheibling, Hennigar & Balch, 1999). The outbreak of disease is the only known mechanism to cause a large-scale shift from urchin-dominated barrens to lush, species-rich kelp forests (Scheibling *et al.* 1999). Productivity of vegetated habitats is greater than in adjacent areas, and the physical structure provided by macrophytes further influences the food web by increasing habitat complexity and enhancing diversity (Snelgrove *et al.* 2000). Thus, parasites may be important in regu-

lating coastal ecosystem functions. Indeed, parasites may be the cause of trophic cascades (Skorping & Högstedt, 2001).

The ingestion of larval parasites during predatory interactions is a frequent event that aids in the comprehension of foraging dynamics and food web structure due to the ubiquity of trophically-transmitted parasites (Lafferty & Morris, 1986). Without consuming much host energy, parasites greatly increase predation rates on their hosts. This may reduce the density of the intermediate host and actually permit the persistence of a predator that otherwise could not support itself, thus affecting the structure of the entire food web (Lafferty & Morris, 1996).

CONCLUSIONS

Consideration of parasites in food web studies leads to alterations in food web properties and dynamics (Huxham *et al.* 1995; Huxham, Beaney & Raffaelli, 1996). Marcogliese & Cone (1997) suggested that the incorporation of parasitological information would assist in the construction and resolution of food webs and they provided justifications to that end. Herein I propose the reverse: that consideration of the structure and dynamics of food webs permits predictions about the distribution and abundance of parasites. Although this review deals exclusively with marine systems, this generalization also applies to fresh waters. Numerous testable hypotheses based on this generalization result from ideas and concepts discussed here. They include: (1) parasite species composition in local demersal fish varies with substrate quality, benthic invertebrate community structure and local food web patterns; (2) the benthic boundary layer is a focal point for parasite exchange among faunas inhabiting different vertical zones in marine habitats; (3) parasite species composition changes in heavily exploited areas to reflect modifications in fish and invertebrate communities; and more specifically, based on ecosystem information; (4) pelagic parasites are more common in slope waters and benthic parasites in shelf waters off Nova Scotia; and (5) pelagic parasites are more common in the North Sea than off Nova Scotia.

ACKNOWLEDGEMENTS

I thank Dr. Jane L. Cook and 2 anonymous referees for providing comments and advice on the manuscript. Robert Poulin is graciously acknowledged for the invitation to contribute to this volume.

REFERENCES

AMBROSE, W. G. JR. (1991). Are infaunal predators important in structuring marine soft-bottom communities? *American Zoologist* **31**, 849–860.

ANDERSON, R. C. (1992). *Nematode Parasites of Vertebrates. Their Development and Transmission.* Wallingford, UK, CAB International.

ARTHUR, J. R. (1997). Recent advances in the use of parasites as biological tags for marine fish. In *Diseases in Asian Aquaculture III.* (ed. Flegel, T. W. & MacRae, I. H.), pp. 141–154. Manila, Fish Health Section, Asian Fisheries Society.

AZNAR, F. J., BALBUENA, J. A. & RAGA, J. A. (1994). Helminth communities of *Pontoporia blainvillei* (Cetacea: Pontoporiidae) in Argentine waters. *Canadian Journal of Zoology* 72, 702–706.

BALBUENA, J. A. & RAGA, J. A. (1993). Intestinal helminth communities of the long-finned pilot whale (*Globicephalus melas*) off the Faroe Islands. *Parasitology* 106, 327–333.

BASFORD, D. J., ELEFTHERIOU, A. & RAFFAELLI, D. (1989). The epifauna of the northern North Sea (56°–61° N). *Journal of the Marine Biological Association of the United Kingdom* 69, 387–407.

BLAYLOCK, R. B., HOLMES, J. C. & MARGOLIS, L. (1998). The parasites of Pacific halibut (*Hippoglossus stenolepis*) in the eastern North Pacific: host-level influences. *Canadian Journal of Zoology* 76, 536–547.

BRABER, L. & DE GROOT, S. J. (1973). The food of five flatfish species (Pleuronectiformes) in the southern North Sea. *Netherlands Journal of Sea Research* 6, 163–172.

BUSH, A. O. (1990). Helminth communities in avian hosts: determinants of pattern. In *Parasite Communities: Patterns and Processes.* (ed. Esch, G. W. Bush, A. O. & Aho, J. M.), pp. 197–232. London, Chapman and Hall.

BUSH, A. O., HEARD, R. W. JR. & OVERSTREET, R. M. (1993). Intermediate hosts as source communities. *Canadian Journal of Zoology* 71, 1358–1363.

CAMPBELL, R. A. (1983). Parasitism in the deep sea. In *The Sea. Vol. 8.* (ed. Rowe, G. T.), pp. 473–552. New York, John Wiley & Sons.

CAMPBELL, R. A., HAEDRICH, R. L. & MUNROE, T. A. (1980). Parasitism and ecological relationships among deep-sea benthic fishes. *Marine Biology* 57, 301–313.

CHEVRIER, A., BRUNEL, P. & WILDISH, D. J. (1991). Structure of a suprabenthic shelf sub-community of gammaridean Amphipoda in the Bay of Fundy compared with similar sub-communities in the Gulf of St. Lawrence. *Hydrobiologia* 223, 81–104.

CHOUDHURY, A. & DICK, T. A. (2000). Richness and diversity of helminth communities in tropical freshwater fishes: empirical evidence. *Journal of Biogeography* 27, 935–956.

COLLIE, J. S., HALL, S. J., KAISER, M. J. & POINER, I. R. (2000). A quantitative analysis of fishing impacts on shelf-sea benthos. *Journal of Animal Ecology* 69, 785–798.

COULL, B. C. (1970). Shallow water meiobenthos of the Bermuda platform. *Oecologia* 4, 325–357.

COULL, B. C. (1990). Are members of the meiofauna food for higher trophic levels? *Transactions of the American Microscopical Society* 109, 233–246.

CURTIS, M. A. (1975). The marine benthos of Arctic and sub-Arctic continental shelves. *Polar Record* 17, 595–626.

CUSHING, D. H. (1988). The flow of energy in marine ecosystems, with special reference to the continental shelf. In *Ecosystems of the World. 27. Continental Shelves.* (ed. Postma, J. & Zijlstra, J. J.), pp. 203–230. Amsterdam, Elsevier.

DAY, J. H., FIELD, J. G. & MONTGOMERY, M. P. (1971). The use of numerical methods to determine the distribution of the benthic fauna across the continental shelf of North Carolina. *Journal of Animal Ecology* 40, 93–125.

DOBSON, A. P. & HUDSON, P. J. (1986). Parasites, diseases and the structure of ecological communities. *Trends in Ecology and Evolution* 1, 11–15.

DOGIEL, V. A. (1961). Ecology of the parasites of freshwater fishes. In *Parasitology of Fishes.* (ed. Dogiel, V. A., Petrushevski, G. K. & Polyanski, Yu. I.), pp. 1–47. Edinburgh, Oliver and Boyd.

ELEFTHERIOU, A. & BASFORD, D. J. (1989). The macrobenthic infauna of the offshore northern North Sea. *Journal of the Marine Biological Association of the United Kingdom* 69, 123–143.

FALK-PETERSEN, S., HOPKINS, C. C. E. & SARGENT, J. R. (1990). Trophic relationships in the pelagic, Arctic food web. In *Trophic Relationships in the Marine Environment.* (ed. Barnes, M. & Gibson, R. N.), pp. 315–333. Aberdeen, UK, Aberdeen University Press.

FIELD, J. G. (1970). The use of numerical methods to determine benthic distribution patterns from dredgings in False Bay. *Transactions of the Royal Society of South Africa* 39, 183–200.

FLACH, E. & THOMSEN, L. (1998). Do physical and chemical factors structure the macrobenthic community at a continental slope in the NE Atlantic? *Hydrobiologia* 375/376, 265–285.

GALAKTIONOV, K. V. (1995). Long-term changes in the helminth fauna of colonial seabirds in the Seven Islands archipelago (Barents Sea, Eastern Murman). In *Ecology of Fjords and Coastal Waters.* (ed. Skjoldal, H. R., Hopkins, C., Erikstad, K. E. & Leinaas, H. P.), pp. 489–496. Amsterdam, Elsevier Scientific B. V.

GALAKTIONOV, K. V. & BUSTNES, J. O. (1999). Distribution patterns of marine bird digenean larvae in periwinkles along the southern coast of the Barents Sea. *Diseases of Aquatic Organisms* 37, 221–230.

GARRISON, L. P. & LINK, J. S. (2000). Fishing effects on spatial distribution and trophic guild structure of the fish community in the Georges Bank region. *ICES Journal of Marine Science* 57, 723–730.

GARTNER, J. V. JR. & ZWERNER, D. E. (1989). The parasite faunas of meso- and bathypelagic fishes of Norfolk Submarine Canyon, western North Atlantic. *Journal of Fish Biology* 34, 79–95.

GEORGE-NASCIMENTO, M. A. (1987). Ecological helminthology of wildlife animal hosts from South America: a literature review and a search for patterns in marine food webs. *Revista Chilena de Historia Natural* 60, 181–202.

GIBSON, R. N. & EZZI, I. A. (1987). Feeding relationships of a demersal fish assemblage on the west coast of Scotland. *Journal of Fish Biology* 31, 55–69.

GOLLASCH, S. & ZANDER, C. D. (1995). Population dynamics and parasitation of planktonic and epibenthic crustaceans in the Baltic Schlei fjord. *Helgoländer Meeresuntersuchungen* 49, 759–770.

GRANOVITCH, A. & JOHANNESSON, K. (2000). Digenetic trematodes in four species of *Littorina* from the west coast of Sweden. *Ophelia* **53**, 55–65.

GRAY, J. S. (1974). Animal-sediment relationships. *Oceanography and Marine Biology Annual Reviews* **12**, 223–261.

GROENEWOLD, S., BERGHAHN, S. & ZANDER, C.-D. (1996). Parasite communities of four fish species in the Wadden Sea and the role of fish discarded by the shrimp fisheries in parasite transmission. *Helgoländer Meeresuntersuchungen* **50**, 69–85.

HACUNDA, J. S. (1981). Trophic relationships among demersal fishes in a coastal area of the Gulf of Maine. *Fishery Bulletin* **79**, 775–788.

HAGEN, N. T. (1992). Macroparasite epizootic disease: a potential mechanism for the termination of sea urchin outbreaks in Northern Norway? *Marine Biology* **114**, 469–478.

HAGEN, N. T. (1996). Sea urchin outbreaks and epizootic disease as regulating mechanisms in coastal ecosystems. In *Biology and Ecology of Shallow Coastal Waters*. (ed. Eleftheriou, A., Ansell, A. D. & Smith, C. J.), pp. 303–308. Fredensborg, Denmark, Olsen & Olsen.

HAIRSTON, N. G. JR. & HAIRSTON, N. G., SR. (1993). Cause-effect relationships in energy flow, trophic structure, and interspecific interactions. *American Naturalist* **142**, 379–411.

HALL, S. J. & RAFFAELLI, D. (1991). Food-web patterns: lessons from a species-rich web. *Journal of Animal Ecology* **60**, 823–842.

HAMERLYNCK, O. & MEES, J. (1991). Temporal and spatial structure in the hyperbenthic community of a shallow coastal area and its relation to environmental variables. *Oceanologica Acta* **11**, 205–212.

HEMMINGSEN, W. & MacKENZIE, K. (2001). The parasite fauna of the Atlantic cod, *Gadus morhua* L. *Advances in Marine Biology* **40**, 1–80.

HESSLER, R. R. & SANDERS, H. L. (1967). Faunal diversity in the deep-sea. *Deep-Sea Research* **14**, 65–78.

HOBERG, E. P. (1985). Nearshore foodwebs and the distribution of acanthocephalan parasites in Antarctic seabirds. *Antarctic Journal of the United States* **20**, 161–162.

HOBERG, E. P. (1986). Aspects of the ecology and biogeography of Acanthocephala in Antarctic seabirds. *Annales de Parasitologie Humaine et Comparée* **61**, 199–214.

HOLMES, J. C. (1990). Helminth communities in marine fishes. In *Parasite Communities: Patterns and Processes*. (ed. Esch, G. W., Bush, A. O. & Aho, J. M.), pp. 101–130. London, Chapman and Hall.

HOUSTON, K. A. & HAEDRICH, R. L. (1986). Food habits and intestinal parasites of deep demersal fishes from the upper continental slope east of Newfoundland, northwest Atlantic Ocean. *Marine Biology* **92**, 563–574.

HUXHAM, M., BEANEY, S. & RAFFAELLI, D. (1996). Do parasites reduce the chances of triangulation in a real food web? *Oikos* **76**, 284–300.

HUXHAM, M., RAFFAELLI, D. & PIKE, A. (1995). Parasites and food web patterns. *Journal of Animal Ecology* **64**, 168–176.

JACKSON, C. J., MARCOGLIESE, D. J. & BURT, M. D. B. (1997). The role of hyperbenthic crustaceans in the transmission of marine helminth parasites. *Canadian Journal of Fisheries and Aquatic Sciences* **54**, 815–820.

JOHNSTON, N. T. & LASENBY, D. C. (1982). Diet and feeding of *Neomysis mercedis* Holmes (Crustacea, Mysidacea) from the Fraser River estuary, British Columbia. *Canadian Journal of Zoology* **60**, 813–824.

KAARTVEDT, S. (1986). Diel activity patterns in deep-living cumaceans and amphipods. *Marine Ecology Progress Series* **30**, 243–249.

KENNEDY, C. R. (1994). The ecology of introductions. In *Parasitic Diseases of Fish*. (ed. Pike, A. W. & Lewis, J. W.), pp. 189–208. Tresaith, UK, Samara Publishing Limited.

KENNEDY, C. R., BUSH, A. O. & AHO, J. M. (1986). Patterns in helminth communities: why are birds and fish different? *Parasitology* **93**, 205–215.

KESTING, V., GOLLASCH, S. & ZANDER, C. D. (1996). Parasite communities of the Schlei Fjord (Baltic coast of northern Germany). *Helgoländer Meeresuntersuchungen* **50**, 477–496.

KESTING, V. & ZANDER, C. D. (2000). Alteration of the metazoan parasite faunas in the brackish Schlei fjord (northern Germany, Baltic Sea). *International Review of Hydrobiology* **85**, 325–340.

KLÖSER, H., PLÖTZ, J., PALM, H., BARTSCH, A. & HUBOLD, G. (1992). Adjustment of anisakid nematode life cycles to the high Antarctic food web as shown by *Contracaecum radiatum* and *C. osculatum* in the Weddell seal. *Antarctic Science* **4**, 171–178.

KØIE, M. (1983). Digenetic trematodes from *Limanda limanda* (L.) (Osteichthyes, Pleuronectidae) from Danish and adjacent waters, with special reference to their life histories. *Ophelia* **22**, 201–228.

KØIE, M. (1985). On the morphology and life-history of *Lepidapedon elongatum* (Lebour, 1908) Nicoll, 1910 (Trematoda: Lepocreadiidae). *Ophelia* **24**, 135–153.

KØIE, M. (2000a). Life cycle and seasonal dynamics of *Cucullanus cirratus* O. F. Muller, 1999 (Nematoda, Ascaridida, Seuratoidea, Cucullanidae) in Atlantic cod, *Gadus morhua* L. *Canadian Journal of Zoology* **78**, 182–190.

KØIE, M. (2000b). The life-cycle of the flatfish nematode *Cucullanus heterochronus*. *Journal of Helminthology* **74**, 323–328.

KOSTADINOVA, A. K. & GIBSON, D. I. (1994). Microphallid trematodes in the amphipod *Gammarus subtypicus* Stock, 1966 from a Black Sea lagoon. *Journal of Natural History* **28**, 37–45.

LAFFERTY, K. D. & MORRIS, A. K. (1996). Altered susceptibility of parasitized killifish increases susceptibility to predation by bird final hosts. *Ecology* **77**, 1390–1397.

LANGTON, R. W. (1982). Diet overlap between Atlantic cod, *Gadus morhua*, silver hake, *Merluccius bilinearis*, and fifteen other Northwest Atlantic finfish. *Fishery Bulletin* **80**, 745–759.

LANGTON, R. W. & BOWMAN, R. E. (1980). *Food of Fifteen Northwest Atlantic Gadiform Fishes*. NOAA Technical Report NMFS SSRF-740. NOAA, NMFS, Rockville, Maryland.

LANGTON, R. W. & WATLING, L. (1990). The fish-benthos connection: a definition of prey groups in the Gulf of

Maine. In *Trophic Relationships in the Marine Environment*. (ed. Barnes, M. & Gibson, R. N.), pp. 424–438. Aberdeen, UK, Aberdeen University Press.

LILE, N. K. (1998). Alimentary tract helminths of four pleuronectid flatfish in relation to host phylogeny and ecology. *Journal of Fish Biology* **53**, 945–953.

LOTZ, J. M., BUSH, A. O. & FONT, W. F. (1995). Recruitment-driven, spatially discontinuous communities: a null model for transferred patterns in target communities of intestinal helminths. *Journal of Parasitology* **81**, 12–24.

MacDONALD, J. S. & GREEN, R. H. (1986). Food resource utilization by five species of benthic feeding fish in Passamaquoddy Bay, New Brunswick. *Canadian Journal of Fisheries and Aquatic Sciences* **43**, 1534–1546.

MacKENZIE, K. (1999). Parasites as pollution indicators in marine ecosystems: a proposed early warning system. *Marine Pollution Bulletin* **38**, 955–959.

MacKENZIE, K., WILLIAMS, H. H., WILLIAMS, B., McVICAR, A. H. & SIDDALL, R. (1995). Parasites as indicators of water quality and the potential use of helminth transmission in marine pollution studies. *Advances in Parasitology* **35**, 85–144.

MADIN, L. P. & MADIN, K. A. C. (1995). Diversity in a vast and stable habitat. *Oceanus* **38**, 20–24.

MARCOGLIESE, D. J. (1995). The role of zooplankton in the transmission of helminth parasites to fish. *Reviews in Fish Biology and Fisheries* **5**, 336–371.

MARCOGLIESE, D. J. (2001*a*). Pursuing parasites up the food chain: implications of food web structure and function on parasite communities in aquatic systems. *Acta Parasitologica* **46**, 82–93.

MARCOGLIESE, D. J. (2001*b*). Implications of climate change for parasitism of animals in the aquatic environment. *Canadian Journal of Zoology* **79**, 1331–1352.

MARCOGLIESE, D. J. & CONE, D. K. (1997). Food webs: a plea for parasites. *Trends in Ecology and Evolution* **12**, 320–325.

MARKLE, D. F., DADSWELL, M. J. & HALLIDAY, R. G. (1988). Demersal fish and decapod crustacean fauna of the upper continental slope off Nova Scotia from LaHave to St. Pierre Banks. *Canadian Journal of Zoology* **66**, 1952–1960.

MARTELL, D. J. & McCLELLAND, G. (1994). Diets of sympatric flatfishes, *Hippoglossoides platessoides*, *Pleuronectes ferrugineus*, *Pleuronectes americanus*, from Sable Island Bank, Canada. *Journal of Fish Biology* **44**, 821–848.

MATTSON, S. (1981). The food of *Galeus melastomus*, *Gadiculus argenteus thori*, *Trisopterus esmarkii*, *Rhinonemus cimbrius*, and *Glyptocephalus cynoglossus* (Pisces) caught during the day with shrimp trawl in a west-Norwegian fjord. *Sarsia* **66**, 109–127.

MAURER, D. & WIGLEY, R. L. (1984). *Biomass and Density of Macrobenthic invertebrates on the U.S. Continental Shelf off Martha's Vineyard, Mass., in Relation to Environmental Factors*. NOAA Technical Report NMFS SSRF-783. NOAA, NMFS, Rockville, Maryland.

MAY, R. M. (1994). Biological diversity: differences between land and sea. *Philosophical Transactions of the Royal Society of London, Series B* **343**, 105–111.

McLUSKY, D. S. & McINTYRE, A. D. (1988). Characteristics of the benthic fauna. In *Ecosystems of the World. 27. Continental Shelves*. (ed. Postma, J. & Zijlstra, J. J.), pp. 131–154. Amsterdam, Elsevier.

MEES, J., DEWICKE, A. & HAMERLYNCK, O. (1993). Seasonal composition and spatial distribution of hyperbenthic communities along estuarine gradients in the Westerschelde. *Netherlands Journal of Aquatic Ecology* **27**, 359–376.

MILLS, E. L. & FOURNIER, R. O. (1979). Fish production and the marine ecosystems of the Scotian Shelf, eastern Canada. *Marine Biology* **54**, 101–108.

MINCHELLA, D. J. & SCOTT, M. E. (1991). Parasitism: a cryptic determinant of community structure. *Trends in Ecology and Evolution* **6**, 250–254.

OJEDA, F. P. & DEARBORN, J. H. (1990). Diversity, abundance, and spatial distribution of fishes and crustaceans in the rocky subtidal zone of the Gulf of Maine. *Fishery Bulletin* **88**, 403–410.

OUG, E. (1977). Faunal distribution close to the sediment of a shallow marine environment. *Sarsia* **63**, 115–121.

PASCUAL, S., GONZALES, A., ARIAS, C. & GUERRA, A. (1996). Biotic relationships of *Illex condetii* and *Todaropsis eblanae* (Cephalopoda, Ommastrephidae) in the Northeast Atlantic: evidence from parasites. *Sarsia* **81**, 265–274.

PAULY, D., CHRISTENSEN, V., FROESE, R. & PALOMARES, M. L. (2000). Fishing down aquatic food webs. *American Scientist* **88**, 46–51.

PETERSEN, G. H. & CURTIS, M. A. (1980). Differences in energy flow through major components of subarctic, temperate and tropical marine shelf ecosystems. *Dana* **1**, 53–64.

POLYANSKI, YU. I. (1961). Ecology of parasites of marine fishes. In *Parasitology of Fishes*. (ed. Dogiel, V. A., Petrushevski, G. K. & Polyanski, Yu. I.), pp. 48–83. Edinburgh, Oliver and Boyd.

POULIN, R., HECKER, K. & THOMAS, F. (1998). Hosts manipulated by one parasite incur additional costs from infection by another parasite. *Journal of Parasitology* **84**, 1050–1052.

PRENA, J., SCHWINGHAMER, P., ROWELL, T. W., GORDON, D. C. JR., GILKINSON, K. D., VASS, W. P. & MCKEOWN, D. L. (1999). Experimental otter trawling on a sandy bottom ecosystem of the Grand Banks of Newfoundland: analysis of trawl bycatch and effects on epifauna. *Marine Ecology Progress Series* **181**, 107–124.

RAKOCINSKI, C. F., BROWN, S. S., GASTON, G. R., HEARD, R. W., WALKER, W. W. & SUMMERS, J. K. (1997). Macrobenthic responses to natural and contaminant-related gradients in northern Gulf of Mexico estuaries. *Ecological Applications* **7**, 1278–1298.

REIMER, L. W. (1995). Parasites especially of piscean hosts as indicators of the eutrophication in the Baltic Sea. *Applied Parasitology* **36**, 124–135.

RHOADS, D. C. & YOUNG, D. K. (1970). The influence of deposit-feeding organisms on sediment stability and community structure. *Journal of Marine Research* **28**, 150–178.

RICHARDS, S. W. (1963). The demersal fish population of Long Island Sound. II. Food of the juveniles from a sand-shell locality (Station I). *Bulletin of the Binghamton Oceanographic Collection* **18**, 33–72.

SANDERS, H. L. (1968). Marine benthic diversity: a comparative study. *American Naturalist* **102**, 243–282.

SCHEIBLING, R. E., HENNIGAR, A. W. & BALCH, T. (1999). Destructive grazing, epiphytism, and disease: the dynamics of sea urchin – kelp interactions in Nova Scotia. *Canadian Journal of Fisheries and Aquatic Sciences* **56**, 2300–2314.

SCHMIDT, G. D. (1985). Development and life cycles. In *Biology of the Acanthocephala*. (ed. Crompton, D. W. T. & Nickol, B. B.), pp. 275–305. Cambridge, Cambridge University Press.

SCHOENER, T. W. (1989). Food webs from the small to the large. *Ecology* **70**, 1559–1589.

SCOTT, J. S. (1975). Incidence of trematode parasites of American plaice (*Hippoglossoides platessoides*) of the Scotian Shelf and Gulf of St. Lawrence in relation to fish length and food. *Journal of the Fisheries Research Board of Canada* **32**, 479–483.

SCOTT, J. S. (1981). Alimentary tract parasites of haddock (*Melanogrammus aeglefinus* L.) on the Scotian Shelf. *Canadian Journal of Zoology* **59**, 2244–2252.

SCOTT, J. S. (1982*a*). Selection of bottom type by groundfishes of the Scotian Shelf. *Canadian Journal of Fisheries and Aquatic Sciences* **39**, 943–947.

SCOTT, J. S. (1982*b*). Digenean parasite communities in flatfishes on the Scotian Shelf and southern Gulf of St. Lawrence. *Canadian Journal of Zoology* **60**, 2804–2811.

SCOTT, J. S. (1985). Occurrence of alimentary tract helminth parasites of pollock (*Pollachius virens* L.) on the Scotian Shelf. *Canadian Journal of Zoology* **63**, 1695–1698.

SCOTT, J. S. & BRAY, S. A. (1989). Helminth parasites of the alimentary tract of Atlantic halibut (*Hippoglossus hippoglossus* L.) and Greenland halibut (*Reinhardtius hippoglossoides* (Walbaum)) on the Scotian Shelf. *Canadian Journal of Zoology* **67**, 1476–1481.

SHERMAN, K., GROSSLEIN, M., MOUNTAIN, D., BUSCH, D., O'REILLY, J. & THEROUX, R. (1988). The continental shelf ecosystem off the northeast coast of the United States. In *Ecosystems of the World. 27. Continental Shelves*. (ed. Postma, J. & Zijlstra, J. J.), pp. 279–337. Amsterdam, Elsevier.

SIBERT, J. R. (1981). Intertidal hyperbenthic populations in the Nanaimo estuary. *Marine Biology* **64**, 259–265.

SKORPING, A. & HÖGSTEDT, G. (2001). Trophic cascades: a role for parasites? *Oikos* **94**, 191–192.

SMETACEK, V. (1988). Plankton characteristics. In *Ecosystems of the World. 27. Continental Shelves*. (ed. Postma, J. & Zijlstra, J. J.), pp. 93–130. Amsterdam, Elsevier.

SMITH, C. R., AUSTEN, M. C., BOUCHER, G., HEIP, C., HUTCHINGS, P. A., KING, G. M., KOIKE, I., LAMBSHEAD, J. D. & SNELGROVE, P. (2000). Global change and biodiversity linkages across the sediment-water interface. *BioScience* **50**, 1108–1120.

SNELGROVE, P. V. R., AUSTEN, M. C., BOUCHER, G., HEIP, C., HUTCHINGS, P. A., KING, G. M., KOIKE, I., LAMBSHEAD, J. D. & SMITH, C. R. (2000). Linking biodiversity above and below the marine sediment-water interface. *BioScience* **50**, 1076–1088.

SNELGROVE, P. V. R. & GRASSLE, J. F. (1995). The deep sea: desert AND rainforest. *Oceanus* **38**, 25–29.

SUCHANEK, T. H. (1994). Temperate coastal marine communities: biodiversity and threats. *American Zoologist* **34**, 100–114.

SULGOWSTOSKA, T., BANACZYK, G. & GRABDA-KAZUBSKA, B. (1987). Helminth fauna of flatfish (Pleuronectiformes) from Gdansk Bay and adjacent areas (south-east Baltic). *Acta Parasitologica Polonica* **31**, 231–240.

SULGOWSTOSKA, T., JERZEWSKA, B. & WICIKOWSKI, J. (1990). Parasite fauna of *Myoxocephalus scorpius* (L.) and *Zoarces viviparus* (L.) from environs of Hel (southeast Baltic) and seasonal occurrence of parasites. *Acta Parasitologica Polonica* **35**, 143–148.

THONEY, D. A. (1991). Population dynamics and community analysis of the parasite fauna of juvenile spot, *Leiostomus xanthurus* (Lacepede), and Atlantic croaker, *Micropogonias undulatus* (Linnaeus), (Sciaenidae) in two estuaries along the middle Atlantic coast of the United States. *Journal of Fish Biology* **39**, 515–534.

THONEY, D. A. (1993). Community ecology of the parasites of adult spot, *Leiostomus xanthurus*, and Atlantic croaker, *Micropogonias undulatus* (Sciaenidae) in the Cape Hatteras Region. *Journal of Fish Biology* **43**, 781–804.

TYLER, A. V. (1972). Food resources division among northern, marine, demersal fishes. *Journal of the Fisheries Research Board of Canada* **29**, 997–1003.

USPENSKAJA, A. V. (1960). Parasitofaune des crustacés benthiques de la mer Barents. *Annales de Parasitologie Humaine et Comparée* **35**, 221–242.

VALTONEN, E. T. & NIINIMAA, A. (1983). Dispersion and frequency distribution of *Corynosoma* spp. (Acanthocephala) in the fish of the Bothnian Bay. *Aquilo Ser Zoologica* **22**, 1–11.

VINOGRADOV, V. I. (1984). Food of silver hake, red hake and other fishes of Georges Bank and adjacent waters, 1968–74. *NAFO Scientific Council Studies* **7**, 87–94.

VIRNSTEIN, R. W. (1977). The importance of predation by crabs and fishes on benthic infauna in Chesapeake Bay. *Ecology* **58**, 1199–1217.

WARWICK, R. M. (1988). Analysis of community attributes of the macrobenthos of Frierfjord/Langesundfjord at taxonomic levels higher than species. *Marine Ecology Progress Series* **46**, 167–170.

WARWICK, R. M. & CLARKE, H. R. (1993). Comparing the severity of disturbance: a meta-analysis of marine macrobenthic community data. *Marine Ecology Progress Series* **92**, 221–231.

WARWICK, R. M. & CLARKE, H. R. (1995). New 'biodiversity' measures reveal a decrease in taxonomic distinctness with increasing stress. *Marine Ecology Progress Series* **129**, 301–305.

WESLAWSKI, J. M. & KWASNIEWSKI, S. (1990). The consequences of climatic fluctuations for the food web in Svalbard coastal waters. In *Trophic Relationships in the Marine Environment*. (ed. Barnes, M. & Gibson, R. N.), pp. 281–295. Aberdeen, UK, Aberdeen University Press.

WESTON, D. P. (1988). Macrobenthos-sediment relationships on the continental shelf off Cape Hatteras, North Carolina. *Continental Shelf Research* **8**, 267–286.

WIGLEY, R. L. & McINTYRE, A. D. (1964). Some quantitative comparisons of offshore meiobenthos and macrobenthos south of Martha's Vineyard. *Limnology and Oceanography* **9**, 485–493.

WILDISH, D. J., WILSON, A. J. & FROST, B. (1992). Benthic boundary layer macrofauna of Browns Banks, northwest Atlantic, as potential prey of juvenile benthic fish. *Canadian Journal of Fisheries and Aquatic Sciences* **49**, 91–98.

WILLIAMS, H. H. & JONES, A. (1994). *Parasitic Worms of Fish*. London, Taylor & Francis.

WILLIAMS, H. H., MACKENZIE, K. & McCARTHY, A. M. (1992). Parasites as biological indicators of the population biology, migrations, diet, and phylogenetics of fish. *Reviews in Fish Biology and Fisheries* **2**, 144–176.

WILSON, W. H. (1991). Competition and predation in marine soft-sediment communities. *Annual Review of Ecology and Systematics* **21**, 221–241.

YIP, S. Y. (1984). Parasites of *Pleurobrachia pileus* Muller, 1776 (Ctenophora), from Galway Bay, western Ireland. *Journal of Plankton Research* **6**, 107–121.

ZANDER, C. D. (1992/1993). The biological indication of parasite life-cycles and communities from the Lubeck Bight, SW Baltic Sea. *Zeitschrift für Angewandte Zoologie* **79**, 377–389.

ZANDER, C. D. (1998). Ecology of host parasite relationships in the Baltic Sea. *Naturwissenschaften* **85**, 426–436.

ZANDER, C. D. (2001). The guild as a concept and a means in ecological parasitology. *Parasitology Research* **87**, 484–488.

ZANDER, C. D., GROENEWOLD, S. & STROHBACH, U. (1994). Parasite transfer from crustacean to fish hosts in the Lubeck Bight, SW Baltic Sea. *Helgoländer Meeresuntersuchungen* **48**, 89–105.

ZANDER, C. D. & KESTING, V. (1996). The indicator properties of parasite communities of gobies (Teleostei, Gobiidae) from Kiel and Lubeck Bight, SW Baltic Sea. *Applied Parasitology* **37**, 186–204.

ZANDER, C. D. & KESTING, V. (1998). Colonization and seasonality of goby (Gobiidae, Teleostei) parasites from the southwestern Baltic Sea. *Parasitology Research* **84**, 459–466.

ZANDER, C. D., REIMER, L. W. & BARZ, K. (1999). Parasite communities of the Salzhaff (Northwest Mecklenburg, Baltic Sea). I. Structure and dynamics of communities of littoral fish, especially small-sized fish. *Parasitology Research* **85**, 356–372.

ZANDER, C. D., REIMER, L. W., BARZ, K., DIETEL, G. & STROHBACH, U. (2000). Parasite communities of the Salzhaff (Northwest Mecklenburg, Baltic Sea). II. Guild communities, with special regard to snails, benthic crustaceans, and small-sized fish. *Parasitology Research* **86**, 359–372.

ZANDER, C. D., STROHBACH, U. & GROENEWOLD, S. (1993). The importance of gobies (Gobiidae, Teleostei) as hosts and transmitters of parasites in the SW Baltic. *Helgoländer Meeresuntersuchungen* **47**, 81–111.

ZDZITOWIECKI, K. (2001). Acanthocephala occurring in intermediate hosts, amphipods, in Admiralty Bay (South Shetland Islands, Antarctica). *Acta Parasitologica* **46**, 202–207.

ZUBCHENKO, A. V. (1981). Parasitic fauna of some Macrouridae in the Northwest Atlantic. *Journal of Northwest Atlantic Fisheries Science* **2**, 67–72.

Parasitism, community structure and biodiversity in intertidal ecosystems

K. N. MOURITSEN *and* R. POULIN*

Department of Zoology, University of Otago, P.O. Box 56, Dunedin, New Zealand

SUMMARY

There is mounting evidence that parasites can influence the composition and structure of natural animal communities. In spite of this, it is difficult to assess just how important parasitism is for community structure because very few studies have been designed specifically to address the role of parasites at the community level, no doubt because it is difficult to manipulate the abundance of parasites in field experiments. Here, we bring together a large amount of published information on parasitism in intertidal communities to highlight the potential influence of parasites on the structure and biodiversity of these communities. We first review the impact of metazoan parasites on the survival, reproduction, growth and behaviour of intertidal invertebrates, from both rocky shores and soft-sediment flats. Published evidence suggests that the impact of parasites on individuals is often severe, though their effects at the population level are dependent on prevalence and intensity of infection. We then put this information together in a discussion of the impact of parasitism at the community level. We emphasize two ways in which parasites can modify the structure of intertidal communities. First, the direct impact of parasites on the abundance of key host species can decrease the importance of these hosts in competition or predator-prey interactions with other species. Second, the indirect effects of parasites on the behaviour of their hosts, e.g. burrowing ability or spatial distribution within the intertidal zone, can cause changes to various features of the habitat for other intertidal species, leading to their greater settlement success or to their local disappearance. Our synthesis allows specific predictions to be made regarding the potential impact of parasites in certain intertidal systems, and suggests that parasites must be included in future community studies and food web models of intertidal ecosystems.

Key words: Altered behaviour, castration, food webs, host survival, rocky shores, soft-sediment flats, trematodes.

INTRODUCTION

Parasitism is now widely recognized as a factor that can influence the composition and structure of natural animal communities (Minchella & Scott, 1991; Combes, 1996; Hudson & Greenman, 1998; Poulin, 1999). The presence of certain species, or their abundance relative to that of other species in a community, may be entirely dependent on the action of parasites. In addition to the numerous studies that have shown that parasites can decrease the survival or reproductive output of their host, there is evidence that the outcome of interspecific competition between hosts (Hudson & Greenman, 1998) and the interaction between predators and prey (Lafferty, 1999) can be modified by parasites. The latter effect often involves parasite-mediated changes in host behaviour that increase their probability of being captured by a predator; these changes, however, can also impact on communities in other ways (Poulin, 1999). Despite this accumulating evidence, it is still difficult to assess just how important parasitism is for community structure for at least two reasons. First, very few studies have been designed specifically to address the role of parasites at the community level. This is no doubt because it is difficult to manipulate the abundance of parasites in field experiments.

Second, the many studies with results that imply a role for parasites in structuring communities come from a wide range of systems. It is more difficult to interpret these isolated reports than it would be if they all came from a single community-type.

Here, we bring together a large amount of published information on parasitism in intertidal communities to highlight the potential influence of parasites on the structure and biodiversity of these communities. Parasites of intertidal animals have been well studied, and an integration of what is currently known can be informative. We first begin by briefly summarising the main features of intertidal ecosystems, and giving an overview of the parasites found in those systems. Then, we examine the known effects of parasites on intertidal hosts. This review focuses only on intertidal macro-invertebrates, and not on the meiofauna or interstitial animals (< 0.5 mm in dimension), for which little information is available regarding parasitism. Finally, we put this information together in a discussion of the impact of parasitism at the community level.

INTERTIDAL SYSTEMS AT A GLANCE

Intertidal areas are at the interface between marine and terrestrial habitats (Nybakken, 1993; Bertness, 1999). Although inhabited almost exclusively by

* Corresponding author: Tel: +64 3 479 7983. Fax: +64 3 479 7584.
E-mail: robert.poulin@stonebow.otago.ac.nz

marine organisms, they are regularly visited by shorebirds and even passerines that feed on intertidal animals. Invertebrates living permanently in the intertidal zone face the greatest temporal variations in environmental conditions of any marine habitat. The action of tides exposes intertidal invertebrates to the air at regular intervals; the tolerance of animals to air exposure and the higher temperatures associated with it can be the major determinant of zonation patterns among the species forming intertidal communities. The upper intertidal, for instance, is inhabited only by the species most resistant to water loss, heat stress, mechanical stress from wave action, and osmotic stress caused by rainfall during low tide. Since many intertidal invertebrates have planktonic larvae, processes controlling larval settlement play a major role in determining the species composition of intertidal communities. The species found in one locality are a subset of the pool of species whose larvae occur in coastal currents. Local conditions, such as wave exposure, the type of substrate and the present benthic community of adults, have a strong influence on whether or not larvae of a given species will establish successfully. Intertidal areas can be classified into different types of habitats according to the predominant substrate; here we will consider rocky shores and soft-sediment flats as the best studied types of intertidal habitats.

Rocky shores

Both the density and diversity of plants and invertebrates reach high levels on rocky shores, because they usually are bathed in nutrient- and plankton-rich waters and provide a good substrate for algal growth (Bertness, 1999). A striking feature on rocky shores is the marked vertical zonation of the major organisms, resulting in often well-defined horizontal bands populated only by the few species restricted or adapted to this particular intertidal zone (Nybakken, 1993; Bertness, 1999). Physical factors, such as the coastal slope and air exposure time, are of prime importance in determining these zonation patterns, but biological processes are also at work in the structuring of rocky shore communities. Interference competition for attachment space on the rocky substrate is usually intense among species of algae as well as between sessile filter-feeding invertebrates such as mussels and barnacles. Predation, e.g. by starfish, crabs, or gastropods, is also an important structuring factor particularly on the more sheltered shores: the action of predators, and their preferences for certain prey, can prevent one prey species from out-competing others, and will affect the distribution and relative abundances of prey species. Similar effects exist regarding the impact of grazers, such as limpets and littorinid snails, on algae. The interactions of physical factors, competition and predation

(or grazing), together with the spatial and temporal variability in larval recruitment, are expected to be the main determinants of community structure and species composition in rocky shore systems (Menge & Sutherland, 1987; Nybakken, 1993; Navarrete & Menge, 1996; Bertness, 1999).

Soft-sediment flats

Unlike rocky shores, soft-sediment intertidal flats can only develop in sheltered areas protected from wave action, such as bays, harbours, lagoons or estuaries. The more protected an area is from the action of waves, the more it tends to accumulate fine-grained sediments and organic matter (Nybakken, 1993). There is thus a continuum among soft-sediment intertidal flats, from mud flats to sand flats, with increasing exposure to waves. Typically, the intertidal slope in these areas is very small, and even at low tide water is retained in interstitial space of the sediment making desiccation a minor problem for soft-bottom invertebrates. For these reasons zonation of organisms is not as pronounced as on the rocky shore. However, the low renewal rate of interstitial water and the action of reducing bacteria in the sediments result in a rapid decrease in oxygen concentration with depth in the sediments, giving rise to anaerobic conditions a few centimetres below the sediment surface. The main plants include a variety of microalgae (diatoms), green and red algae and, usually at lower tidal levels, various seagrasses. Most invertebrates of soft-sediment shores live either freely burrowed into the sediments or construct more or less permanent tubes, relying mainly on surface water for oxygen. They include deposit and suspension feeding polychaetes, bivalves, and crustaceans as well as endobenthic predatory polychaetes and nemertines. Snails that graze algal mats are also common on the surface, in addition to invertebrate predators such as crabs, shrimps and whelks. Because of the accumulation of organic matter on intertidal flats, the density of many of these organisms is often very high. Exploitative competition, however, does not appear to play a major role in the structure of soft-sediment intertidal communities, though it can have more limited effects (Wilson, 1991; Bertness, 1999). Predation, on the other hand, can be very important, especially by fish (rays, flatfish and gobies), epibenthic invertebrates (crabs, shrimps) and shorebirds. Many bird species are migratory and only seasonal visitors of intertidal flats, but during those brief annual visits they can dramatically decrease the densities of many key invertebrates (see e.g. Nybakken, 1993 and references therein). Finally, disturbances that disrupt or destabilise the sediments also have major impacts on soft-sediment communities, whether these disturbances are of abiotic (e.g. storms) or biotic (e.g. burrowing) origins (Bertness, 1999). On the other hand, sediment stabilisation through the construction of permanent or semi-permanent burrows and tubes, by polychaetes and amphipods for instance, may also significantly influence the character of the surrounding benthic community (see e.g. Mouritsen, Mouritsen & Jensen, 1998 and references therein).

Parasites of intertidal organisms

The present review will focus only on the metazoan parasites of intertidal macro-invertebrates. There are, of course, a wide variety of viruses, bacteria and protozoans that infect intertidal invertebrates (see Sindermann, 1990) and that may impact on community structure; cases involving metazoan parasites, however, should suffice to illustrate the potential importance of parasitism.

Among the metazoans, trematodes are the most common parasites of intertidal animals. Their complex life cycle almost always involves a gastropod, less frequently a bivalve, as first intermediate host, in which larval stages (sporocysts or rediae) engage in the asexual production of cercariae, the next stage in the life cycle. Prevalence of infection in the mollusc first intermediate host is notoriously variable in time and space, and usually relatively low ($< 25\%$). However, much higher values have regularly been reported (see review in Sousa, 1991; Curtis, this supplement). It must be emphasized that trematode infections in gastropods are ubiquitous: given a large enough sample covering a species' range of habitats, dissections of intertidal gastropods almost always reveal the presence of one or more trematode species. After their release from the first intermediate host, trematode cercariae penetrate and encyst in a second intermediate host, usually another mollusc, a crustacean, a polychaete or a fish. Prevalence in the second intermediate host is usually much higher than in the first intermediate host, often approaching 100% (see Sousa, 1991). The life cycle of the parasite is completed when the second intermediate host is eaten by a suitable definitive host, most often a shorebird or a fish.

Nematodes are also common parasites of intertidal invertebrates; they use them as intermediate host or as their only host, depending on the parasite species. Larval cestodes and acanthocephalans often use intertidal crustaceans as their intermediate host, their life cycle requiring the ingestion of the crustacean by a suitable fish or bird. Finally, decapod crustaceans in intertidal systems are often infected by parasites that either castrate them or greatly reduce their reproductive output, e.g. nematomorphs, nemertean egg parasites, rhizocephalans, and parasitic isopods (see below). Very few quantitative studies have been performed on these parasites of decapods, but it is clear from the available evidence that they can have dramatic impact on individual host fitness and on host population dynamics.

IMPACT OF PARASITISM ON INTERTIDAL HOSTS

Survival

By definition any parasite will damage its host to a certain degree, with effects ranging from minor metabolic changes to severe tissue destruction (see Lauckner, 1980, 1983; Price, 1980). Under normal environmental conditions the pathology inflicted may be insignificant in terms of mortality, but when conditions approach the limit to which the host is adapted, reduced survivorship may be the rule. This can be expected to apply in particular to intertidal species that exist at the edge of 'possible', exposed to wide short- as well as long-term fluctuations in temperature, osmotic stress, desiccation and (in soft-bottom habitats) oxygen conditions. Accordingly, several species of intertidal or estuarine snails show decreased resistance to extreme abiotic conditions when infected by trematodes. The survival of infected periwinkles *Littorina littorea* is significantly reduced in comparison to control snails when exposed to high temperatures within the natural thermal range (McDaniel, 1969; Lauckner, 1987a), and a similar effect has been demonstrated in the mud whelk *Nassarius reticulatus* regarding both high and low temperatures (Tallmark & Norrgren, 1976). Lower resistance to freezing, osmotic stress, desiccation and/or low oxygen conditions has likewise been shown in an array of trematode-infected hard- as well as soft-bottom snail species, including *L. obtusata*, *L. saxatilis*, *Cerithidea californica*, *Hydrobia neglecta* and *H. ulvae* (Berger & Kondratenkov, 1974; Sergievsky, Granovich & Mikhailova, 1986; Sousa & Gleason, 1989; Jensen, Latama & Mouritsen, 1996). The synergistic effect of parasites and extreme environmental conditions can also be inferred from a more long-term field experiment on *L. littorea* (Huxham, Raffaelli & Pike, 1993; Fig. 1). In addition to the unequivocal overall lower survival rate of infected specimens, it appears that the higher mortality among infected snails is largely confined to the winter and summer months, where the abiotic factors also reach their extremes. In addition, Jensen & Mouritsen (1992) were able to ascribe a 40% decline in the density of the mud snail *H. ulvae* in Denmark to the combined effect of infection by microphallid trematodes and unusually high ambient temperatures. The density of infected snails was reduced from more than 10000 to less than 1000 individuals m^{-2} over a mere five weeks. That trematodes can affect the population dynamics of their snail host significantly is also tentatively supported by data on *L. littorea* (Lauckner, 1987a) and demonstrated in a field experiment in the case of *C. californica* (Lafferty, 1993).

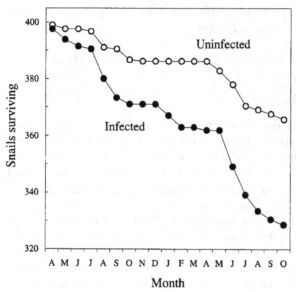

Fig. 1. Mortality of uninfected and trematode-infected periwinkles *Littorina littorea* caged *in situ*. Infected specimens were identified on the basis of foot-colour. Subsequent dissections revealed that this method was correct in about 75 % of the cases. (After Huxham *et al.* 1993.)

Compromised survival of infected individuals is not restricted to gastropods. Bivalves acting either as first or second intermediate host of trematodes experience reduced condition, reduced byssus thread production (in the case of *Mytilus edulis*), and may be easier to open by predators (Lauckner, 1983; Zwarts, 1991; Calvo-Ugarteburu & McQuaid, 1998). Shell-size/infection intensity relationships in *Cerastoderma edule*, *M. edulis* and *Tapes philippinarum* also suggest increased mortality among both recruits and adults infected by various macroparasites (trematodes, but also cestodes, turbellarians and copepods) (Montaudouin *et al.* 2000). Similarly, Goater (1993) showed a peaking pattern in mean abundance and in the level of aggregation of the trematode *Meiogymnophallus minutus* as a function of cockle size (*C. edule*), suggesting parasite-induced mortality in the older more heavily infected specimens. Jonsson & André (1992) could even ascribe an event of mass mortality in a population of *C. edule* to infection by the digenean trematode *Cercaria cerastodermae* I. Experimental work on *Cerastoderma* supports the above observations, showing significantly reduced survivorship in cockles infected by digenean trematodes (echinostomatids) particularly under environmental stress, such as hypoxia (Lauckner, 1987b; Wegeberg & Jensen, 1999).

Crustaceans are also commonly subject to significant parasite-induced mortality. The small and often very abundant amphipod of north Atlantic intertidal mud flats, *Corophium volutator*, is the second intermediate host to a range of trematode species of which several have been shown to elevate substantially the mortality rate of the amphipod under

laboratory conditions (Mouritsen & Jensen, 1997; Jensen, Jensen & Mouritsen, 1998; McCurdy, Forbes & Boates, 1999a; Meissner & Bick, 1999). That these observations are relevant also *in situ* can be inferred from field studies in both tidal and non-tidal coastal habitats showing impacts of microphallid trematodes on *Corophium* populations ranging from subtle regulation (Bick, 1994; Meissner, 2001; see Thomas *et al.* 1995 for a *Gammarus* example), to major reductions in abundance (Meissner & Bick, 1997) and even local extinction (Jensen & Mouritsen, 1992, Fig. 2A). The latter effect coincided with high ambient temperatures that triggered a mass release of the infective cercarial stage from the sympatric population of first intermediate snail hosts. Interestingly, because *C. volutator* can, under certain circumstances, stabilise the substrate by its tube-building activity, the disappearance of the amphipods released sediment erosion that significantly changed the particle size composition of the substrate as well as the topography of the tidal flat (Mouritsen *et al.* 1998, Fig. 2B–D). Since sediment characteristics are important distributional factors for many infaunal invertebrates (Snelgrove & Butman, 1994), the changes in the substrate may in part have been responsible for the significant changes in abundance of other macrofaunal species following the amphipod die-off (Jensen & Mouritsen, 1992; K. N. Mouritsen & K. T. Jensen, unpublished data).

Other crustaceans also feel the effects of parasites. On rocky shores, the combined effect of infection by microphallid trematodes (*Maritrema arenaria*) and desiccation has also been suggested to be responsible for observations of sudden shoreward declines in the abundance of the encrusting barnacle *Semibalanus balanoides* (Carrol, Montgomery & Hanna, 1990). Prevalence of *M. arenaria* in encrusting barnacles varies among sites but may approach 100 % in the uppermost fringe of the intertidal zone (Irwin & Irwin, 1980; Mitchell & Dessi, 1984; Carrol *et al.* 1990). Among the larger epibenthic predators, crabs are commonly infected by parasites that inflict mortality, in particular at the juvenile stage. For instance, juvenile shore crabs *Carcinus maenas* appear very sensitive to infection by microphallid trematodes (Lauckner, 1987b), and a directly transmitted parasitic rhizocephalan has been shown to cause an 8-fold increase in mortality in infected juvenile *Rhithropanopeus* crabs (Alvarez, Hines & Reaka-Kudla, 1995). Whereas significant intertidal predators, such as the shore crab (*C. maenas*) and blue crabs (*Callinectes* spp.), are rarely found to be infected by rhizocephalans in the intertidal zone, prevalences between 10 and 20 % among adults are common in subtidal samples, occasionally approaching 50 or even 100 % (Wardle & Tirpak, 1991; Alvarez & Calderon, 1996; Larzaro-Chavez, Alvarez & Rosas, 1996; Glenner & Hoeg, 1997; Mathieson,

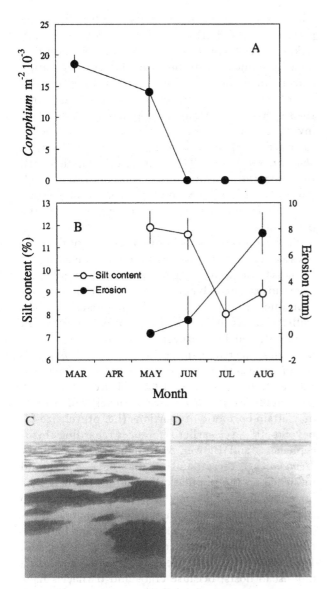

Fig. 2. Extermination of a dense intertidal amphipod population by microphallid trematodes and changes in sediment characteristics and flat topography that followed. (A) Mean density (±S.E.) of *Corophium volutator*. (B) Mean silt content (i.e. particle diameter < 63 μm) (○) and substrate erosion (●; the first data point represents first measurement, i.e. the reference level) (±S.E.). (C) The topography of the *Corophium*-bed prior to the die-off. The amphipods occurred mainly on the emerged areas. (D) The topography of the flat a few months after the disappearance of amphipods. (After Mouritsen *et al.* 1998.)

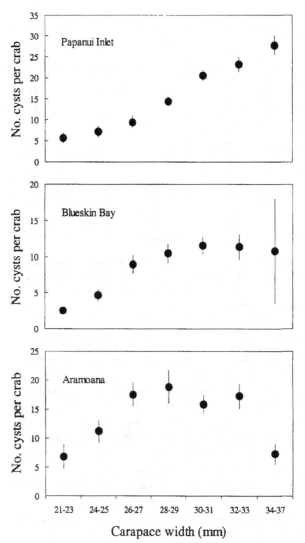

Fig. 3. Acantocephalan-induced mortality in the mud crab *Macrophthalmus hirtipes*. Mean larvae intensity (±S.E.) as a function of size-classes from three different intertidal populations of the South Island, New Zealand. The levelling-out and decrease of mean parasite load in the larger size-classes in Blueskin Bay and at Aramoana, respectively, suggest parasite-induced mortality among the larger and more heavily infected crabs. (Data from Latham & Poulin, unpublished.)

populations of the mud crab *Macrophthalmus hirtipes* in New Zealand (Fig. 3).

Reproduction

Although parasitism, by increasing mortality, may reduce the host population significantly, its effect on host fecundity may be even more severe in terms of population dynamics. Infection by parasites commonly reduces the reproductive output of the host, and in the case of important intertidal genera of molluscs acting as first intermediate hosts to trematodes (e.g. *Mytilus*, *Cerastoderma*, *Macoma*, *Hydrobia*, *Littorina*, *Cerithidea*, *Ilyanassa*, *Nucella*), infection inevitably results in partial or complete castration (Feral *et al.* 1972; Lauckner, 1983; Sousa,

Berry & Kennedy, 1998; Alvarez *et al.* 1999). Assuming high mortality among infected juveniles, the rhizocephalans may significantly depress the subtidal population of crabs which, in turn, may affect the abundance also in the adjacent intertidal habitat. Little information is available on the effects of other types of parasites on crab survival, but these can be non-negligible. For instance, the relationship between intensity of infection and crab size suggests that acanthocephalans cause mortality in certain

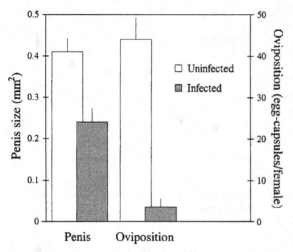

Fig. 4. Parasitic castration of *Hydrobia ulvae*. Average penis size and oviposition (±s.e.) of uninfected (white columns) and trematode infected (shaded columns) snails. (After Mouritsen & Jensen, 1994.)

Table 1. Results from statistical analysis testing the difference in growth rate between uninfected and trematode infected hydrobiids from several different snail populations sampled in the White Sea, Denmark and northern Germany. + denotes significantly higher growth rates among infected than uninfected snails ($P < 0.05$). NS denotes no statistically significant difference between growth rates. +/NS denotes that the result varied between investigated populations. Growth rate was measured as shell-increment from last growth (winter) interruption line and therefore represent *in situ* growth rate. (After Gorbushin, 1997.)

Type of infection	Snail species	
	Hydrobia ulvae	*Hydrobia ventrosa*
Notocotylus sp.		NS
Bonocotyle progenetica	NS	NS
Himasthla sp.	+/NS	
Cryptocotyle sp.	+	+
Maritrema subdolum	+	
Microphallus claviformis	+	
Microphallus pirum	+	

1983; Pearson & Cheng, 1985; Lim & Green, 1991; Coustau *et al.* 1993; Huxham *et al.* 1993; Mouritsen & Jensen, 1994, see Fig. 4). Since the prevalence of trematode infections usually increases with the size/age of the molluscan host, often approaching 100 % in the largest size classes, the impact of the parasites on recruitment may be substantial because the largest individuals in the population also contribute the most to the pool of larvae (Køie, 1975; Hughes & Answers, 1982; Lauckner, 1980, 1987a; Huxham *et al.* 1993; Curtis, 1997). For host species with pelagic larvae, it is unknown whether this

affects the density of recruits locally, but data presented by Lafferty (1993) on *C. californica* and Sokolova (1995) on *L. saxatilis* strongly suggest that parasite-induced castration can depress population density in molluscan host species with direct development. Since uninfected individuals of some snail species can engage in copulation activity (i.e. investing time and energy) with infected and hence castrated conspecifics (e.g. *L. littorea* and *Ilyanassa obsoleta*; see Saur, 1990; Curtis, this supplement), the impact of castration on recruitment may reach beyond its immediate influence and be more than just proportional to the prevalence of infection in the population.

Trematode infections in molluscs are not the only ones associated with host castration. For instance, infections by nemertean egg parasites can be highly prevalent in certain crab populations, and result in lower reproductive output in infected individuals (Torchin, Lafferty & Kuris, 1996). More importantly, adult crabs infected by rhizocephalans and adult sessile barnacles infected by the epicarid isopod *Hemioniscus balani* are also found to be completely castrated (e.g. Blower & Roughgarden, 1988; Alvarez *et al.* 1995 and references therein). Since the level of parasitism may attain epizootic proportions (i.e. prevalence far beyond 50 %, see earlier references regarding rhizocephalans, and Blower & Roughgarden, 1989 for *Hemioniscus*), these parasites may have a profound influence on the population dynamics of their host in certain areas. This may be particularly relevant to the sessile barnacles that require internal cross-fertilisation, and where the risk of infection and barnacle fecundity increase with size (Blower & Roughgarden, 1987, 1988). As also mentioned above regarding molluscan hosts, because crabs and barnacles have pelagic larvae it may be difficult to identify an impact of the parasites on local host abundance if other regional populations are not affected by parasitism. However, in the case of the direct-developing soft-bottom amphipod, *C. volutator*, the apparently trematode-induced lower brood size of infected females is likely to result in lower abundance locally or reduce the rate of colonization in adjacent habitats under high infection intensities (Bick, 1994; McCurdy *et al.* 1999a). The mechanism by which the trematodes compromise reproduction in corophiids is unknown, but old females that were newly infected by *Gynaecotyla adunca* have been observed to abort their young (McCurdy *et al.* 1999a).

Growth

Parasites may or may not affect the growth rate of their host. In short-lived intertidal snails, such as *Hydrobia* spp., castration by trematodes usually results in gigantism (Mouritsen & Jensen, 1994; Gorbushin, 1997; Gorbushin & Levakin, 1999; Probst & Kube, 1999; Table 1), whereas no effect on

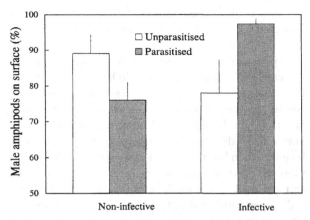

Fig. 5. Parasitic manipulation of male *Corophium volutator*. Surface activity (mean percent ± S.E.) of unparasitised (white columns) and parasitised (shaded columns) amphipods, the latter harbouring respectively non-infective and infective larval stages of the nematode *Skrjabinoclava morrisoni*. (After McCurdy *et al.* 1999*b*.)

growth or reduced growth rates have been observed in longer-lived host species, such as *Littorina* spp., *C. californica* and *I. obsoleta* (Sousa, 1983; Lafferty, 1993; Curtis, 1995; Gorbushin & Levakin, 1999; Mouritsen, Gorbushin & Jensen, 1999). The qualitative and quantitative effects of trematode infections on snail growth seem to depend on a range of factors other than just the life history characteristics of the host. Most important are the species of trematode involved (i.e. the inflicted level of pathology), the abundance of food and the intensity of exploitative competition experienced by the host (see Mouritsen & Jensen, 1994; Gorbushin, 1997; Mouritsen *et al.* 1999). The above statements may apply also to bivalves acting as hosts to digenean trematodes (see Bowers, 1969; Lauckner, 1983 and references therein; Machkevskij, 1988; Lim & Green, 1991; Calvo-Ugarteburu & McQuaid, 1998). Castration by rhizocephalans and epicarid isopods usually results in reduced growth rates in their crustacean host. In rhizocephalan infections, the number of moults and the size increment at each moult may be reduced in infected crabs, eventually causing infected individuals to be significantly smaller than uninfected ones (see Alvarez *et al.* 1995 and references therein). As a consequence of the more or less interrupted moulting activity, crabs infected by rhizocephalans may support a higher abundance or diversity of fouling organisms (e.g. barnacles and serpulid polychaetes) on their cuticles (Thomas *et al.* 1999; K. N. Mouritsen & T. Jensen, unpublished data).

Behaviour

Parasites commonly alter the behaviour of their hosts either as a side effect of infection or as an adaptive manipulation by the parasite with the purpose of facilitating transmission. In any case, the changed behaviour may result in increased predation pressure on the infected host or affect the way the host species interacts with the remaining benthic community of plants and animals due to its changed spatial distribution. Behavioural effects of infection believed to result from adaptive manipulation include the changed zonation or microhabitat selection of the periwinkles *L. saxatilis* and *L. obtusata* infected by microphallid metacercariae, and the dogwhelk *Nucella lapillus* infected by *Parorchis acanthus* (Philophthalmidae), that might increase predation by definitive hosts such as gulls and oystercatchers (Feare, 1971; Mikhailova, Granovich & Sergievsky, 1988; Granovich & Sergievsky, 1989; McCarthy, Fitzpatrick & Irwin, 2000). Another example comes from parasite-specific changes in the zonation of *I. obsoleta* that facilitate cercarial transmission to various species of second intermediate amphipod hosts inhabiting different intertidal zones (Curtis, 1987, 1990; McCurdy, Boates & Forbes, 2000). Surfacing behaviour of otherwise buried soft-bottom hosts, known or assumed to facilitate trophic transmission to shorebirds, has been observed in the mud snail *H. ulvae* (Huxham, Raffaelli & Pike, 1995*a*), in the cockles *C. edule* (Lauckner, 1984; Jonsson & André, 1992) and *Austrovenus stutchburyi* (Thomas & Poulin, 1998), and in the amphipod *C. volutator* (Mouritsen & Jensen, 1997; McCurdy *et al.* 1999*a*). And although the crawling behaviour of the tellinid bivalve *Macoma balthica* on the sediment surface appears natural in this species, trematode infections might enhance the activity (Mouritsen, 1997). The reverse position, closer to the sediment surface, in the bivalve *Venerupis aurea* infected by *Meiogymnophallus fossarum* seems, on the other hand, directly parasite induced (Bartoli, 1974, 1976). Among behavioural alternations of hosts following infection that may be simply side effects, there are reports of reduced mobility, changed direction of locomotion and altered zonation in gastropods. Good examples include trematode-infected periwinkles *L. littorea* (Lambert & Farley, 1968; Williams & Ellis, 1975), the mud snail *H. ulvae* (Mouritsen & Jensen, 1994; Huxham *et al.* 1995*a*), the top shells *Diloma subrostrata* (Miller & Poulin, 2001) and the mud snail *I. obsoleta* (Curtis, this supplement). Finally, the mere presence of parasites in the habitat can also have impacts on snail behaviour; for example, the movements of *L. littorea* on the substrate are influenced by the presence of bird droppings and its content of trematode eggs, with the snails apparently trying to avoid infection (Davies & Knowles, 2001).

Parasites other than trematodes can also affect the behaviour of intertidal invertebrates. For instance, the nematode *Skrjabinoclava* alters the crawling behaviour of the amphipod *C. volutator* in ways that make it more susceptible to predation by shorebirds (McCurdy, Forbes & Boates, 1999*b*; Fig. 5). Acanthocephalans are known as master manipulators of their crustacean intermediate hosts, as became apparent from numerous studies on freshwater amphipods and isopods (Moore, 1984). Similar results are likely to be obtained with intertidal species. Some effects on either behaviour or colouration have been observed in crabs, the only intertidal crustaceans in which the impact of acanthocephalans have been investigated (Pulgar *et al.* 1995; Haye & Ojeda, 1998; Latham & Poulin, 2001).

PARASITISM AND INTERTIDAL COMMUNITIES

What is the impact of these direct effects on host behaviour, growth, reproduction or survival, on the overall structure and diversity of intertidal communities? On the one hand, the effects of parasites on given hosts may have little impact on local larval recruitment, i.e. on the number of larvae of the host species that settle per unit time. High levels of infection may reduce the production of host larvae at one locality, but if for some reason the parasite is not prevalent in adjacent localities, the pool of planktonic larvae available for recruitment will be mostly unaffected. On the other hand, following larval establishment, parasitism could result in reductions in host population size, and in altered distributions of host body sizes. Very few studies have specifically addressed the role of parasitism at the level of community organisation, whether in intertidal systems or in other ecosystems. Here, we will discuss two of the ways, based on whether they have numerical or functional effects on their hosts, in which parasites may affect intertidal communities, trying to integrate the information presented above into scenarios that are plausible without being too speculative.

The first way in which parasites can influence intertidal communities is via their *direct* effects on host density, mediated by reductions in host survival or reproductive output. If the local abundance of a species declines, we should expect that this would have consequences for its prey, its predators or its competitors. For instance, classical removal/exclusion experiments (e.g. Paine, 1966; Dayton, 1975; Menge, 1976) have shown that the local abundance of a single, 'keystone' species can be the main determinant of the relative abundance and diversity of other organisms in intertidal systems. Key predators can prevent one or more prey species from becoming over-abundant and driving other species to local extinction, especially when they compete for attachment space on rocky substrates. A decrease in predator abundance can allow a few prey species to profit at the expense of others, with the outcome being a reduction in local diversity. The same argument applies to key grazers and the algae on which they feed. Many key predators and grazers in intertidal systems are infected by parasites that reduce their survival or reproduction, and reductions

in the local abundance of these key species should be expected to have the sort of consequences outlined above. Clearly, if the species affected by parasitism is a prey instead of a predator, we may also expect that a decrease in its abundance would have consequences for other species: if it becomes scarce, predators would switch to other prey species that would otherwise not be selected. The actual abundance of a host species does not necessarily need to be reduced by parasitism for impacts on other species to manifest themselves: the parasites may simply reduce the activity levels of the host, or cause it to move towards a different microhabitat. As long as the perceived abundance of the host is altered, we can expect community-wide impacts.

The second major way in which parasites may affect intertidal communities is *indirectly*, via their coincidental effects on habitat variables that matter to other organisms. The process by which certain organisms modulate the availability or quality of resources for other species, by physically changing some component of the habitat, is referred to as ecosystem engineering (Jones, Lawton & Shachak, 1997). Parasites can either have impacts on existing ecosystem engineers, or act as engineers themselves (Thomas *et al.* 1999). A prime example of the interaction between parasitism and ecosystem engineering is provided by the marked changes in sediment composition observed on a Danish mud flat following a massive die-off of the amphipod *C. volutator* caused by trematode infections (see Fig. 2). Under normal circumstances, the amphipods help to stabilise the substrate with their tube-building activity. Their sudden extinction from the site allowed rapid sediment erosion that significantly changed the particle size composition of the mud flat (Mouritsen *et al.* 1998). Because the characteristics of the sediment are major factors determining whether various species of infaunal invertebrates will settle and survive (Snelgrove & Butman, 1994), the changes in the substrate appear in part responsible for the significant changes in the diversity and relative abundance of other macrofaunal species following the disappearance of the amphipods (Jensen & Mouritsen, 1992; K. N. Mouritsen & K. T. Jensen, unpublished data). This is a dramatic illustration of the far-reaching consequences of parasitism in one species mediated by coincidental changes in the physical features of the habitat.

There are other examples of this nature. In soft-sediment communities, many trematodes impair the burrowing ability of their molluscan host (see above). A parasite-induced increase in the density of bivalves exposed at the surface of the sediments could alter water flow and sediment deposition, thus changing the characteristics of the sediments and the infaunal invertebrates that can become established (see Cummings *et al.* 1998; Ragnarsson & Raffaelli, 1999). In addition, the presence of molluscs at the surface of

PARASITES ABSENT   PARASITES PRESENT

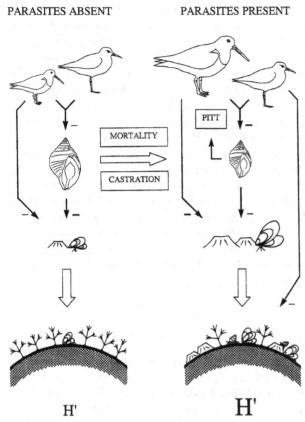

MORTALITY

CASTRATION

PITT

H'   H'

Fig. 6. The predicted appearance of the benthic intertidal community on a hypothetical *Nucella* dominated rocky shore in the absence and presence of a castrating and pathogenic trematode. Icons from above: oystercatcher and sandpiper; dogwhelk (*Nucella*); *Balanus* spp. and *Mytilus* sp.; macroalgae and littorinids. Arrows with a minus denote negative impact on abundance. Open arrow means 'results in'. Text boxes represent parasite-mediated processes. PITT: parasite increased trophic transmission. H' denote species diversity. Larger or smaller symbols indicate respectively increase or decrease in the given process, value or abundance. See text (scenario I) for other details.

the sediments provides a hard substrate for various plants and invertebrates that would otherwise have no such substrate for attachment in soft-sediment habitats. For example, heavy infections of the trematode *Curtuteria australis* prevent the cockle *A. stutchburyi* from burrowing 2–3 cm into the sediments, as healthy cockles do (Thomas & Poulin, 1998). Several epibiotic, or fouling, organisms use cockle shells as attachment substrates and even compete for space in areas where they are abundant. There is evidence that a species of limpet, which is normally out-competed for space on burrowed cockles by a sea anemone, has a preference for surface cockles, and that its coexistence with the anemone is possible mainly because of the parasite-mediated availability of surface cockles (Thomas *et al.* 1998). Thus the fouling community of the host is not independent of the action of its parasites. This

may be true whether or not burrowing is involved. For instance, the proportion of the shell covered by epiphytic algae is greater in snails, *H. ulvae*, parasitised by trematodes than in uninfected snails (Mouritsen & Bay, 2000); a possible explanation could be that the defences of snails against fouling algae are compromised by infection. Since hydrobiids, through their action as substrate for germlings, seem important for the development of green algae mats in some soft-bottom intertidal habitats (Schories *et al.* 2000), and because algal mats significantly affect the abundance and composition of the benthic fauna (Raffaelli, Raven & Poole, 1998), an interaction between parasitism and fouling defences of the snails may have consequences also for the benthic community surrounding these hosts. As a final example of an indirect effect of parasitism on fouling organisms, the infrequent moulting of crabs infected by rhizocephalans allows greater numbers and a higher diversity of fouling organisms (e.g. barnacles and serpulid polychaetes) to establish on their carapace (Thomas *et al.* 1999; K. N. Mouritsen & T. Jensen, unpublished data). In this case, the parasite increases the permanence of a habitat that would normally be too ephemeral to support many epibionts. The point of all these examples is that the impact of parasites on the growth or behaviour of intertidal invertebrates can have a myriad of consequences for the infauna, the epibiotic organisms, or other components of the community; these effects can include major increases as well as decreases in local biodiversity, but they have not been emphasised before simply because they have not been studied.

Based on documented processes acting to shape intertidal communities, and on available information on the impact of parasites on species involved in these processes, it is possible to make specific predictions about what certain well-studied intertidal communities would look like in the presence or absence of parasites.

Scenario I (Fig. 6)

Consider the mid-intertidal level of a semi-exposed rocky shore recently cleared of sessile organism by ice scouring, for instance. Whelks (*Nucella*) have, together with a few patches of mussels (*Mytilus*) and acorn barnacles (e.g. *Balanus* spp.), survived the winter in crevices and rock pools and now occur abundantly throughout the mid-intertidal. These predatory snails feed preferentially on barnacles and mussels and are in high densities capable of significantly diminishing or even completely eradicating their prey populations (Nybakken, 1993; Carroll & Highsmith, 1996; Navarrete, 1996; Noda, 1999). Barnacles and mussels themselves are superior to macroalgae as space competitors, but in their absence, the algae will spread into the unoccupied space that eventually develops into a mono-cultural

PARASITES ABSENT   PARASITES PRESENT

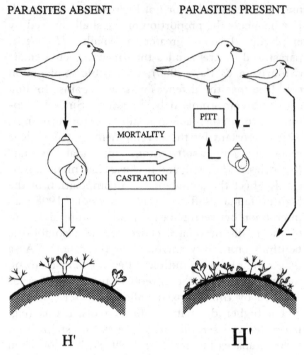

Fig. 7. The predicted appearance of the encrusting algal community on a hypothetical periwinkle (*Littorina littorea*) dominated rocky shore in the absence and presence of castrating and pathogenic trematodes. Icons from above: Gull and sandpiper; *L. littorea*; low diversity algal community (left); high diversity algal community and various herbivorous snails other than *L. littorea* (right). See Fig. 7 and text (scenario II) for additional explanations and details.

stand (Nybakken, 1993 and references therein). During this process purple sandpipers and a few oystercatchers visit the site feeding on both the whelks (see Feare, 1970, 1971) and the patches of older mussels left.

Let us introduce a parasite, for instance the castrating trematode *Parorchis acanthus* that utilizes *Nucella* as first intermediate host and oystercatchers as final host, the latter reached by trophic transmission (see Feare, 1970, 1971). Since *Nucella* is ovoviviparous (Hughes, 1986), the parasitic castration will result in a decline in local whelk abundance, reinforced by apparently direct parasite-induced mortality under the physical stress that prevails on the rock face, and parasite-induced behavioural changes increasing the predation pressure from oystercatchers (Feare, 1971). Released from predatory exclusion due to lower density of whelks, barnacles and mussels now manage to populate the site, which prevents the community succession to proceed into a pure algae culture (see Nybakken, 1993). Because the prevalence of the parasite is moderate (as it is usually the case in first intermediate host populations), and in concert with the increasing number of oystercatchers that visit the site to feed, not only on the now more easily accessible snails but also the more numerous mus-

sels, the abundance of whelks is still sufficiently high to keep barnacles and mussels from monopolising the shore. As the dominant space competitors, the mussels will eventually take over the site if not kept in check (see Nybakken, 1993 and references therein). The whelks' predation on the increasing population of barnacles also leaves empty shells scattered throughout the site, paving the way for herbivorous littorinids that, in the case of larger species (e.g. *L. saxatilis* and *L. littorea*), use them as nursery ground for juveniles, or, in the case of smaller species (e.g. *L. neritoides* and *L. neglecta*), entirely rely on them as shelter in the mid-intertidal zone (Hughes, 1986; Granovitch & Sergievsky, 1989). The appearance of the littorinids may in turn keep the sandpipers, known to consume large quantities of the smaller specimens (Faller-Fritsch & Emson, 1985), attracted to the area.

Scenario II (Fig. 7)

The herbivorous periwinkle *L. littorea* has been shown to control the species diversity of algae on rock surfaces in a density-dependent manner (Lubchenco, 1978). Under high snail density the algal diversity is lower than when *L. littorea* is scarce. Moreover, herbivorous snails on rocky shores are likely to face intra- and interspecific exploitative or interference competition that depresses the abundance or, at times, results in exclusion of the inferior species/size class (Underwood, 1976, 1978; Hylleberg & Christensen, 1978; Brenchley & Carlton, 1983; Faller-Fritsch & Emson, 1985; Petraitis, 1987; Yamada & Mansour, 1987; Chow, 1989). Then consider the mid-intertidal area on a rocky shore, cobble beach or stone jetty that supports an abundant population of, say, *L. littorea*. As a consequence of the intensive grazing pressure only the most grazing-resistant algae species persist, and because of the high level of combined interference and exploitative competition only few con- as well as heterospecific snail-recruits manage to become established on the site, almost completely dominated by larger *L. littorea* specimens. A few scattered gulls are found to feed particularly on the smaller snails available. However, by introducing a range of digenean trematode species using *Littorina* as first intermediate host and shorebirds as their final host, the snail density decreases due to parasite-induced mortality. Since *L. littorea* has pelagic larvae, the impact of the inevitable parasitic castration on the abundance of the local snails will at first be undetectable. Later, as the parasites spread along the coastline, the number of recruits that settle on the site may decrease. Because the parasites have made infected snails more vulnerable to bird predation, the gulls in the area have been little affected by the decline in the abundance of *L. littorea*. For the same reason, however, the number of algal species on the

PARASITES ABSENT   PARASITES PRESENT

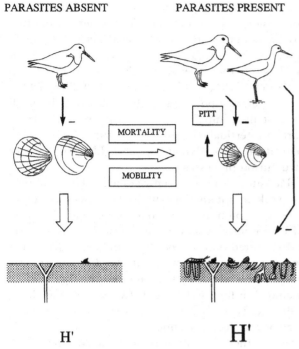

H'   H'

Fig. 8. The predicted appearance of the zoobenthic community on a hypothetical soft-sediment flat dominated by the cockle *Austrovenus stutchburyi* or *Cerastoderma edule* in the absence and presence of a behaviour-manipulating trematode. Icons from above: oystercatcher and stilt; *C. edule* (left) and *A. stutchburyi* (right); burrows from larger and smaller species of polychaetes and amphipods, herbivorous snails and limpets, tellinid bivalve and epibenthic shrimp. See Fig. 7 and text (scenario III) for additional explanations and details.

shore has increased significantly, accompanied by a more diverse assemblage of herbivorous snails (e.g. more *L. littorea* recruits, *L. saxatilis*, *L. obtusata*, or *I. obsoleta*), profiting either from increased algal diversity used for browsing (see Hughes, 1986; Bertness, 1999), or directly from relaxed competition with *L. littorea*. As these changes happen, the site becomes more attractive also to smaller bird species (e.g. purple sandpiper and rock pipit) feeding on the increasing number of smaller littorinids (Faller-Fritsch & Emson, 1985).

Scenario III (Fig. 8)

The two cockle species, *C. edule* in the northern hemisphere and *A. stutchburyi* in the south, occur abundantly on tidal flats. In terms of occupied space, they may literally monopolise the upper sediment strata at lower tidal levels (personal observation). Albeit suspension-feeding infaunal organisms, they frequently move, even over longer distances, resulting in substantial disturbance of the upper sediment layer (Richardson, Ibarrola & Ingham, 1993; Flach, 1996; Whitlatch *et al.* 1997; K. N. Mouritsen, unpublished data). As a consequence of this biotur-

bation, both species have been shown to depress significantly the density of co-occurring infaunal invertebrates or change their relative abundances (Flach, 1996; Whitlatch *et al.* 1997). Both cockle species are preyed upon by oystercatchers, and are in addition haunted by a range of trematode species that can be expected to affect the cockles' survival rate, fecundity and/or behaviour (see earlier sections and references therein). Among those trematodes *Himasthla elongata* (and allies), in the case of the European cockle, and *C. australis*, in the case of the New Zealand cockle, are of particular interest. Both are echinostomids that utilize intertidal snails as first intermediate host, cockles as second intermediate host, and cockle-eating shorebirds as final hosts (e.g. Allison, 1979; Lauckner, 1983). In cockles, they specifically infest the foot-tissue causing the foot to become immobilised and stunted which impairs the cockles' ability to move and re-burrow if dislodged to the surface (Lauckner, 1983; Thomas & Poulin, 1998; Mouritsen, this supplement). Hence, it can be expected that the introduction of echinostomid trematodes to a dense population of cockles will result in increasing densities of other infaunal organisms that profit from (1) unoccupied substrate brought about by the cockles' greater tendency to be dislodged to the surface and an increasing predation pressure from shorebirds feeding preferentially on the surfaced specimens (Lauckner, 1983; Thomas & Poulin, 1998), and (2), reduced bioturbation from the increasingly immobilised infected population of cockles. This should in turn promote an increasing variety of epifauna, for instance limpets, as previously mentioned, or shrimps, smaller crabs, and benthic fish that feed on the increasingly abundant infaunal polychaetes and amphipods (e.g. Reise, 1985). Eventually, other and smaller species of shorebirds, such as sandpipers, stilts and plovers, may find it worthwhile to feed on the growing population of smaller and more soft-bodied benthic organisms.

Scenario IV

Through a multitude of processes including grazing, bioturbation, fertilization, and direct interference competition, the largely deposit-feeding mud snail *I. obsoleta* has been shown to impact profoundly on the intertidal benthic community wherever it occurs in reasonable densities. Depending on density, the mud snail may either increase or decrease microalgal productivity and alter the relative abundance of functional groups, which in turn also affect the abundance of nematodes relying on these algae (Nichols & Robertson, 1979; Connor & Edgar, 1982; Connor, Teal & Valiela, 1982). *Ilyanassa* has also been demonstrated to diminish significantly the density of co-existing oligochaetes, polychaetes, gastropods, bivalves and harpacticoid copepods

(Levinton & Stewart, 1982; Hunt, Ambrose & Peterson, 1987; Dunn, Mullineaux & Mills, 1999). In particular, the snails can almost completely displace their exploitative competitor *Hydrobia totteni* to the upper- and lowermost fringes of the intertidal zone, where *I. obsoleta* usually occur in low densities during the productive season (Levinton, Stewart & Dewitt, 1985). *Ilyanassa* also acts as first intermediate host to trematodes that change the behaviour of the snail. Seemingly dependent on the vertical distribution of potential second intermediate host on the tidal flat, the infected snails move either toward the high water line or toward the low water line in search for the next host in the parasites' life cycle (Curtis, 1987, 1990; McCurdy *et al.* 2000). By doing so, the snails' substantial impact on the benthic community can be envisaged to follow in their trails, transforming it accordingly throughout a larger intertidal range than otherwise would have been expected if no parasites were present in the system.

Although the above scenarios necessitate assumptions regarding the physical conditions, the strength of biotic interactions, and the composition of the initial benthic community (for instance, an abundant starfish population under scenario I would likely undermine the ability of whelks to control their common prey population [see Navarrete & Menge, 1996; Bertness, 1999]), scenarios I–III clearly suggest that parasites retain the potential of increasing the species diversity of the tidal community on all trophic levels. Regarding scenario IV, it is more difficult to make specific predictions. However, it is reasonably safe to expect that the structure of the benthic community will change wherever the infected snails appear. And should such sites be strongly dominated e.g. by amphipod second intermediate hosts, the sediment re-working activity of *I. obsoleta* may allow for immigration of other infaunal organisms that cause the diversity to increase.

Whether parasites affect intertidal communities directly or indirectly, the magnitude of their impact is unlikely to be constant in time or space. Seasonal fluctuations in recruitment of parasites by intertidal invertebrates have been documented (see review by Lauckner, 1980, 1983). Similarly, the abundance of parasites also varies in space (see Carrol *et al.* 1990; Sousa, 1991; Curtis, this supplement), for a variety of reasons including the local availability of other essential hosts in the parasite's life cycle and physical features of the habitat such as wave exposure. Often, a parasite with a marked effect on host populations in one area will be almost completely absent from another geographical area. There is also spatial variation in parasite infections on much smaller scales. For instance, the precise position of a snail on a mud flat, relative to where shorebirds aggregate and defaecate, can determine its probability of being infected by larval trematodes (e.g. Smith, 2001),

creating a mosaic of infection foci in the two-dimensional snail habitat. Also, there are often gradients in the intensity or prevalence of trematode infection in more or less sessile invertebrates from the lower to the upper shore (Carrol *et al.* 1990; Lim & Green, 1991; Poulin, Steeper & Miller, 2000). This spatiotemporal variability does not diminish the potential importance of parasitism in the structure of intertidal communities; it simply means that parasitism is a dynamic factor like most other structuring processes.

Parasite species are deeply embedded in a complex network of interactions with the other species in intertidal communities, and ignoring them in studies of community structure can be misleading (Sousa, 1991). Traditional food web models have often focused on energy flows along the different trophic links in the web. Parasites are very rarely, if at all, included in food web models (Marcogliese & Cone, 1997; Raffaelli, 2000). One practical reason for this may be that ecologists find it easier simply to exclude them; a more scientific reason may be that often energy flows from hosts to parasites are trivial compared to those from prey to predators. The single study that has incorporated parasites in an intertidal food web model has shown that including them changes estimates of various statistics used to summarize food web processes, such as food chain length and linkage density (Huxham, Raffaelli & Pike, 1995*b*). These parameters are often linked with the stability of food webs and ecosystems and their resilience to perturbations, and Huxham *et al.*'s (1995*b*) study thus emphasizes the importance of parasitism at the community level. Parasites may not themselves take part in significant energy flows in the food web, but they have huge functional effects on other energy flows and on the species composition of the food web. Their main roles are to directly mediate other trophic or competitive interactions in the systems, or indirectly cause changes to the availability of physical resources (e.g. attachment space) for various organisms. These are the roles documented thus far, but there are likely others. The key importance of other types of symbiotic interactions for community structure has also been recently highlighted (e.g. Stachowicz, 2001), and the time has come to take these suggestions seriously. In the past, the experimental approach has proven fruitful in community studies of intertidal systems, mainly because it is logistically easy to carry out replicated, small-scale, short- or long-term manipulative experiments in these systems (Raffaelli, 2000). Manipulating the abundance of parasites, e.g. by excluding them from certain areas, is logistically more challenging, but it is not unthinkable (Sousa, 1991). Clever solutions to these experimental difficulties, tailored to specific systems, are likely to pave the way toward a better understanding of parasitism and community structure.

Parasitism in intertidal systems

ACKNOWLEDGEMENTS

We are grateful to Dr Andrey Granovitch, St Petersburg State University, who translated key parts of several of the Russian papers, and to David Latham for access to unpublished data on acanthocephalan infections in crabs. We also wish to acknowledge the referees for useful comments to an earlier draft. The work was supported by The Marsden Foundation (New Zealand) and The Danish Natural Science Research Council (K.N. Mouritsen).

REFERENCES

ALLISON, F. R. (1979). Life cycle of *Curtuteria australis* n. sp. (Digenea: Echinostomatidae: Himasthlinae), intestinal parasite of the South Island pied oystercatcher. *New Zealand Journal of Zoology* **6**, 13–20.

ALVAREZ, F. & CALDERON, J. (1996). Distribution of *Loxothylacus texanus* (Cirripedia: Rhizocephala) parasitising crabs of the genus *Callinectes* in the southwestern Gulf of Mexico. *Gulf Research Report* **9**, 205–210.

ALVAREZ, F., GARCIA, A., ROBLES, R. & CALDERON, J. (1999). Parasitization of *Callinectes rathbunae* and *Callinectes sapidus* by the rhizocephalan barnacle *Loxothylacus texanus* in Alvarado Lagoon, Veracruz, Mexico. *Gulf Research Report* **11**, 15–21.

ALVAREZ, F., HINES, A. H. & REAKA-KUDLA, M. L. (1995). The effect of parasitism by the barnacle *Loxothylacus panopaei* (Gissler) (Cirripedia: Rhizocephala) on growth and survival of the crab *Rhithropanopeus harrisii* (Gould) (Brachyura: Xanthidae). *Journal of Experimental Marine Ecology and Biology* **192**, 221–232.

BARTOLI, P. (1974). Recherches sur les Gymnophallidae F. N. Morozov, 1955 (Digenea), parasites d'oiseaux des côtes de Camargue: Systématique, Biologie et Écologie. Ph.D. Dissertation, University of Aix-Marseille, Aix-Marseille, France.

BARTOLI, P. (1976). Modification de la croissance et du comportement de *Venerupis aurea* parasité par *Gymnophallus fossarum* P. Bartoli, 1965 (Trematoda, Digenea). *Haliotis* **7**, 23–28.

BERGER, V. G. & KONDRATENKOV, A. P. (1974). The effect of infection of *Hydrobia ulvae* with larvae of trematodes and its resistance to drying and water-freshening. *Parazitologiya* **8**, 563–564 (In Russian).

BERTNESS, M. D. (1999). *The Ecology of Atlantic Shorelines.* Sunderland, U.K. Sinauer Associates Inc.

BICK, A. (1994). *Corophium volutator* (Corophiidae: Amphipoda) as an intermediate host of larval Digenea – an ecological analysis in a coastal region of the southern Baltic. *Ophelia* **40**, 27–36.

BLOWER, S. M. & ROUGHGARDEN, J. (1987). Population dynamics and parasitic castration: a mathematical model. *American Naturalist* **129**, 730–754.

BLOWER, S. M. & ROUGHGARDEN, J. (1988). Parasitic castration: host species preferences, size-selectivity and spatial heterogeneity. *Oecologia* **75**, 512–515.

BLOWER, S. M. & ROUGHGARDEN, J. (1989). Population dynamics and parasitic castration: test of a model. *American Naturalist* **134**, 848–858.

BOWERS, E. A. (1969). *Cercaria bucephalopsis haimeana* (Lacaze-Duthiers, 1854) (Digenea: Bucephalidae) in cockles, *Cardium edule* L. in South Wales. *Journal of Natural History* **3**, 409–422.

BRENCHLEY, G. A. & CARLTON, J. T. (1983). Competitive displacement of native mud snails by introduced periwinkles in the New England intertidal zone. *Biological Bulletin (Woods Hole)* **165**, 543–558.

CALVO-UGARTEBURU, G. & MCQUAID, C. D. (1998). Parasitism and invasive species: effects of digenetic trematodes on mussels. *Marine Ecology Progress Series* **169**, 149–163.

CARROL, H., MONTGOMERY, W. I. & HANNA, R. E. B. (1990). Dispersion and abundance of *Maritrema arenaria* in *Semibalanus balanoides* in north-east Ireland. *Journal of Helminthology* **64**, 151–160.

CARROLL, M. L. & HIGHSMITH, R. C. (1996). Role of catastrophic disturbance in mediating *Nucella–Mytilus* interactions in the Alaskan rocky intertidal. *Marine Ecology Progress Series* **138**, 125–133.

CHOW, V. (1989). Intraspecific competition in a fluctuating population of *Littorina plena* Gould (Gastropoda: Prosobranchia). *Journal of Experimental Marine Biology and Ecology* **130**, 147–165.

COMBES, C. (1996). Parasites, biodiversity and ecosystem stability. *Biodiversity and Conservation* **5**, 953–962.

CONNOR, M. S. & EDGAR, R. K. (1982). Selective grazing by the mud snail *Ilyanassa obsoleta*. *Oecologia* **53**, 271–275.

CONNOR, M. S., TEAL, J. M. & VALIELA, I. (1982). The effect of feeding by mud snails, *Ilyanassa obsoleta* (Say), on the structure and metabolism of a laboratory benthic algal community. *Journal of Experimental Marine Biology and Ecology* **65**, 29–45.

COUSTAU, C., ROBBINS, I., DELAY, B., RENAUD, F. & MATHIEU, M. (1993). The parasitic castration of the mussel *Mytilus edulis* by the trematode parasite *Prosorhynchus squamatus*: specificity and partial characterization of endogenous and parasite-induced anti-mitotic activities. *Comparative Biochemical Physiology* **104**, 229–233.

CUMMINGS, V. J., THRUSH, S. F., HEWITT, J. E. & TURNER, S. J. (1998). The influence of the pinnid bivalve *Atrina zelandica* (Gray) on benthic macroinvertebrate communities in soft-sediment habitats. *Journal of Experimental Marine Biology and Ecology* **228**, 227–240.

CURTIS, L. A. (1987). Vertical distribution of an estuarine snail altered by a parasite. *Science* **235**, 1509–1511.

CURTIS, L. A. (1990). Parasitism and the movements of intertidal gastropod individuals. *Biological Bulletin (Woods Hole)* **179**, 105–112.

CURTIS, L. A. (1995). Growth, trematode parasitism, and longevity of a long-lived marine gastropod (*Ilyanassa obsoleta*). *Journal of the Marine Biological Association of the United Kingdom* **75**, 913–925.

CURTIS, L. A. (1997). *Ilyanassa obsoleta* (Gastropoda) as a host for trematodes in Delaware estuaries. *Journal of Parasitology* **83**, 793–803.

DAVIES, M. S. & KNOWLES, A. J. (2001). Effects of trematode parasitism on the behaviour and ecology of a common marine snail (*Littorina littorea* (L.)). *Journal of Experimental Marine Biology and Ecology* **260**, 155–167.

DAYTON, P. K. (1975). Experimental evaluation of ecological dominance in a rocky intertidal algal community. *Ecological Monographs* **45**, 137–159.

DUNN, R., MULLINEAUX, L. S. & MILLS, S. W. (1999). Resuspension of postlarval soft-shell clams *Mya arenaria* through disturbance by the mud snail *Ilyanassa obsoleta*. *Marine Ecology Progress Series* **180**, 223–232.

FALLER-FRITSCH, R. J. & EMSON, R. H. (1985). Causes and patterns of mortality in *Littorina rudis* (Maton) in relation to intraspecific variation: a review. In *The Ecology of Rocky Coasts* (eds. Moore, P. G. & Seed, R.), pp. 157–177. London; Hodder & Stoughton.

FEARE, C. J. (1970). Aspects of the ecology of an exposed shore population of dogwhelks *Nucella lapillus* (L.). *Oecologia* **5**, 1–18.

FEARE, C. J. (1971). Predation of limpets and dogwhelks by oystercatchers. *Bird Study* **18**, 121–129.

FERAL, C., BRETON, J.-LE. & STREIFF, W. (1972). New observations on parasitic castration in some gastropods. *Annales de l'Insitut Michel Pacha* **5**, 28–40.

FLACH, E. C. (1996). The influence of the cockle *Cerastoderma edule*, on the macrozoobenthic community of tidal flats in the Wadden Sea. *Marine Ecology* **17**, 87–98.

GLENNER, H. & HOEG, J. T. (1997). Rhizocephalan parasites and their decapod hosts. *Journal of Shellfish Research* **16**, 319.

GOATER, C. P. (1993). Population biology of *Meiogymnophallus minutus* (Trematoda: Gymnophallidae) in cockles from the Exe estuary. *Journal of the Marine Biological Association of the United Kingdom* **73**, 163–177.

GORBUSHIN, A. M. (1997). Field evidence of trematode-induced gigantism in *Hydrobia* spp. (Gastropoda: Prosobranchia). *Journal of the Marine Biological Association of the United Kingdom* **77**, 785–800.

GORBUSHIN, A. M. & LEVAKIN, I. A. (1999). The effect of trematode parthenitae on the growth of *Onoba aculeus*, *Littorina saxatilis* and *L. obtusata* (Gastropoda: Prosobranchia). *Journal of the Marine Biological Association of the United Kingdom* **79**, 273–279.

GRANOVICH, A. I. & SERGIEVSKY, S. O. (1989). The use of acorn barnacles' settlements by molluscs *Littorina saxatilis* (Gastropoda, Prosobranchia) depending on their infestation with parthenites of trematodes. *Zoologicheskii Zhurnal* **68**, 39–47 (In Russian).

HAYE, P. A. & OJEDA, F. P. (1998). Metabolic and behavioural alterations in the crab *Hemigrapsus crenulatus* (Milne-Edwards 1837) induced by its acanthocephalan parasite *Profilicollis antarcticus* (Zdzitowiecki 1985). *Journal of Experimental Marine Biology and Ecology* **228**, 73–82.

HUDSON, P. & GREENMAN, J. (1998). Competition mediated by parasites: biological and theoretical progress. *Trends in Ecology & Evolution* **13**, 387–390.

HUGHES, R. N. (1986). *A Functional Biology of Marine Gastropods*. London, Croom Helm Ltd.

HUGHES, R. N. & ANSWERS, P. (1982). Growth, spawning and trematode infection of *Littorina littorea* (L.) from an exposed shore in North Wales. *Journal of Molluscan Studies* **48**, 321–330.

HUNT, J. H., AMBROSE, W. G. JR & PETERSON, C. H. (1987). Effects of the gastropod, *Ilyanassa obsoleta* (Say), and the bivalve, *Mercenaria mercenaria* (L.), on larval settlement and juvenile recruitment of infauna. *Journal of Experimental Marine Biology and Ecology* **108**, 229–240.

HUXHAM, M., RAFFAELLI, D. & PIKE, A. (1993). The influence of *Cryptocotyle lingua* (Digenea: Plathyhelminthes) infections on the survival and fecundity of *Littorina littorea* (Gastropoda: Prosobranchia); an ecological approach. *Journal of Experimental Marine Biology and Ecology* **168**, 223–238.

HUXHAM, M., RAFFAELLI, D. & PIKE, A. (1995 a). The effect of larval trematodes on the growth and burrowing behaviour of *Hydrobia ulvae* (Gastropoda: Prosobranchia) in the Ythan estuary, north-east Scotland. *Journal of Experimental Marine Biology and Ecology* **185**, 1–17.

HUXHAM, M., RAFFAELLI, D. & PIKE, A. (1995 b). Parasites and food web patterns. *Journal of Animal Ecology* **64**, 168–176.

HYLLEBERG, J. & CHRISTENSEN, J. T. (1978). Factors affecting the intra-specific competition and size distribution of the periwinkle *Littorina littorea* (L.). *Natura Jutlandica* **20**, 193–202.

IRWIN, S. W. B. & IRWIN, B. C. (1980). The distribution of the metacercariae of *Maritrema arenaria* (Digenea: Microphallidae) in the barnacle *Balanus balanoides* at three sites on the east coast of Northern Ireland. *Journal of the Marine Biological Association of the United Kingdom* **60**, 959–962.

JENSEN, K. T., LATAMA, G. & MOURITSEN, K. N. (1996). The effect of larval trematodes on the survival rates of two species of mud snails (Hydrobiidae) experimentally exposed to desiccation, freezing and anoxia. *Helgoländer Meeresuntersuchungen* **50**, 327–335.

JENSEN, K. T. & MOURITSEN, K. N. (1992). Mass mortality in two common soft-bottom invertebrates, *Hydrobia ulvae* and *Corophium volutator* – the possible role of trematodes. *Helgoländer Meeresuntersuchungen* **46**, 329–339.

JENSEN, T., JENSEN, K. T. & MOURITSEN, K. N. (1998). The influence of the trematode *Microphallus claviformis* on two congeneric intermediate host species (*Corophium*): infection characteristics and host survival. *Journal of Experimental Marine Biology and Ecology* **227**, 35–48.

JONES, C. G., LAWTON, J. H. & SHACHAK, M. (1997). Positive and negative effects of organisms as physical ecosystem engineers. *Ecology* **78**, 1946–1957.

JONSSON, P. R. & ANDRE, C. (1992). Mass mortality of the bivalve *Cerastoderma edule* on the Swedish west coast caused by infestation with the digenean trematode *Cercaria cerastodermae* I. *Ophelia* **36**, 151–157.

KØIE, M. (1975). On the morphology and life history of *Opechona bacillaris* (Molin, 1859) Looss, 1907 (Trematoda, Lepocreadiidae). *Ophelia* **13**, 63–86.

LAFFERTY, K. D. (1993). Effect of parasitic castration on growth, reproduction and population dynamics of the marine snail *Cerithidea californica*. *Marine Ecology Progress Series* **96**, 229–237.

LAFFERTY, K. D. (1999). The evolution of trophic transmission. *Parasitology Today* **15**, 111–115.

LAMBERT, T. C. & FARLEY, J. (1968). The effect of parasitism by the trematode *Cryptocotyle lingua* (Creplin) on zonation and winter migration of the common periwinkle, *Littorina littorea* (L.). *Canadian Journal of Zoology* **46**, 1139–1147.

LARZARO-CHAVEZ, E., ALVAREZ, F. & ROSAS, C. (1996). Records of *Loxothylacus texanus* (Cirripedia: Rhizocephala) parasitising the blue crab *Callinectes sapidus* in Tamiahua Lagoon, Mexico. *Journal of Crustacean Biology* **16**, 105–110.

LATHAM, A. D. M. & POULIN, R. (2001). Effect of acanthocephalan parasites on the behaviour and coloration of the mud crab *Macrophthalmus hirtipes* (Brachyura: Ocypodidae). *Marine Biology* **139**, 1147–1154.

LAUCKNER, G. (1980). Diseases of Mollusca: Gastropoda. In *Diseases of Marine Animals*, vol. 1 (ed. Kinne, O.), pp. 311–424. Hamburg, Biologische Anstalt Helgoland.

LAUCKNER, G. (1983). Diseases of Mollusca: Bivalvia. In *Diseases of Marine Animals*, vol. 2 (ed. Kinne, O.), pp. 477–961. Hamburg, Biologische Anstalt Helgoland.

LAUCKNER, G. (1984). Impact of trematode parasitism on the fauna of a North Sea tidal flat. *Helgoländer Meeresuntersuchungen* **37**, 185–199.

LAUCKNER, G. (1987a). Ecological effects of larval trematode infestations on littoral marine invertebrate populations. *International Journal for Parasitology* **17**, 391–398.

LAUCKNER, G. (1987b). Effects of parasites on juvenile Wadden Sea invertebrates. In *Proceedings of the 5th International Wadden Sea Symposium* (eds. Tougaard, S. & Asbirk, S.), pp. 103–121. Esbjerg, The National Forest and Nature Agency and the Museum of Fisheries and Shipping.

LEVINTON, J. S. & STEWART, S. (1982). Marine succession: the effect of two deposit-feeding gastropod species on the population growth of *Paranais litoralis* Müller 1784 (Oligochaeta). *Journal of Experimental Marine Biology and Ecology* **59**, 231–241.

LEVINTON, J. S., STEWART, S. & DEWITT, T. H. (1985). Field and laboratory experiments on interference between *Hydrobia totteni* and *Ilyanassa obsoleta* (Gastropoda) and its possible relation to seasonal shifts in vertical mudflat zonation. *Marine Ecology Progress Series* **22**, 53–58.

LIM, S. L. & GREEN, R. H. (1991). The relationship between parasite load, crawling behaviour, and growth rate of *Macoma balthica* (L.) (Mollusca, Pelecypoda) from Hudson Bay, Canada. *Canadian Journal of Zoology* **69**, 2202–2208.

LUBCHENCO, J. (1978). Plant species diversity in a marine intertidal community: importance of herbivore food preference and algal competitive ability. *American Naturalist* **112**, 23–39.

MACHKEVSKIJ, V. K. (1988). Effect of *Proctoeces maculatus* parthenitae on the growth of *Mytilus galloprovincialis*. *Parazitologiya* **22**, 341–344 (In Russian).

MARCOGLIESE, D. J. & CONE, D. K. (1997). Food webs: a plea for parasites. *Trends in Ecology & Evolution* **12**, 320–325.

MATHIESON, S., BERRY, A. J. & KENNEDY, S. (1998). The parasitic rhizocephalan barnacle *Sacculina carcini* in

crabs of the Forth Estuary, Scotland. *Journal of the Marine Biological Association of the United Kingdom* **78**, 665–667.

MCCARTHY, H. O., FITZPATRICK, S. & IRWIN, S. W. B. (2000). A transmissible trematode affects the direction and rhythm of movements in a marine gastropod. *Animal Behaviour* **59**, 1161–1166.

MCCURDY, D. G., BOATES, J. S. & FORBES, M. R. (2000). Spatial distribution of the intertidal snail *Ilyanassa obsoleta* in relation to parasitism by two species of trematodes. *Canadian Journal of Zoology* **78**, 1137–1143.

MCCURDY, D. G., FORBES, M. R. & BOATES, J. S. (1999a). Testing alternative hypotheses for variation in amphipod behaviour and life history in relation to parasitism. *International Journal for Parasitology* **29**, 1001–1009.

MCCURDY, D. G., FORBES, M. R. & BOATES, J. S. (1999b). Evidence that the parasitic nematode *Skrjabinoclava* manipulates host *Corophium* behavior to increase transmission to the sandpiper, *Calidris pusilla*. *Behavioral Ecology* **4**, 351–357.

MCDANIEL, J. S. (1969). *Littorina littorea*: lowered heat tolerance due to *Cryptocotyle lingua*. *Experimental Parasitology* **25**, 13–15.

MEISSNER, K. (2001). Infestation patterns of microphallid trematodes in *Corophium volutator* (Amphipoda). *Journal of Sea Research* **45**, 141–151.

MEISSNER, K. & BICK, A. (1997). Population dynamics and ecoparasitological surveys of *Corophium volutator* in coastal waters in the Bay of Mecklenburg (southern Baltic Sea). *Diseases of Aquatic Organisms* **29**, 169–179.

MEISSNER, K. & BICK, A. (1999). Mortality of *Corophium volutator* (Amphipoda) caused by infestation with *Maritrema subdolum* (Digenea, Microphallidae) – laboratory studies. *Diseases of Aquatic Organisms* **35**, 47–52.

MENGE, B. A. (1976). Organization of the New England rocky intertidal community: role of predation, competition and environmental heterogeneity. *Ecological Monographs* **46**, 355–393.

MENGE, B. A. & SUTHERLAND, J. P. (1987). Community regulation: variation in disturbance, competition, and predation in relation to environmental stress and recruitment. *American Naturalist* **130**, 730–757.

MIKHAILOVA, N. A., GRANOVICH, A. I. & SERGIEVSKY, S. O. (1988). Effect of trematodes on the microbiotopical distribution of molluscs *Littorina obtusata* and *L. saxatilis*. *Parazitologiya* **22**, 398–407 (In Russian).

MILLER, A. A. & POULIN, R. (2001). Parasitism, movements and distribution of the snail *Diloma subrostrata* (Trochidae) in a soft-sediment intertidal zone. *Canadian Journal of Zoology* **79**, 2029–2035.

MINCHELLA, D. J. & SCOTT, M. E. (1991). Parasitism: a cryptic determinant of animal community structure. *Trends in Ecology & Evolution* **6**, 250–254.

MITCHELL, J. B. & DESSI, J. (1984). A note on the distribution of metacercariae of *Maritrema arenaria* in *Balanus balanoides* at a site on the north-east coast of England. *Journal of the Marine Biological Association of the United Kingdom* **64**, 734–735.

MONTAUDOUIN, X. DE, KISIELEWSKI, I., BACHELET, G. & DESCLAUX, C. (2000). A census of macroparasites in an intertidal bivalve community, Archachon Bay, France. In *National Programme on the Determinism of Recruitment*, vol. 23 (eds. Bhaud, M., Nival, P. & Bachelet, G.), pp. 453–468. Paris, Elsevier.

MOORE, J. (1984). Altered behavioural responses in intermediate hosts: an acanthocephalan parasite strategy. *American Naturalist* 123, 572–577.

MOURITSEN, K. N. (1997). Crawling behaviour in the bivalve *Macoma balthica*: the parasite-manipulation hypothesis revisited. *Oikos* 79, 513–520.

MOURITSEN, K. N. & BAY, G. M. (2000). Fouling of gastropods: a role for parasites? *Hydrobiologia* 418, 243–246.

MOURITSEN, K. N., GORBUSHIN, A. & JENSEN, K. T. (1999). Influence of trematode infections on *in situ* growth rates of *Littorina littorea*. *Journal of the Marine Biological Association of the United Kingdom* 79, 425–430.

MOURITSEN, K. N. & JENSEN, K. T. (1994). The enigma of gigantism: effect of larval trematodes on growth, fecundity, egestion and locomotion in *Hydrobia ulvae* (Pennant) (Gastropoda: Prosobranchia). *Journal of Experimental Marine Biology and Ecology* 181, 53–66.

MOURITSEN, K. N. & JENSEN, K. T. (1997). Parasite transmission between soft-bottom invertebrates: temperature mediated infection rates and mortality in *Corophium volutator*. *Marine Ecology Progress Series* 151, 123–134.

MOURITSEN, K. N., MOURITSEN, L. T. & JENSEN, K. T. (1998). Changes of topography and sediment characteristics on an intertidal mud-flat following mass-mortality of the amphipod *Corophium volutator*. *Journal of the Marine Biological Association of the United Kingdom* 78, 1167–1180.

NAVARRETE, S. A. (1996). Variable predation: effects of whelks on a mid-intertidal successional community. *Ecological Monographs* 66, 301–321.

NAVARRETE, S. A. & MENGE, B. A. (1996). Keystone predation and interaction strength: interactive effects of predators on their main prey. *Ecological Monographs* 66, 409–429.

NICHOLS, J. A. & ROBERTSON, J. R. (1979). Field evidence that the eastern mud snail, *Ilyanassa obsoleta*, influences nematode community structure. *Nautilus* 93, 44–46.

NODA, T. (1999). Within- and between-patch variability of predation intensity on the mussel *Mytilus trossulus* Gould on a rocky intertidal shore in Oregon, USA. *Ecological Research* 14, 193–203.

NYBAKKEN, J. W. (1993). *Marine Biology: An Ecological Approach*. 3rd edn. New York, Harper Collins.

PAINE, R. T. (1966). Food web complexity and species diversity. *American Naturalist* 100, 65–75.

PEARSON, E. J. & CHENG, T. C. (1985). Studies on parasitic castration: occurrence of a gametogenesis-inhibiting factor in extract of *Zoogonus lasius* (Trematoda). *Journal of Invertebrate Pathology* 46, 239–246.

PETRAITIS, P. S. (1987). The effect of the periwinkle *Littorina littorea* (L.) and of intraspecific competition on growth and survivorship of the limpet *Notoacmea testudianalis* (Mueller). *Journal of Experimental Marine Biology and Ecology* 125, 99–115.

POULIN, R. (1999). The functional importance of parasites in animal communities: many roles at many

levels? *International Journal for Parasitology* 29, 903–914.

POULIN, R., STEEPER, M. J. & MILLER, A. A. (2000). Non-random patterns of host use by the different parasite species exploiting a cockle population. *Parasitology* 121, 289–295.

PRICE, P. W. (1980). *Evolutionary Biology of Parasites*. Princeton, Princeton University Press.

PROBST, S. & KUBE, J. (1999). Histopathological effects of larval trematode infections in mudsnails and their impact on host growth: what causes gigantism in *Hydrobia ventrosa* (Gastropoda: Prosobranchia). *Journal of Experimental Marine Biology and Ecology* 238, 49–68.

PULGAR, J., ALDANA, M., VERGARA, E. & GEORGE-NASCIMENTO, M. (1995). La conducta de la jaiba estuarina *Hemigrapsus crenulatus* (Milne-Edwards 1837) en relación al parasitismo por el acantocefalo *Profilicollis antarcticus* (Zdzitowiecki 1985) en el sur de Chile. *Revista Chilena de Historia Natural* 68, 439–450.

RAFFAELLI, D. (2000). Trends in research on shallow water food webs. *Journal of Experimental Marine Biology and Ecology* 250, 223–232.

RAFFAELLI, D. G., RAVEN, J. A. & POOLE, L. J. (1998). The ecological impact of macroalgal blooms. *Oceanographic and Marine Biology Annual Review* 36, 97–125.

RAGNARSSON, S. Á. & RAFFAELLI, D. (1999). Effect of the mussel *Mytilus edulis* L. on the invertebrate fauna of sediments. *Journal of Experimental Biology and Ecology* 241, 31–43.

REISE, K. (1985). *Tidal Flat Ecology: An Experimental Approach to Species Interactions*. Berlin, Springer-Verlag.

RICARDSON, C. A., IBARROLA, I. & INGHAM, R. J. (1993). Emergence pattern and spatial distribution of the common cockle *Cerastoderma edule*. *Marine Ecology Progress Series* 99, 71–81.

SAUR, M. (1990). Mate discrimination in *Littorina littorea* (L.) and *L. saxatilis* (Olivi) (Mollusca: Prosobranchia). *Hydrobiologia* 193, 261–270.

SCHORIES, D., ANIBAL, J., CHAPMAN, A. S., HERRE, E., ISAKSSON, I., LILLEBØ, A. I., PIHL, L., REISE, K., SPRUNG, M. & THIEL, M. (2000). Flagging greens: hydrobiid snails as substrate for the development of green algal mats (*Enteromorpha* spp.) on tidal flats of North Atlantic coasts. *Marine Ecology Progress Series* 199, 127–136.

SERGIEVSKY, S. O., GRANOVICH, A. I. & MIKHAILOVA, N. A. (1986). Effect of trematode infection on survival of periwinkles *Littorina obtusata* and *L. saxatilis* under the conditions of extremely low salinity. *Parazitologiya* 20, 202–207 (In Russian).

SINDERMANN, C. J. (1990). *Principal Diseases of Marine Fish and Shellfish*, Vol. 2. New York, Academic Press.

SMITH, N. F. (2001). Spatial heterogeneity in recruitment of larval trematodes to snail intermediate hosts. *Oecologia* 127, 115–122.

SNELGROVE, P. V. R. & BUTMAN, C. A. (1994). Animal-sediment relationships revisited: cause versus effect. *Oceanography and Marine Biology: An Annual Review* 32, 111–177.

SOKOLOVA, I. M. (1995). Influence of trematodes on the demography of *Littorina saxatilis* (Gastropoda:

Prosobranchia: Littorinidae) in the White Sea. *Diseases of Aquatic Organisms* **21**, 91–101.

SOUSA, W. P. (1983). Host life history and the effect of parasitic castration on growth: a field study of *Cerithidea californica* Haldemann (Gastropoda: Prosobranchia) and its trematode parasites. *Journal of Experimental Marine Biology and Ecology* **73**, 273–296.

SOUSA, W. P. (1991). Can models of soft-sediment community structure be complete without parasites? *American Zoologist* **31**, 821–830.

SOUSA, W. P. & GLEASON, M. (1989). Does parasitic infection compromise host survival under extreme environmental conditions? The case for *Cerithidea californica* (Gastropoda: Prosobranchia). *Oecologia* **80**, 456–464.

STACHOWICZ, J. J. (2001). Mutualism, facilitation, and the structure of ecological communities. *BioScience* **51**, 235–246.

TALLMARK, B. & NORRGREN, G. (1976). The influence of parasitic trematodes on the ecology of *Nassarius reticulatus* (L.) in Gullmar Fjord (Sweden). *Zoon* **4**, 149–154.

THOMAS, F. & POULIN, R. (1998). Manipulation of a mollusc by a trophically transmitted parasite: convergent evolution or phylogenetic inheritance? *Parasitology* **116**, 431–436.

THOMAS, F., POULIN, R., DE MEEUS, T., GUEGAN, J.-F. & RENAUD, F. (1999). Parasites and ecosystem engineering: what roles could they play? *Oikos* **84**, 167–171.

THOMAS, F., RENAUD, F., DE MEEUS, T. & POULIN, R. (1998). Manipulation of host behaviour by parasites: ecosystem engineering in the intertidal zone? *Proceedings of the Royal Society of London B* **265**, 1091–1096.

THOMAS, F., RENAUD, F., ROUSSET, F., CEZILLY, F. & DE MEEUS, T. (1995). Differential mortality of two closely related host species induced by one parasite. *Proceedings of the Royal Society of London B* **260**, 349–352.

TORCHIN, M. E., LAFFERTY, K. D. & KURIS, A. M. (1996). Infestation of an introduced host, the European green crab, *Carcinus maenas*, by a symbiotic nemertean egg predator, *Carcinonemertes epialti*. *Journal of Parasitology* **82**, 449–453.

UNDERWOOD, A. J. (1976). Food competition between age-classes in the intertidal Neritacean *Nerita atramentosa* Reeve (Gastropoda: Prosobranchia). *Journal of Experimental Marine Biology and Ecology* **23**, 145–154.

UNDERWOOD, A. J. (1978). An experimental evaluation of competition between three species of intertidal prosobranch gastropods. *Oecologia* **33**, 185–202.

WARDLE, W. J. & TIRPAK, A. J. (1991). Occurrence and distribution of an outbreak of infection of *Loxothylacus texanus* (Rhizocephala) in blue crabs in Galveston Bay, Texas, with special reference to size and coloration of the parasite's external reproductive structures. *Journal of Crustacean Biology* **11**, 553–560.

WEGEBERG, A. M. & JENSEN, K. T. (1999). Reduced survivorship of *Himasthla* (Trematoda, Digenea)-infected cockles (*Cerastoderma edule*) exposed to oxygen depletion. *Journal of Sea Research* **42**, 325–331.

WHITLATCH, R. B., HINES, A. H., THRUSH, S. F., HEWITT, J. E. & CUMMINGS, V. (1997). Benthic faunal responses to variations in patch density and patch size of a suspension-feeding bivalve. *Journal of Experimental Biology and Ecology* **216**, 171–189.

WILLIAMS, I. C. & ELLIS, C. (1975). Movements of the common periwinkle, *Littorina littorea* (L.), on the Yorkshire coast in winter and the influence of infection with larval Digenea. *Journal of Experimental Biology and Ecology* **17**, 47–58.

WILSON, W. H. (1991). Competition and predation in marine soft-sediment communities. *Annual Review of Ecology and Systematics* **21**, 221–241.

YAMADA, S. B. & MANSOUR, R. A. (1987). Growth inhibition of native *Littorina saxatilis* (Olivi) by introduced *L. littorea* (L.). *Journal of Experimental Marine Ecology and Biology* **105**, 187–196.

ZWARTS, L. (1991). Seasonal variation in body weight of the bivalves *Macoma balthica*, *Scrobicularia plana*, *Mya arenaria* and *Cerastoderma edule* in the Dutch Wadden Sea. *Netherlands Journal of Sea Research* **28**, 231–245.

Parasitism at the ecosystem level in the Baltic Sea

C. D. ZANDER[1] *and* L. W. REIMER[2]

[1] *Zoological Institute and Museum, University, Martin-Luther-King-Platz 3, D-20146 Hamburg, Germany*
[2] *Am Bahnhof Minden Stadt 4, D-32423 Minden, Germany.*

SUMMARY

The Baltic Sea is characterized by organisms that can tolerate brackish water. Because of the Sea's history during glacial times, its flora and fauna (and also their parasites) can be traced to marine, freshwater and genuine brackish elements beside glacial relics. Snails, planktonic copepods, benthic amphipods and isopods are important intermediate hosts of diverse helminths; in addition polychaetes, bivalves and fishes may also act as final hosts. The most important final hosts, beside fishes, were seals and birds; these were able to disperse the parasites over the whole of the Baltic. Decreasing salinity from west to east limits the distribution of many parasites. Several marine and genuine brackish water species have almost spread over the whole Baltic. Freshwater species, however, have a lower tolerance than marine species and are only rarely found in the western part. A serious problem in the Baltic is eutrophication which can lead to massive abundances of generalist parasites, in host populations as well as host individuals. The final stage of this influence can cause a general decrease of host abundance and, as a consequence, of all kinds of parasites, due to oxygen deficiency. In comparison with the species spectrum of other brackish waters in Europe, the Baltic presents some endemic parasites as well as sharing parasite species with the Mediterranean and even the Black Sea.

Key words: Brackish water, distribution, tolerance, eutrophication, species spectrum.

INTRODUCTION

The Baltic Sea is of very recent origin; its status as a brackish water body goes back no longer than 3500 years. Therefore, evolutionary processes did not have sufficient time for spectacular specialisation. The Baltic is a shallow sea enclosed by land and has only a restricted connection from the Kattegat to the North Sea. From there salinity decreases from 30–20‰ to only 3–1‰ in the Bothnian Bay and the Gulf of Finland in the east (Fig. 1). A sharp decrease in salinity follows in the areas of the Belt Sea, the Sound and the south-west Baltic (Kiel Bight 20–14‰, Mecklenburg Bight 14–10‰), up to the shallow Bank of Darss (19 m deep). This is the limit between the Western and the huge area of the Central Baltic with an almost uniform surface salinity of 8–6‰. Salinity increases again in the depths and, together with the decline in temperature, creates a boundary layer. Therefore, an exchange of surface and deep layers is severely restricted, which causes oxygen deficits in the depths. Only sporadically (last time in 1994) can water bodies from the North Sea overcome the Bank of Darss and provide the deep Central Baltic with oxygen-saturated water for a short time period.

The aim of this article is to examine parasite–host relationships in the Baltic Sea as an extreme environment. Whereas changes of salinity are natural

phenomena to which organisms can adapt over a long time, man-made eutrophication has only been in effect since the last century and has become a great problem in the Sea. The influences of salinity and eutrophication will be dealt with in the context of parasite abundance. The specialists among the parasites deserve special attention, because their occurrence in the Baltic Sea is expected to be restricted, in comparison to other waters. We follow the concept of Holmes & Price (1986), furthered by Zander (1998 a, 2001), which discerns parasite populations and communities at different hierarchical levels: individual hosts (infrapopulation or infracommunity), host populations (component), group of ecologically similar host species (guild) and all stages of a species or all organisms of an ecosystem, respectively (suprapopulation or supracommunity). Because not all existing hosts (as well as parasites) may be recorded, the supracommunity can barely be analyzed and, therefore, can only be considered theoretically. The best approach may be to compile as many analyzed guilds as possible in order to characterize the ecosystem level. This is done in regard to possible processes of co-evolution in the greatest brackish sea of northern Europe.

PARASITE FAUNA AND ITS ORIGIN

Like all other organisms of the Baltic (Remane, 1958), the parasite communities are also composed of marine, brackish and freshwater elements. This composition does not mean that marine hosts are only infected by marine parasites, freshwater hosts only by freshwater parasites, and so on. The fact is that in the Baltic all groups of hosts and parasites are

Corresponding author: Dr. C. Dieter Zander, Zoologisches Institut und Zoologisches Museum, Martin-Luther-King-PI, 3 D-20146 Hamburg, Germany. Tel: +49 40 428 38 3876. Fax: +49 40 428 38 3937. E-mail: cedezet@zoologie.uni-hamburg.de

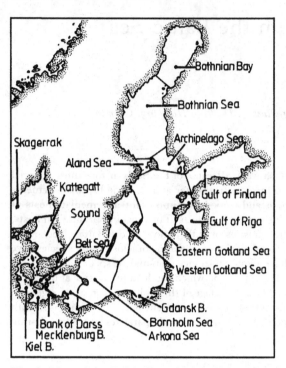

Fig. 1. Map of the Baltic Sea which is divided into several main areas according to salinity. Beginning in the Kattegat (30–20‰ S) salinity decreases until the Bank of Darss rapidly. Arkona Sea, Bornholm Sea, Western and Eastern Gotland Sea comprise the huge Central Baltic Sea with a salinity of 8–6‰. These low values decrease to a freshwater level at the end of the Bothnian Bay and the Gulf of Finland.

thoroughly mixed. Consequently, one can find many generalists and only some specialists among the parasites of this sea.

The marine elements are represented by many helminths and copepods. Several species of Digenea which are common in the Atlantic and North Sea also tolerate brackish waters. The intertidal fish parasite *Podocotyle atomon* (James, 1970) lives in the sublittoral of the Baltic. *Cryptocotyle lingua* infects fish as second intermediate host and birds as the final host (Stunkard, 1930). The digenean group of Hemiurida, (e.g. *Brachyphallus crenatus*) are obvious by their planktobenthic way of life (see below), which is also observed in the brackish Baltic (Reimer, 1970). Abundant cestodes of marine origin include *Bothriocephalus* spp. which have become specialized to several fish as final hosts (Renaud & Gabrion, 1984). Characteristic marine representatives among nematodes include *Anisakis simplex* from whales and *Hysterothylacium aduncum* from fishes. Common Acanthocephala are *Echinorhynchus gadi* in fish or *Corynosoma strumosum* in birds (Reimer, 1969). *Lernaeocera branchialis* is an example of a marine copepod found in Kiel Bight; the larvae may even be found in the Mecklenburg Bight (Arntz, 1972; Lüthen, 1989).

The number of species of freshwater origin is clearly lower because their existence depends on the

occurrence of distinct intermediate hosts or, alternatively, the capture of replacement hosts (Holmes, 1990). The bird digenean *Diplostomum spathaceum*, widespread in fresh water causing blindness in many fish intermediate hosts, depends also in brackish water on the occurrence of the snail *Radix ovata* (Zander *et al.* 2000). The bird cestode *Schistocephalus solidus* has become specialized to its second intermediate host, the stickleback *Gasterosteus aculeatus*. Parasitic freshwater nematodes include *Raphidascaris acus* which is widespread in freshwater fish (Moravec, 1994) and regularly found in the Baltic, in flatfish, eels or sprats (Reimer & Walter, 2000). *Pomphorhynchus laevis* (Acanthocephala) matures in freshwater fish (Hine & Kennedy, 1974) as well as in several fish of the Baltic (Lüthen, 1989). *Caligus lacustris* (Copepoda) was found in the Schlei fjord (Kesting & Zander, 2000), Arkona Sea (Kozikowska, 1957; Schwarz 1960) and Gdansk Bight (Grabda, 1962). This parasite is also known from the Black and Kaspian Sea (Petrushewskij, 1957).

Genuine brackish water parasites are difficult to distinguish from marine parasites. One of their characteristics is a more abundant presence than closely-related species of marine origin. The digenean *Cryptocotyle concavum* is common in the Baltic and other brackish waters; its first and preferred second intermediate hosts (the snail, *Hydrobia stagnalis* and the fish, *Pomatoschistus microps*) are also genuine brackish water species. The digeneans *Aphalloides timmi*, also in *P. microps*, and *Acanthostomum balthicum* in *Syngnathus typhle* may also be representatives of this category (Reimer, 1971; Reimer *et al.* 1996; Zander, Reimer & Barz, 1999), as well as the copepod *Thersitina gasterostei* living in the gill chamber of *Gasterosteus aculeatus* (Hildebrand & Scharberth, 1992; Zander & Westphal, 1991).

The nematode *Ascarophis arctica* may be a glacial relic (Fagerholm & Berland, 1988) which must have immigrated into the Baltic, together with its main host, the sculpin *Myoxocephalus quadricornis*, from the Barents Sea. The most peculiar non-native parasite of the Baltic is the eel nematode *Anguillicola crassus* which uses the brackish milieu as a bridge between the rivers of continental Europe and those of Scandinavia on its way of dispersal (Reimer *et al.* 1994).

Several parasites of the Baltic or closely-related species can also be found in other European brackish waters, e.g. off the French Atlantic and Mediterranean coasts as well as in the Black and Asov Seas (Bauer, 1987; Reimer *et al.* 1996). Certain species found in the Baltic Sea also occur in the Black Sea (e.g. *Asymphylodora demeli*), and other species occur also in France (e.g. *Brachyphallus crenatus*). The distribution patterns may possibly be influenced by the history of the Tethys Sea (Gibson & Køie, 1991; Reimer *et al.* 1996) as well as by the Parathethys in its extension during the Oligocene, when the regions of the Black and North Seas as well as the Baltic Sea were connected. The parasitic copepod, *Thersitina gasterostei*, was not only found in the Black, Caspian, White Seas but also off the coasts of Canada (Yamaguti, 1963). Endemic species of the Baltic Sea like *Aphalloides timmi*, *Acanthostomum balthicum* or possibly *Magnibursatus caudofilomentosa* may be the result of speciation events after the isolation of this sea, followed decreasing salinities as a natural barrier.

Regarding faunistic patterns, the composition of parasitic species of the Baltic does not differ principally from that of free living organisms with a majority of marine elements in both groups. The category of genuine brackish water species has strong taxonomic affinities to those of other European brackish waters which is in accordance with non-parasitic species which can even be distributed world-wide (Remane, 1958). Albeit its short existence the Baltic Sea has evolved some endemic parasite species which all are genuine brackish water species.

CHARACTERISTIC LIFE CYCLES IN THE BALTIC SEA

Here, one should consider the functional types of hosts in the brackish Baltic Sea; these guarantee the existence of parasites. Marine mammals, birds and fish are highly mobile hosts which contribute to the dispersion of parasites. According to Holmes & Price (1986) parasites of birds are allogenic ones which are transferred from the terrestrial to the aquatic environment. If fish are final hosts, the parasites are autogenic elements. Fish as well as invertebrates can also be intermediate hosts.

The helminth parasites present characteristic cycles which need several intermediate hosts (Fig. 2). For transmission to the next host it is very important whether the host lives in open water or on the bottom. *Podocotyle atomon* presents a characteristic benthic cycle of digeneans which starts in the snail *Littorina saxatilis*, continues in benthic crustaceans like gammarids and isopods which are actively infected by creeping cercariae and ends in benthic fish like gobies or flatfishes. The digenean group of Hemiurida show a planktobenthic cycle from snails to planktonic copepods to pelagic fish or fish larvae which later may have a bottom-dwelling way of life. The digenean, *Cryptocotyle* spp., finally attain birds which are infected by consuming fish where the metacercariae encyst normally on the skin and fins. An exception is the microhabitat of *C. concavum* in the common goby, *Pomatoschistus microps*: the cercariae penetrate the gills and encyst in the kidney (Zander *et al.* 1984; Kreft, 1991).

Deviations of the general cycle are numerous in the Baltic Sea. *Asymphylodora demeli* which infect generally snails, mussels or polychaetes as inter-

mediate hosts and fishes as final hosts, can shorten the life cycle by maturing in polychaetes (Reimer, 1973); the same phenomenon was found by Vaes (1974) in brackish waters of the Belgian coast. In a similar way, *Proctoeces maculatus* can already mature in mussels instead of flatfish, *Bunocotyle cingulata* in *Hydrobia ventrosa* (Reimer, 1961). *Parvatrema affinis* which normally finish its life cycle in birds do not leave its first intermediate host, *Macoma balthica* (Fig. 2). *Aphalloides timmi* regularly ends its life cycle in the second host, the common goby, *Pomatoschistus microps*. Microphallids of brackish waters present a great variability in life cycles: they must reach birds as final hosts – normally by benthic crustaceans as second host, but *Microphallus claviformis* and *M. pygmaeus* may infect birds directly by snails, where metacercariae can encyst; *M. claviformis* may include additionally fish as third host (Reimer, 1963; Kesting, Gollasch & Zander, 1996) (Fig. 2). This alteration enlarges the spectrum of final hosts from crustacean-eating to snail- and fish-eating birds.

Other helminth groups have often simpler life cycles. Cestoda present 2- or 3-host cycles, terminating in fish or birds; but probably only birds are infected by benthic crustaceans in the Baltic (Fig. 2). Tapeworms of freshwater origin, e.g. the bird parasite *Schistocephalus*, infect fish species of marine origin like gobies and cod (Reimer & Walter, 1993; Zander & Kesting, 1996) though the second host are in fresh water solely three-spined sticklebacks, *Gasterosteus aculeatus*. Nematodes need either 2 or 3 hosts (Fig. 2). Remarkably, a more complicated web prevails in the nematode *Hysterothylacium*, which can be directly transferred to fish by benthic crustaceans but also by smaller fish, or several invertebrates like ophiurids, copepods, polychaetes, ctenophores or chaetognaths (Køie, 1993). In similar ways eels are infected by *Anguillicola crassus* via planktonic copepods or by fish of which the pipefish, *Syngnathus typhle*, is the most frequent second intermediate host in the Baltic Sea (Reimer *et al.* 1994). No acanthocephalan is known from the Baltic, which deviates from the characteristic 2- (e.g. *Echinorhynchus*) or 3-host cycles (e.g. *Corynosoma*) (Fig. 2).

Remarkable are some cycles by which fish are infected by fish; besides the already mentioned nematodes, the digenean *Acanthostomum balthicum* is unique because it is transferred from small fish like gobies to the pipefish, *Syngnathus typhle* (Fig. 2).

Regarding patterns of ontogeny, successful parasites of the Baltic Sea often show great flexibility in altering life cycles (microphallids) which may even lead to cycles without vertebrate hosts (*Asymphylodora demeli*), and use a wide spectrum of paratenic hosts (*Hysterothylacium*). Parasites of extreme environments such as brackish water have better possibilities of existence when alternative ways of on-

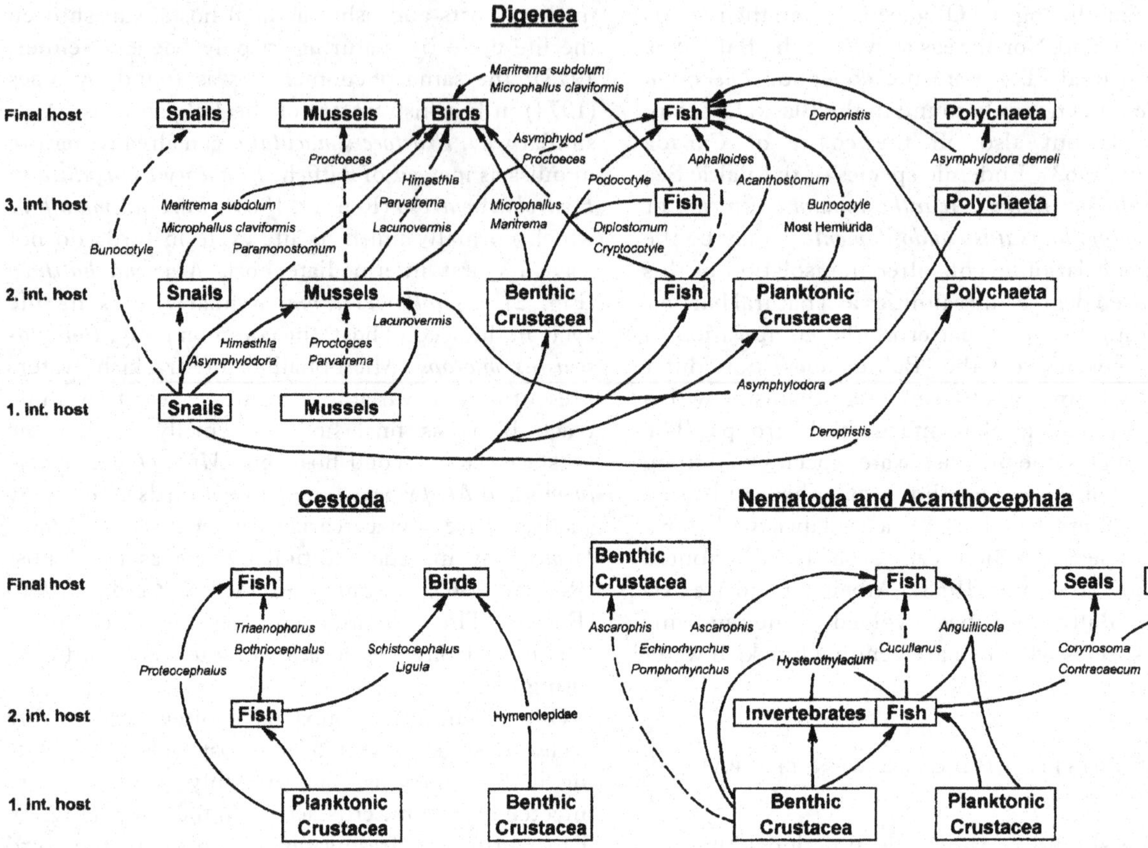

Fig. 2. Life cycles of parasites from the 4 most important helminth groups. The ways of digeneans are highly inhomogeneous because brackish water can lead to omission of second, to infect third intermediate hosts or to remain and mature in paratenic hosts (dashed lines).

togenetic development are adopted. Galaktianov (1985) reported *Microphallus* spp. from the White Sea which metacercariae encyst already in their first hosts, *Littorina rudis*, obviously as an adaptation of the short ice-free season in the year.

INFLUENCE OF HOSTS

Gastropods are first and, also sometimes, second intermediate hosts of most digeneans (Fig. 2). In the Baltic, the number of snail species is clearly lower than in the North Sea and is associated with a lower number of parasite species. However, the fresh water gastropods *Radix ovata* (up to 8) and *Theodoxus fluviatilis* (up to 15‰ salinity) can survive in brackish water. These species harbour not only fresh water digeneans like *Diplostomum spathaceum* but also brackish water species like *Asymphylodora demeli* (Zander *et al.* 2000). As many as 18 digenean species can live in *Hydrobia ulvae* and *H. ventrosa* from the Baltic Sea (Reimer, 1971).

In the brackish Baltic, digenean parasites are less specialised to snail species than in marine and freshwater environments. *Psilochasmus oxyuris* can infect as many as 6 snail species as intermediate hosts, *Asymphylodora demeli* is found in as many as 8 intermediate hosts, *Microphallus papillorobustus* in 4 (Reimer 1973; Kesting *et al.* 1996; Strohbach, 1999; Zander *et al.* 2000) (Table 1). The bird

parasite, *Cryptocotyle lingua*, normally develops in the periwinkle, *Littorina littorina*, but was in the Baltic Sea also found in *L. saxatilis* and 3 *Hydrobia* species (Reimer, 1970, 1995). This broad host spectrum may induce cases of host capture (Holmes, 1990), also known from fish hosts. The fish digenean, *Podocotyle atomon*, which infects *Littorina saxatilis* in the North Sea, is also found in *Littorina littorea* in the Baltic, e.g. in the Schlei Fjord, Kiel Bight (Kesting *et al.* 1996) (Table 1). Remarkably, we did not find in the non-native (neozoon) *Pomatopyrgus jenkinsi* from New Zealand, a relative of *Hydrobia*, within many thousand specimens as infected one (Kesting & Zander, 2000).

Bivalves may be first and second intermediate as well as final hosts. Four host species are common in the Baltic: *Mytilus edulis*, *Macoma balthica*, *Cerastoderma edule* and *C. lamarcki*, whereas the non-native *Mya arenaria* is very rarely infected (Markowski, 1936; Pekkarinen, 1988; Reimer & Pohl, 1989; Junge, 1993). The digeneans, *Asymphylodora demeli*, *Himasthla* spp. and *Lacunovermis macomae* are very conspicuous because of their high prevalences in *Macoma balthica* or *Cerastoderma lamarcki*. Two turbellarians may occur, *Urostoma cyprina* in *Mytilus edulis* and *Paravortex cardii* in *Cerastoderma* spp. and *Macoma balthica*.

Annelids are parasitized by protozoans and by

Table 1. Presence of digeneans in gastropods from 5 Baltic localities: B = Blank Eck, F = West Fehmarn, SC = Schlei (Kiel Bight); D = Dahmeshöved, S = Salzhaff (Mecklenburg Bight)
(According to Kesting, Gollasch & Zander, 1996; Strohbach, 1999; Zander *et al.* 2000.)

Parasites	Snail hosts										Number of hosts
	Littorina saxatilis	*Littorina littorea*	*Littorina obtusata*	*Hydrobia ulvae*	*Hydrobia ventrosa*	*Hydrobia neglecta*	*Zippora membranacea*	*Turboella conspicuosa*	*Limnaea ovata*	*Theodoxus fluviatilis*	
Cryptocotyle concavum	B, D			S, F, D	S, F, D						2
C. lingua				S							2
Aphalloides timmi				S, F	F						2
Acanthostomum balthicum				S	S						2
Psilochasmus oxyuris	SC	SC	SC	S, F	S, F		S				6
Bunocotyle cingulata				S							1
Magnibursatus caudofilamentosa							S				1
Asymphylodora demeli	S			S, F, D	S, F, D	F	S	F	S	S	8
Paramonostomum alveolatum				S	S						2
Podocotyle atomon	S, SC, B	SC		S, SC, F, D	S, SC, F, D	SC, F					2
Maritrema subdolum				S, SC, F	S, SC, F	SC					3
Microphallus claviformis				F, D	S, SC, F	F		F			3
M. papillorobustum						F	S				4
M. pygmaeus	S, B, D								S		2
Cotylurus cornutus									S	S	2

Table 2. Number of fish hosts of some abundant parasites of the
German coasts of the Baltic Sea.
(According to Palm, Klimpel & Bucher (1999).)

Group	Species	Origin	Number of host spp.
Digenea	*Podocotyle atomon*	marine	27
	Brachyphallus crenatus	marine	18
	Diplostomum spathaceum	freshwater	15
Cestoda	*Bothriocephalus scorpii*–plerocercoids	marine	19
Nematoda	*Hysterothylacium aduncum*	marine	19–26
Acanthocephala	*Echinorhynchus gadi*	marine	20
	Pomphorhynchus laevis	freshwater	16

digeneans. *Asymphylodora demeli* can infect *Hediste dervisicolor*, *Alkmaria romijni* and *Spirorbis granulatus*, which act as second intermediate hosts. Additionally, *H. diversicolor* can also serve as final host of this digenean in the Baltic Sea (Reimer, 1973), and therefore replaces fishes (Fig. 2). *Neanthes virens* is the only second intermediate host of the eel parasite *Deropristis inflata*. Metacercariae of *Himasthla* sp. were found in *Arenicola marina* and also in *H. diversicolor*. Again, a non-native species of the Baltic, *Marenzelleria viridis*, remained uninfected (n = 109).

Planktonic crustaceans are very important intermediate hosts of digeneans of the group Hemiurida, cestodes and nematodes (Fig. 2). Strohbach (1999) found several calanoid copepods (*Acartia*, *Centropages*, *Eurytemora*, *Paracalanus* and *Temora* spp.) to be hosts of tapeworm and nematode larvae in Kiel and Mecklenburg Bight. The very abundant cyclopoid, *Oithona*, was only infected in the Schlei Fjord but not in the open Kiel and Lübeck Bight (Kesting *et al.* 1996). Copepods with lower abundance seem to be more heavily infected than common ones. They are also intermediate hosts for the important fish parasite *Ichthyophonus hoferi* (Fungi).

Benthic crustaceans are also important intermediate hosts of several helminth parasites. In particular, species of gammarids (Amphipoda), idotheids and sphaeromids (Isopoda) – being the preferred prey of fish or birds – can be heavily infected by larvae of helminth parasites (Skroblies, 1998). *Carcinus maenas* is the only host of the bird digenean *Microphallus primas*. Other specializations may not have evolved in parasites of benthic crustaceans, though preference of single species usually changes from locality to locality. *Gammarus salinus* seems to be less infected by cercariae of *Podocotyle atomon* or microphallids than other *Gammarus* species. *Maritrema subdolum* is an abundant parasite of *Idothea chelipes* or *I. granulosa* whereas *I. balthica* is hardly infected (Skroblies, 1998; Zander *et al.* 2000). Procercoids of the bird tapeworm family Hymenolepidae were also found in *Gammarus* spp. (Reimer, 1969; Zander & Döring, 1989; Gollasch & Zander, 1995; Strohbach,

1999). Acanthellae of acanthocephalans regularly infect gammarids and isopods (Gollasch & Zander, 1995; Strohbach, 1999). In the Baltic Sea, *Crangon crangon* is an occasional host of digenean metacercariae though metazoan parasites were never found in North Sea specimens (Gollasch *et al.* 1996).

Fish may be hosts of helminths but also of parasitic protozoans, crustaceans and of an annelid (*Piscicola*). Besides genuine fish parasites, the Baltic Sea also includes many parasites of birds and marine mammals which are transferred by fish as intermediate or paratenic hosts. The component community of small-sized fish from the Salzhaff (Mecklenburg Bight) like *Pomatoschistus microps* (Gobiidae) can attain as much as 16 parasite species, *Gasterosteus aculeatus* 20, and *Pungitius pungitius* (Gasterosteidae) 18 (Zander *et al.* 2000).

The number of generalist parasites is numerous on the German Baltic coasts of Schleswig-Holstein and Mecklenburg-Vorpommern, e.g. *Podocotyle atomon* can infect as much as 27 different fish hosts (Palm, Klimpel & Bucher, 1999) (Table 2). Recent investigations revealed the low presence or even absence of this parasite in *Gobius niger*, obviously because metacercariae cysts are not digested when these are ingested together with the second intermediate hosts, benthic crustaceans (Zander & Kesting, 1996; Zander *et al.* 1999).

Certain fish species may be the main hosts for a particular parasite and other fish species can serve as occasional hosts which are infected in clearly lower abundance probably due to different levels of co-evolution (Zander, 2001) (Table 3). Whereas *Aphalloides timmi* is restricted to gobies, the gill-inhabiting *Magnibursatus caudofilomentosa* is not only present in sticklebacks but also in gobies and in the brood pouches of *Syngnathus typhle* (Zander, unpublished). Obviously, among the small-sized fish of the Salzhaff (Mecklenburg Bight) *Proteocephalus percae* is present in gobies whereas *P. filicollis* is restricted to sticklebacks (Zander *et al.* 1999).

The specific parasites of smaller and larger fish hosts (Table 4) comprise in the Baltic Sea 7 digeneans, 8 cestodes, 2 nematodes and 4 copepods.

Table 3. Prevalences (%) of 8 selected specialist parasites in small-sized fish hosts of the Salzhaff, Mecklenburg Bight. (The main hosts are underlined. According to Zander (2001).)

Parasites	Hosts								Number of hosts	Specifity index (Rohde, 1993)
	Gobius niger	*Gobiusculus flavescens*	*Pomatoschistus minutus*	*Pomatoschistus microps*	*Gasterosteus aculeatus*	*Pungitius pungitius*	*Spinachia spinachia*	*Syngnathus typhle*		
Apatemon gracilis		2		35 / 52					2	0·973
Aphalloides timmi			7						2	0·941
Acanthostomum balthicum				0·6	8			59	4	0·877
Bunocotyle cingulata		2					7		2	0·964
Magnibursatus caudofilimentosa					38	26			2	0·987
Schistocephalus solidus	1		1	0·6	3	1			4	0·705
Schistocephalus pungitii						12			1	1
Thersitina gasterostei					71	8	10	1	4	0·911

Table 4. Parasite specialists of several fish species from the Baltic Sea

Hosts	Parasites			
	Digenea	Cestoda	Nematoda	Copepoda
Pomatoschistus microps	*Aphalloides timmi*			
Pomatoschistus microps	*Apatemon gracilis*, metacercaria			
Gasterosteus aculeatus	*Magnibursatus caudofilamentosa*	*Schistocephalus solidus*, plerocercoid		
Pungitius pungitius		*Schistocephalus pungitii*, plerocercoid		
Spinachia spinachia		*Diplocotyle olriki*		
Syngnathus typhle	*Acanthostomum balthicum*			*Thersitina gasterostei*
Myoxocephalus scorpius		*Bothriocephalus scorpii*		
Myoxocephalus scorpius		*Prosorhynchus squamatus*		
Zoarces viviparus			*Hysterothylacium auctum*	
Anguilla anguilla	*Deropristis inflata*	*Bothriocephalus claviceps*	*Anguillicola crassus*	*Ergasilus gibbus*
Anguilla anguilla		*Proteocephalus macrocephalus*		
Pleuronectes flesus	*Aporocotyle simplex*			*Acanthochondria cornuta*
Pleuronectes flesus				*Lemaeocera branchialis*, larva
Pleuronectes limanda	*Proctoeces maculatus*			
Psetta maxima		*Bothriocephalus cf. gregarius*		

Table 5. Specifity of helminth parasites in bird hosts. S = number of species; x/y = number of parasite species found only in the respective bird group/number of parasite species found only in 1 host species; Anatidae I = Anserini, Tadornini, Anatini; Anatidae II = Aythyini, Mergini, Somaterini

		Birds													
Parasites	S	Gavii- dae 1	Podici- pedidae 3	Phalacro- coracidae 1	Ardei- dae 1	Anati- dae I 6	Anati- dae II 9	Chara- driidae 2	Scolo- pacidae 7	Laridae 4	Sterni- dae 1	Ralli- dae 1	Species restricted species	Group restricted species	Infection stage or host
Digenea															
Diplostomatidae	7	1/0	1/0	1/0	1/0	1/0	0/1			0/1	1/0		5	2	Fish
Strigeidae	5	1/0			1/0		0/2			2/0			3	2	Molluscs and fish
Cyathocotylidae	1						1/0						1		Fish
Schistosomatidae	1									1/0			1		Cercariae
Renicolidae	4			1/0			0/1			0/1			1	2	Fish
Plagiorchidae	2					1/0				0/1			1	1	Chironomid-larvae
Gymnophallidae	4			1/1			0/2			1/0			1	2	Mussels
Microphallidae	9					1/0	1/2			?		1/0	3	3	Crustaceans
Eucotylidae	4					1/0	0/1		1/0			1/0	3	1	Molluscs
Notocotylidae	5					2/0	0/1					1/0	3	1	Metacercarial cysts on substrate
Psilostomatidae	2						0/1						2	1	Molluscs
Echinostomatidae	12		0/1	2/0	1/0	1/0	1/0			1/1			6	2	Molluscs and fish
Opisthorchiidae	1			1/0									1		Fish
Heterophyidae	4				1/0								1		Fish
Cestoda															
Pseudophylloidea	3	1/0		1/0									2		Fish
Tetrabothriidea	2	1/0											2		Fish
Dilepididae	6			1/0	1/0				1/0	1/0			5		Fish
Davaneidae	1									2/0					?
Hymenolepidae	19		1/0			4/0	2/4	0/1		4/0			11	5	Oligochaetes, Molluscs, Crustacea
Amabiliidae	1		0/1										1	1	Copepods, Dragonfly-larv.
Nematoda															
Amidostomatidae	3						0/2	1/0					1	2	Larvae
Heterakidae	1						0/1						1	1	Eggs, (Earthworms)
Anisakidae	3		1/0							0/1			1	1	Fish
Toxocaridae	1							1/0					1		Earthworms
Tetrameridae	1						1/0						1		Crustaceans
Streptocaridae	1						0/1						1	1	
Acuariidae	2		1/0				1/0			1/0			2	1	Amphipoda, (fish)
Ancyranthidae	2									1/0			2	1	
Capillariidae	6						2/1			1/0			3	1	Larvae, (Earthworms)
Acanthocephala															
Polymorphinae	4			1/0			0/1					1/0	2	1	Crustacea, Fish
Centrorhynchinae	1								1/0				1		
Plagiorhynchinae	1								0/1					1	

The nematode *Anguillicola crassus* is a non-native species which was imported from Asia to Europe in the late 1970s. Though adults are restricted to eels its larvae can occur in several intermediate fish hosts (Reimer *et al.* 1994).

Birds are important final hosts in the life cycles of several helminths (Fig. 2). Thirty six species out of 12 families of aquatic birds of the Baltic area turned out to be infected by 119 helminth species (Reimer, compiled in Reimer, 2002) (Table 5). Birds are highly mobile hosts that can collect their parasites in other climatic zones or even on other continents, resulting in a diverse mix of different elements in their parasite faunas. Digeneans of freshwater origin are mainly the Diplostomatidae, Strigeidae and Echinostomatidae, of which several species are found in fish as in intermediate hosts from the Mecklenburg and Kiel Bight (Reimer, 1970; Kesting *et al.* 1996; Kesting & Zander, 2000; Zander *et al.* 2000). Renicolidae and many species of Microphallidae and Gymnophallidae are of entirely marine, and Eucotylidae, of terrestrial origin. The cestodes, nematodes and acanthocephalans that were found are mostly limited to fresh waters. Dilepididae, Diplostomatidae (Digenea) or Polymorphinae (Acanthocephala) seem to be specialists of host families (Table 5). The majority of parasites are generalists which infect a lot of bird hosts. Among the helminths found the microphallids, *Levinseniella brachysoma* (16 bird host species) and *Maritrema subdolum* (13 bird species), as well as the heterophyid *Cryptocotyle concavum* (20 bird species) display the greatest host spectra. Several parasites with only 1 host are found among the Tetrabrothiidae, Dilepididae and Hymenolepidae (Cestoda) as well as the Acuariidae (Nematoda) (Table 5). It is not clear in every case whether these results are a consequence of co-evolution or of strict preference of distinct food (snails, crustaceans or fish) by the respective hosts. Most larvae of the above mentioned bird parasites are found in areas of lower salinity and can complete therefore their life cycle in brackish waters.

Regarding the patterns of biotic factors (hosts), Baltic parasites display loss or reductions of specialisation in a greater extent in comparison to marine and freshwater habitats. Even species which are not infected in other areas can become hosts in the Baltic Sea (*Crangon crangon*) which make host capture evident (Holmes, 1990). Non-native parasites (*Anguillicola crassus*) can find native hosts without difficulties. But in contrast, non-native hosts are not or only very seldom infected by Baltic parasites (*Pomatopyrgus jenkinsi*, *Mya arenaria* and *Marenziella viridis*).

INFLUENCE OF SALINITY

The brackish condition of the Baltic Sea is an obvious stress factor to all organisms, including parasites. In the central Baltic with salinity of only 6–8‰, the lowest number of species ('species minimum', Remane, 1958) is found. Generally, parasites depend on the salt tolerance of their hosts, the most restricted ones confine to the distribution of the parasite.

The restriction of parasites to certain areas of the Baltic depends mainly on the limits of the distribution of benthic, less mobile intermediate hosts, as well as on the tolerance of their own free larval stages. The cercariae of the bird digenean, *Cryptocotyle lingua* can exist down to salinities of 2‰ at temperatures below 10 °C, but higher temperatures reduce salinity tolerance to 8‰ and the life span to a maximum of 4 days (Möller, 1978). The distribution of intermediate hosts is probably more important for the presence of parasites than salinity. Thus, several snails of marine and fresh water origin are only found in parts of the Baltic, restricting the distribution of many digeneans (Table 6).

In other cases, the limits of the distribution of the first hosts can be further restricted but also enlarged by the next hosts. The first larval stages of the cosmopolitan *Derogenes varicus* are bound to *Natica* snails up to the Kiel Bight but may be exported with the help of currents to the Mecklenburg Bight in their second hosts, copepods, where they may be ingested by benthic fish (Table 6). Several migrating fish species are capable of transferring parasites like *Hemiurus luehei*, *Lecithaster confusus*, *Deropristis inflata* or *Diplostomum spathaceum* to areas where the salinity cannot be tolerated by the first hosts. The nematode, *Anisakis simplex*, was found repeatedly in fishes of the Baltic though euphausiids, their intermediate hosts, cannot exist below salinities of the Kattegat (Köhn & Gosselck, 1989). *Clupea harengus* migrate from feeding grounds in the North Sea where they prey also on infected krill to spawning grounds off Rügen or in the Oder Bight, Arkona Sea (Grabda, 1974). However, benthic fish like *Zoarces viviparus* in the Kiel Bight (Bucher, 1998) or 5 goby species in the Lübeck Bight were found to be infected by *A. simplex* (Zander, Strohbach & Groenewoldt, 1993; Zander & Kesting, 1996). The garfish, *Belone belone*, which migrates from the Atlantic Sea to spawn in the Central Baltic, harboured parasites like the marine cestodes *Ptychobothrium belones* and *Lacistorhynchus tenuis* as well as *Anisakis simplex* (Grabda, 1981).

The influence of salinity may be best recognized by the comparison of host populations from different localities along the salt gradient. Whereas highly mobile fish species are not suited to such analyses, resident benthic fish like the eelpout, *Zoarces viviparus*, are very suitable (Table 7). The population of the Lübeck Bight possesses the most marine parasite species (Zander, 1991). Gdansk Bight is the other extreme, with the most fresh water species (Markowski, 1938; Sulgostowska, Jerzewska &

Table 6. Distribution limits of some digeneans in the Baltic Sea according to the distribution of their hosts (According to diverse authors.)

Parasites	First host	Restricted distribution of first host	Second host	Restricted distribution of second host	Second or Third host beyond distribution limits	Locality of record
Marine species						
Derogenes varicus	Natica spp.	Kiel Bight	Calanoid copepods		Gobiidae, Zoarces	Lübeck Bight
Deropristis inflata	Bittium reticulatum	Lübeck Bight	Neanthes virens	Kiel Bight	Anguilla anguilla	Gdansk Bight
Hemiurus luehei	Philine denticulata	Lübeck Bight	Calanoid copepods		Gadus morhua	Bornholm
H. communis	Retusa truncatula	Bornholm	Calanoid copepods			
Lecithaster confusus	Odostomia rissoides	Mecklenburg Bight	Calanoid copepods		Clupea harengus	Arkona Sea
L. gibbosus	Odostomia spp.	Mecklenburg Bight	Calanoid copepods			
Magnibursatus caudofilamentosa	Zippora membranacea	Arkona Sea	Calanoid copepods			
Brachyphallus crenatus	Retusa obtusa	Gotland	Calanoid copepods			
Podocotyle atomon	Littorina saxatilis	Bornholm	Amphipoda, Isopoda			
Cryptocotyle lingua	Littorina littorea	Bornholm	Fish			
Microphallus primas	Hydrobia ulvae	Bothnian Bay	Carcinus maenas	Bank of Darss		
M. pygmaeus	Littorina saxatilis	Bornholm				
Freshwater species						
Diplostomum spathaceum	Radix ovata	8‰, Arkona Sea	Fish		Gadus morhua	Öresund
Cotylurus cornutus	Radix ovata	8‰, Arkona Sea	Theodoxus fluviatilis	15‰, Lübeck B.		

Table 7. Prevalences (%) and origin of helminth parasites of *Zoarces viviparus* from 5 areas of the Baltic Sea

Parasite	Salinity / Authors / N hosts (origin)	Kiel Fjord 14–18‰ Möller, 1975 (1112)	Lübeck Bight 10–14‰ Zander, 1991 (36)	Salzhaff 10–12‰ Reimer & Walter, 1998 (191)	Gdansk Bight 7‰ Markowski, 1938 (550)	Gdansk Bight 7‰ Sulgostowska et al. 1990 (280)	Riga Bight 5–6‰ Vismanis et al. 1980 (191)
Gyrodactylus medius	freshwater				21		22
Diplostomum spathaceum M.	freshwater			25		50	4
Cardiocephalus longicollis M.	freshwater			11			10
Cryptocotyle concavum M.	brackish	50		4			
Acanthostomum sp. M.	brackish			3			
Podocotyle atomon	marine	20	72	46			
Derogenes varicus	marine		3				
Eubothrium sp.	freshwater				3		0·5
Bothriocephalus sp. Plerocercoid	freshwater		28		11	0·7	1
Triaenophorus nodulosus	freshwater				5		
Caryophyllaeus sp.	freshwater						
Hysterothylacium auctum	marine	71	94	82	54	50	38
Ascarophis arctica	brackish	0·1		17	2		1
Metabronema canadense	freshwater						6
Raphidascaris sp.	freshwater				12		
Neoechinorhynchus rutili	freshwater		5		0·4		
Echinorhynchus gadi	marine	14	38	3	2	7	1
Echinorhynchus salmonis	freshwater				0·2		
Pomphorhynchus laevis	freshwater	0·6	3	8	9	48	0·5
Corynosoma semerme Acanthella	marine		30	5	2		33
Corynosoma strumosum Acanthella	marine						
Sum of species	marine	3	5	4	3	2	3
	brackish	2	0	3	1	0	1
	freshwater	1	3	3	8	3	7

Wicikoski, 1990), the Salzhaff having an intermediate position with the most brackish parasite species (Reimer & Walter, 1998). The most marginal localities, Kiel Fjord (Möller, 1975) and Riga Bight (Vismanis *et al.* 1980), may be influenced by special factors and are not as persuasive as the other localities.

The same phenomena observed over large spatial scales can also be acting over short distances in isolated waters of the Baltic, e.g. in the Schlei Fjord (Kiel Bight) which is characterized by great freshwater influx (Kesting *et al.* 1996). The numbers of metazoan parasite species change from the innermost station in the Fjord (salinity 3‰) with 2 marine, 1 brackish and 7 freshwater species, through a middle station (6‰, 'species minimum') with 4, 3 and 6 species respectively, to the outer Station (15‰) near fjord entrance with 5 marine, 4 brackish and 3 fresh water species (Kesting & Zander, 2000). Here, also prevalence and intensity of the digenean *Cryptocotyle concavum* clearly increase with increasing salinities (Zander *et al.* 1984). This effect may be influenced and varied by hydrographical, topographical and biotic factors (Reimer, 1970).

Regarding the patterns of the main abiotic factor, salinity, in the Baltic Sea, the distribution of parasites is limited along the gradient from west to east. This phenomenon depends on the respective tolerance of their hosts and is found to be least in parasites which are generalists. The rule of 'species minimum' which is claimed for free living species at salinities of 6–8‰ (Remane, 1958) seems to be not absolutely valid for parasites, probably due to their ability to switch to several hosts.

INFLUENCE OF EUTROPHICATION

Eutrophication is an anthropogenic stress factor, caused by an increased influx of organic wastes, especially nitrate and phosphate which stimulates plant growth. This favours production throughout the trophic chain, and parasites can also profit from the increased abundance of their hosts. The Orther Bight off west Fehmarn (Kiel Bight) is a very shallow water with a maximal salinity of 11–12‰. Although a great stock of benthic algae provide high assimilation rates, low oxygen saturations of 50–60% on the bottom in summer document the high deposition rate of organic detritus. Benthic crustaceans like *Idothea chelipes* attain densities of 28 individuals/m² on the bottom and more than 250 on a *Fucus* plant in summer. These can be infected at prevalences as high as 85% and at intensities as high as 202 per host, with a mean of 88 metacercariae by the digenean *Maritrema subdolum* (Skroblies, 1998). This creates a high infection potential for birds that feed on crustaceans. The Orther Bight, therefore, has been designated as a habitat serving as a centre of epidemics (Lauckner, 1994). Because of the re-

stricted mobility of infected isopods, their availability for birds is greatly increased.

Three localities in the western Baltic, characterized by clear differences in oxygen deficiencies, could be compared with respect to the digenean *Cryptocotyle concavum* which encysts in the kidneys of the goby, *Pomatoschistus microps* (Zander, 1998*b*). The lowest values of oxygen content were measured in the Orther (Kiel) Bight, followed by Salzhaff, then by Dahmeshöved (Mecklenburg Bight). The highest infection levels were found in the Salzhaff, attaining a maximum of 2300, and a mean of 430 metacercariae per fish, whereas in the Orther Bight, these values were 347 and 112 metacercariae, and in Dahmeshöved even lower (Zander, 1998*b*). These infection rates were promoted by different levels of eutrophication which influence the densities of the herbivorous first intermediate host, *Hydrobia* spp.

The localities Blank Eck in the Kiel Bight (15‰ salinity) and Dahmeshöved in the Lübeck Bight (11‰ salinity) differ in terms of eutrophication, as indicated by oxygen saturation: in Blank Eck oxygen saturation only drops below 100% in autumn but in Dahmeshöved it is below 100% over almost the whole year. The parasite fauna of 4 goby species from Blank Eck was a little poorer in species numbers and showed clearly lower prevalence and intensities, but attained higher diversity indices and evenness values than in Dahmeshöved (Zander & Kesting 1996) (Table 8). Also, the prevalence and parasite density of benthic crustaceans of Dahmeshöved by far surpassed the values from Blank Eck, whereas infection of planktonic crustaceans hardly differed between localities (Strohbach, 1999). This results from better conditions for herbivores and generalist parasites which may increase species richness and parasite density but decrease diversity indices. Only in *Littorina saxatilis* was the opposite result found, with periwinkles in Blank Eck attaining higher prevalences and parasite densities than in Dahmeshöved, which is explained by their smaller component community than in mud-snails.

Another way of using parasites as indicators of environmental influences is to calculate the regression of parasite prevalences versus host specificity (Zander, 1993). The idea is that specific parasites are favoured in more pristine habitats where their number is higher than in affected habitats where the generalists with many hosts prevail. Regarding the respective prevalences (y-axis in Fig. 3), parasites with only 1 host may prevail in pristine habitats over parasites with more hosts (x-axis in Fig. 3). In eutrophic waters, where generalists dominate over specialists, the reverse may be the result. The regression line in oligotrophic waters is therefore, negative, whereas in eutrophic habitats the slope of the regression line increases and indicates the level of eutrophication. The comparison of the habitats in Kiel and Lübeck Bights resulted in a clearly stronger

Table 8. Comparison of levels of parasite infection in 4 host groups from Blank Eck (Kiel Bight, less eutrophicated) and Dahmeshöved (Lübeck Bight, eutrophicated). Hs = Shannon-Wiener index, J = index of equibility. (According to Strohbach (1999), Zander & Kesting (1996).)

Parameter	*Littorina saxatilis*		Planktonic copepods		4 *Gammarus* spp.		4 gobiid fishes	
	Blank Eck	Dahmeshöved	Blank Eck	Dahmeshöved	Blank Eck	Dahmeshöved	Blank Eck	Dahmeshöved
Mean prevalence (%)	4·1	0·4	5·5	6·1	1·2-5·0	5·6-16·2	0·2-21·0	0·6-61·0
Mean intensity (N)			1·0	1·0	1·0-1·6	1·0-5·0	1·0-23·4	1·0-16·8
Parasite density (N/m²)	3·1	0·4	78	55	59	1921		3·8
Species Numbers (S)	3	2	11	12	14	9	4-9	3-10
Diversity (Hs)	0·17	0·09	1·95	2·08	1·67	1·52	0·78-1·74	0·55-1·62
Evenness (J)	0·15	0·13	0·81	0·83	0·63	0·69	0·48-0·89	0·45-0·70

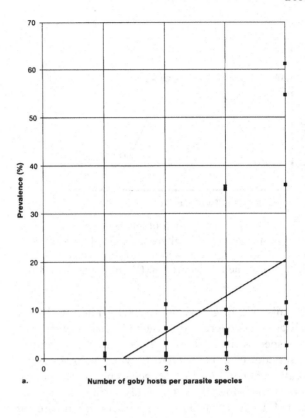

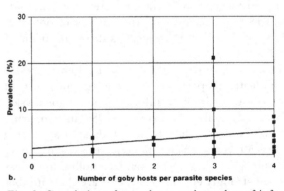

Fig. 3. Correlation of prevalence and number of infected goby (Gobiidae, Pisces) hosts at the sites of Lübeck Bight (a) and Kiel Bight (b), Germany. (From Zander, 1998 *a*).

positive regression in Dahmeshöved than in Blank Eck (Fig. 3), which corresponds to what was expected based on prevalence values (Zander & Kesting, 1996).

Long-term changes by eutrophication were studied in mud snails, *Hydrobia ventrosa*, from the Wismar Bight (Mecklenburg Bight) and the Strela Sound (Arkona Sea) over nearly 30 years (Reimer, 1995). In particular the digeneans, *Microphallus claviformis* and *Paramonostomum alveatum* but also *Cryptocotyle concavum*, clearly increased in prevalence during this time of increasing eutrophication. Other parasites like *Asymphylodora demeli* or *Maritrema subdolum* showed minor changes, or none.

A faunal impoverishment over a period of 15 years was also found in the Schlei Fjord, which is strongly polluted in its inner parts, where lower salinity

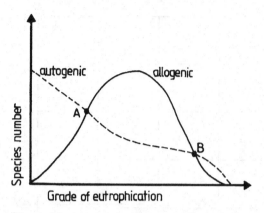

Fig. 4. Model of the relation of allogenic and autogenic parasite species numbers along the scale of increasing eutrophication. A and B = points of dominance alteration of the two categories. (From Zander, 1998a).

prevails (Kesting & Zander, 2000). The number of fish parasites decreased in Selk (2‰ salinity) from 8 to 4 species, in Haddeby (4‰) from 8 to 6 and in Missunde (6‰) from 11 to 6, whereas in the outer locality of Olpenitz (15‰) species numbers remained almost the same (10 or 11). Only during the 3 years from 1991 to 1994, did the numbers of parasites of snail, crustacean and fish hosts decrease in Missunde (from 18 to 8 species) with the number of potential host species among these groups dropping from 21 to 14.

The last example from the Schlei Fjord stresses the indicator properties of the parasite community of an ecosystem. Similar results attained Sulgostowska, Banaczyk & Grabda-Kabzubska (1987) in comparing parasite faunas of flatfish from Gdansk Bay during almost 50 years where the releaser was not only eutrophication but also chemical pollution. Clear signs of strongly eutrophic systems are: low number of parasite species, because less tolerant species vanish; predominance of generalists, because special hosts are extincted; predominance of planktonic parasites like tapeworms and hemiurid digeneans over benthic ones, because free water offers better oxygen supply than the bottom; predominance of parasites with direct development – like copepods or monogeneans – over those with complex cycles, because these are only dependent to one or few hosts (Overstreet & Howse, 1977; Kesting, 1996; Zander, 1998a). In oligotrophic waters, autogenic parasite species (see above) dominate. The more a water body is eutrophic, the more herbivores and carnivores profit from the increased plant growth and attract birds to rest and prey on them. The single hosts may be more heavily infected when the density increases, as consequence of better possibilities of parasite larvae to meet their hosts (Fig. 4). Finally, the number of allogenic parasite species decreases in extremely eutrophic waters because the environment becomes unattractive for birds, as a consequence of overall species impoverishment.

Regarding the patterns of eutrophication, parasites can profit from it in the first stage as indicated by increasing prevalences and intensities of hosts. In the second stage, oxygen deficiency influences hosts and parasites, at first on the bottom and thereafter also in the free water (Schlei Fjord). When a sufficient oxygen supply is guaranteed, as in very shallow habitats where exchange with the air is easy, higher levels of eutrophication may have an increased promoting effect not only on plants, but also on herbivores and parasites (Salzhaff, Orther Bight). This process differs from the effect of chemical pollutants which reduce parasite prevalence and intensity already with the start of their influence leading finally to lethality (Khan & Thulin, 1991). Only low doses of chemical pollutants can sometimes result in higher infection rates as a consequence of immunosuppression in the hosts.

CONCLUSIONS

The composition of parasite communities in the Baltic Sea is influenced by a complex of static and dynamic factors. The brackish condition with decreasing salinities from west to east means a strong stress to hosts as well as parasites, which excludes stenohaline marine and freshwater species. Beside some relic elements, the host species differ hardly from those of the North Sea or the adjacent freshwaters, respectively. Several non-native invertebrates which had immigrated from other continents were not infected by Baltic parasites whereas non-native parasites find native species as hosts without difficulty. These reactions may be evidence for the lack of co-evolution between the partners. The parasite communities of the Baltic Sea comprise especially euryhaline species of marine and freshwater origin. In contrast to hosts, several parasites are found outside the Baltic Sea only in other European brackish waters or have developed to endemic species.

Because brackish water is an extreme environment in which organisms tolerate lower or changing salinities, parasites have developed special adaptations which guarantee their successful existence in such an extreme environment. Therefore, some generalisations from effect of brackish water can be made (Zander 1998a): (1) Relaxation of host specificity. Parasites which infect only 1 host in marine or fresh water, infect several host species in brackish habitats. This is valid for many snails but also other hosts. (2) Exploitation of new hosts, e.g. *Crangon crangon* which is not infected at all in the adjacent North Sea. (3) Reduction or elongation of life cycles, as seen in several species of Microphallidae (Digenea) (Fig. 2). (4) Adaptation to genuine brackish-water species as hosts which is best explained by the rare co-evolution in the Baltic Sea of the digenean, *Cryptocotyle concavum*, with the gobiid fish, *Pomato-*

schistus microps, where the parasite captures a new microhabitat.

In summary, the structure of the parasite community in the Baltic Sea is the result of the tolerance of parasites to low salinities, of the distribution and the dispersion of their hosts, of the possibility to alter the life cycles and sometimes of the co-evolution with hosts. The results from the brackish Baltic Sea can be a model for other environments where extreme intensities of abiotic factors prevail.

ACKNOWLEDGEMENTS

We wish to thank Dr. Robert Poulin and two unknown reviewers for constructive criticism, Andy Godfrey, Dr. Jacob Parzefall and Dr. Robert Poulin for revising the English text.

REFERENCES

ARNTZ, W. E. (1972). Über das Auftreten des parasitischen Copepoden *Lemaeocera branchialis* in der Kieler Bucht und seine Bedeutung als biologische Markierung. *Archiv für Fischereiwissenschaften* **23**, 72–74.

BAUER, O. N. (1987). *Opredelitel parasitov presnovodnych Ryb Fauny SSSR*. Tom 3. Leningrad, Izdatelstvo Nauka, 583 pp. (In Russian)

BUCHER, C. (1998). Parasiten von Fischen der Kieler Förde. M.Sc. thesis, University of Kiel, 111 pp.

FAGERHOLM, H.-P. & BERLAND, B. (1988). Description of *Ascarophis arctica* Poljanski, 1952 (Nematoda, Cysticolidae) in Baltic Sea fishes. *Systematic Parasitology* **11**, 151–158.

GALAKTIONOV, K. V. (1985). Special features of the infection of the mollusk, *Littorina rudis* (Maton, 1797), with parthenitae of *Microphallus pygmaeus* (Levinsen, 1881) nec Odhner, 1905 and *M. piriformis* (Odhner, 1905) Galationov, 1980 (Trematoda: Microphallidae) from the White Sea. In *Parasitology and Pathology of Marine Organisms of the World Oceans* (ed. Hargis, W. J. Jr) p. 111, NOAA Technical Report NMFS 25, U.S. Department of commerce, Washington.

GIBSON, D. I. & KØIE, M. (1991). *Magnibursatus caudofilamentosa* (Reimer, 1971) n. comb. (Digenea: Derogenidae) from the stickleback *Gasterosteus aculeatus* L. in Danish waters: a zoogeographical anomaly? *Systematic Parasitology* **20**, 221–228.

GOLLASCH, S., STROHBACH, U., WINKLER, G. & ZANDER, C. D. (1996). Digene Parasiten der Nordseegarnele, *Crangon crangon* (L., 1758) (Decapoda, Crustacea) aus der westlichen Ostsee. *Seevögel* **17**, 3–4.

GOLLASCH, S. & ZANDER, C. D. (1995). Population dynamics and parasitation of planktonic and epibenthic crustaceans in the Baltic Schlei Fjord. *Helgoländer Meeresuntersuchungen* **49**, 759–770.

GRABDA, J. (1962). Pasozytnicze *Caligus lacustris* STP. et LÜTK. w Polsce. *Prace Moskiego Instytut Rybactwa w Gdyni* **11/A**, 275–286.

GRABDA, J. (1974). The dynamics of the nematode larvae, *Anisakis simplex* (Rud.) invasion in the south-western Baltic herring (*Clupea harengus* L.). *Acta Ichthyologica et Piscatoria* **4**, 3–21.

GRABDA, J. (1981). Parasitic fauna of garfish *Belone belone* (L.) from the Pomerian Bay (Southern Baltic). *Acta Ichthyologica et Piscatoria* **11**, 75–85.

HILDEBRAND, A. & SCHARBERTH, D. (1992). Die Parasitierung der Kleinfische des Salzhaffs (westliche Ostsee). Staatsexamensarbeit, Universität Rostock

HINE, P. M. & KENNEDY, C. R. (1974). Observations on the distribution, specifity and pathogenecity of the acanthocephalan *Pomphorhynchus laevis*. *Journal of Fish Biology* **6**, 521–535.

HOLMES, J. C. (1990). Helminth communities in marine fishes. In *Parasite Communities: Pattern and Processes* (ed. Esch, G. W., Bush, A. O. & Aho, J. M.), pp. 101–130. London, Chapman and Hall.

HOLMES, J. C. & PRICE, P. W. (1986). Communities of parasites. In *Community Biology: Pattern and Processes* (ed. Anderson, D. J. & Kikkawa, J.), pp. 187–213. Oxford, Blackwell.

JAMES, B. L. (1970). Host selection and Ecology of marine digenean larvae. Fourth European Biology Symposium (ed. CRISP, D. J.), pp 179–196. Cambridge University Press.

JUNGE, A. (1993). Parasiten der sich eingrabenden Muscheln aus der Wismarer Bucht. *Hausarbeit Universität Rostock*, 70 pp.

KESTING, V. (1996). Untersuchungen zur Parasitenfauna von Fischen und wirbellosen Zwischenwirten aus Uferbereichen der Ostseeförde Schlei. *Thesis Universität Hamburg*, 200 pp.

KESTING, V., GOLLASCH, S. & ZANDER, C. D. (1996). Parasite communities of the Schlei fjord (Baltic coast of northern Germany). *Helgoländer Meeresuntersuchungen* **50**, 477–496.

KESTING, V. & ZANDER, C. D. (2000). Alteration of the metazoan parasite fauna in the brackish Schlei fjord (Northern Germany, Baltic Sea). *International Review of Hydrobiology* **85**, 325–340.

KHAN, R. A. & THULIN, J. (1991). Influence of pollution on parasites of aquatic animals. *Advances in Parasitology* **30**, 201–238.

KÖHN, J. & GOSSELCK, F. (1989). Bestimmungsschlüssel der Malakostraken der Ostsee. *Mitteilungen aus dem Zoologischen Museum in Berlin* **65**, 3–114.

KØIE, M. (1993). Aspects of the life cycle and morphology of *Hysterothylacium aduncum* (Rudolph, 1802) (Nematoda, Ascarioidea, Anisakidae). *Canadian Journal of Zoology* **71**, 1289–1296.

KOZIKOWSKA, Z. (1957). Crustaces parasites des poissons de la Pologne. III. Resultats des exploration sur les poissons de la cote meridionale de la Mer Baltique. *Polish Archiv Hydrobiologia* **13**, 97–104.

KREFT, K.-A. (1991). Befalls- und Populationsdynamik ausgewählter digener Trematoden und ihrer Wirte in der Schlei. Thesis Universität Hamburg, 289 pp.

LAUCKNER, G. (1994). Parasiten als bestandsregulierender Faktor. In *Warnsignale aus dem Wattenmeer* (ed. Lozan, J. L., Rachor, E., Reise, K., v. Westernhagen, H. & Lenz, W.), pp. 144–149. Berlin, Blackwell.

LÜTHEN, K. (1989). Fischkrankheiten und Parasiten von Flunder, Scholle, Kliesche und Steinbutt aus den Küstengewässern der DDR. *Thesis Pädagogische Hochschule Güstrow*, 186 pp.

MARKOWSKI, S. (1936). Über die Trematodenfauna der baltischen Mollusken aus der Umgebung der

Halbinsel Hel. *Bulletin international de l'Académie Polonaise des Sciences et des Lettres (Ser. B, II)* **1936**, 285–317.

MARKOWSKI, S. (1938). Über die Helminthenfauna der Baltischen Aalmutter *Zoarces viviparus*. *Zoologica Poloniae* **1938**, 89–104.

MÖLLER, H. (1975). Parasitological investigations on the European eelpout (*Zoarces viviparus* L.) in a fjord in the western Baltic. *Berichte der deutschen wissenschaftlichen Kommission für Meeresforschung* **23**, 267–272.

MÖLLER, H. (1978). The effect of salinity and temperature on the development and survival of fish parasites. *Journal of Fish Biology* **12**, 311–323.

MORAVEC, F. (1994). *Parasitic Nematodes of Freshwater Fishes of Europe*. Dorbrecht, Boston. London, Kluwer Academic Publications

OVERSTREET, R. M. & HOWSE, H. D. (1977). Some parasites and diseases of estuarine fishes in polluted habitats of Missisippi. *Annals of the New York Academy of Sciences* **298**, 437–462.

PALM, H. W., KLIMPEL, S. & BUCHER, C. (1999). Checklist of metazoan fish parasites of German coastal waters. *Berichte des Instituts für Meereskunde der Universität Kiel* **307**, 148 pp.

PEKKARINEN, M. (1988). Gymnophallid trematodes parasites of the Baltic clam *Macoma balthica* (L). of the southwest coast of Finland. Thesis University Helsinki, 7–51.

PETRUSHEWSKIJ, G. K. (1957). The parasite fauna of clupeids from the Black Sea. *Izw. VSES. Nauchno-Istvestiya Vsesoyuznye Nauchno-Issledovatel'skiy Institut Ozernogo i Rechnogo Rybnogo Khozyaistva* **42**, 304–314.

REIMER, L. (1961). Die Stufen der Progenesis bei dem Fischtrematoden *Bunocotyle cingulata* Odhner, 1928. *Wiadomosci Parazytologiczne* **7**, 843–849.

REIMER, L. W. (1963). Zur Verbreitung der Adulti und Larvenstadien der Familie Microphallidae VIANA, 1924 (Trematoda, Digenea) in der mittleren Ostsee. *Zeitschrift für Parasitenkunde* **23**, 253–273.

REIMER, L. W. (1969). Helminthen von Kormoranen von Brutkolonien der Deutschen Demokratischen Republik. *Wissenschaftliche Zeitschrift der Universität Greifswald* **18**, 129–135.

REIMER, L. W. (1970). Digene Trematoden und Cestoden der Ostseefische als natürliche Fischmarken. *Parasitologische Schriftenreihe* **20**, 1–144.

REIMER, L. W. (1971). Neue Cercarien der Ostsee mit einer Diskussion ihrer möglichen Zuordnung und einem Bestimmungsschlüssel. *Parasitologische Schriftenreihe* **21**, 125–149.

REIMER, L. W. (1973). Das Auftreten eines Fischtrematoden der Gattung *Asymphylodora* LOOSS, 1899, bei *Nereis diversicolor* O. F. MÜLLER als Beispiel für einen Alternativzyklus. *Zoologischer Anzeiger* **191**, 187–196.

REIMER, L. W. (1995). Parasites especially of piscean hosts as indicators of the eutrophication in the Baltic Sea. *Applied Parasitology* **36**, 124–135.

REIMER, L. W. (2002). Parasitische Würmer (Helminthen) von Seevögeln der Ostseeküste. *Seevögel* (in press).

REIMER, L. W., HILDEBRAND, A., SCHARBERTH, D. & WALTER, U. (1994). *Anguillicola crassus* in the Baltic Sea: field data supporting transmission in brackish waters. *Diseases of Aquatic Organisms* **18**, 77–79.

REIMER, L. W., HILDEBRAND, A., SCHARBERTH, D. & WALTER, U. (1996). Trematodes of the brackisch waters of the Baltic Sea and their distribution together with that of related species in other European areas. *Applied Parasitology* **37**, 177–185.

REIMER, L. W. & POHL, M. (1989). Untersuchungen der Parasitenfauna der Miesmuschel *Mytilus edulis* L. von Standorten der Ostseeküste der DDR. *Wissenschaftliche Zeitschrift der Pädagogischen Hochschule Güstrow* **1989**, 155–161.

REIMER, L. W. & WALTER, U. (1993). Zur Parasitierung von *Gadus morhua* in der südlichen Ostsee. *Applied Parasitology* **34**, 181–186.

REIMER, L. W. & WALTER, U. (1998). Zur Parasitierung der Aalmutter *Zoarces viviparus* (L.) in der Wismar-Bucht. *Mitteilungen der Landesforschungsanstalt für Fischerei in Mecklenburg-Vorpommem* **17**, 114–120.

REIMER, L. W. & WALTER, U. (2000). Veränderungen im Parasitenbefall der Aale der südlichen Ostseeküste in den 90er Jahren. *Mitteilungen der Landesforschungsanstalt für Fischerei in Mecklenburg-Vorpommem* **22**, 160–181.

REMANE, A. (1958). Ökologie des Brackwassers. In *Die Biologie des Brackwassers* (ed. Remane, A. & Schlieper, C.). *Die Binnengewässer* **12**, 1–216.

RENAUD, F. & GABRION, G. (1984). Polymorphisme enzymatique de populations du groupe *Bothriocephalus scorpii* (Müller, 1776) (Cestoda, Pseudophyllidea). Etude des parasites de divers Téléostéens de côtes du Finistère. *Bulletin de la Societé Française de Parasitologie* **2**, 95–98.

SCHWARZ, S. (1960). Zur Crustaceenfauna der Brackwassergebiete Rügens und des Darss. *Hydrobiologia* **16**, 239–300.

SKROBLIES, M. (1998). Populationsdynamische und ökoparasitologische Untersuchungen an benthischen Crustacea der Orther Bucht (Fehmarn, südwestliche Ostsee. Diplom thesis Universität Hamburg, 116 pp.

STROHBACH, U. (1999). Vergleichende Untersuchungen zur Populationsdynamik und Parasitenfauna ausgewählter benthischer und planktischer Crustaceen sowie Gastropoden im Bereich der Kieler – und Lübecker Bucht (SW-Ostsee). Thesis Universität Hamburg, 196 pp.

STUNKARD, H. W. (1930). Some larval trematodes from the coast in the region of Roscoff, Finistère. *Parasitology* **24**, 321–343.

SULGOSTOWSKA, T., BANACZYK, G. & GRABDA-KAZUBSKA, B. (1987). Helminth fauna of flatfish (Pleuronectiformes) from Gdansk Bay and adjacent areas (south-east Baltic). *Acta Parasitologica Polonica* **31**, 231–240.

SULGOSTOWSKA, T., JERZEWSKA, B. & WICIKOSKI, J. (1990). Parasite fauna of *Myoxocephalus scorpius* L. and *Zoarces viviparus* (L.) from environs of Hel (south-east Baltic) and seasonal occurrence of parasites. *Acta Parasitologica Polonica* **35**, 143–148.

VAES F. (1974) A new type of trematode life-cycle: an invertebrate as final host. III. *International Congress of Parasitology, München, Proceedings* **1**, 351.

VISMANIS, K., PETRINA, Z., EGLITE, R., VOLKOVA, A. & SABLE, B. (1980). Materialien über Parasitenfauna einiger Objekte des Fischfangs im Rigaer Meerbusen (in Russian). *Latvijas PSR Bezmugurkaulnieka fauna un Ecologija (Riga)* **1980**, 5–12.

YAMAGUTI, S. (1963). Parasitic Copepoda and Branchiura of fishes. New York, London, Sydney. Interscience publisher.

ZANDER, C. D. (1991). Akkumulation von Helminthen-Parasiten in Aalmuttern *Zoarces viviparus* (L.) (Teleostei) der SW Ostsee. *Seevögel* **12**, 70–73.

ZANDER, C. D. (1993). The biological indication of parasite life cycles and communities from the Lübeck Bight, SW Baltic Sea. *Zeitschrift für angewandte Zoologie* **79**, 377–389.

ZANDER, C. D. (1998*a*). Ecology of host parasite relationships in the Baltic Sea. *Naturwissenschaften* **85**, 426–436.

ZANDER, C. D. (1998*b*). Parasitengemeinschaften bei Grundeln (Gobiidae, Teleostei) der südwestlichen Ostsee. *Verhandlungen der Gesellschaft für Ichthyologie* **1**, 241–252.

ZANDER, C. D. (2001). The guild as a concept and a means in ecological parasitology. *Parasitology Research* **87**, 484–488.

ZANDER, C. D. & DÖRING, W. (1989). The role of gobies (Gobiidae, Teleostei) in the food web of shallow habitats of the Baltic Sea. *Proceedings of the 21st European Marine Biology Symposium, Gdansk* **1986**, 499–508.

ZANDER, C. D. & KESTING, V. (1996). The indicator properties of parasite communities of gobies (Teleostei, Gobiidae) from Kiel and Lübeck Bight. *Applied Parasitology* **37**, 186–204.

ZANDER, C. D., KOLLRA, H.-G., ANTHOLZ, B., MEYER, W. & WESTPHAL, D. (1984). Small-sized euryhaline fish as intermediate hosts of the digenetic trematode *Cryptocotyle concavum*. *Helgoländer Meeresuntersuchungen* **37**, 433–443.

ZANDER, C. D., REIMER, L. W. & BARZ, K. (1999). Parasite communities of the Salzhaff (Northwest Mecklenburg, Baltic Sea). I. Structure and dynamics of communities of littoral fish, especially small-sized fish. *Parasitology Research* **85**, 356–372.

ZANDER, C. D., REIMER, L. W., BARZ, K., DIETEL, G. & STROHBACH, U. (2000). Parasite communities of the Salzhaff (Northwest Mecklenburg, Baltic Sea). II. Guild communities, with special regard to snails, benthic crustaceans, and small-sized fish. *Parasitology Research* **86**, 359–372.

ZANDER, C. D., STROHBACH, U. & GROENEWOLD, S. (1993). The importance of gobies (Gobiidae, Teleostei) as hosts and transmitters of parasites in the SW Baltic. *Helgoländer Meeresuntersuchungen* **47**, 81–111.

ZANDER, C. D. & WESTPHAL, D. (1991). Kleinfischparasiten der Ostseeförde Schlei und ihre Einbindung in die Nahrungskette. *Seevögel* **12**, 4–8.

Parasites and marine invasions

M. E. TORCHIN[1]*, K. D. LAFFERTY[2] and A. M. KURIS[1]

[1] *Marine Science Institute and Department of Ecology, Evolution and Marine Biology, University of California, Santa Barbara, CA 93106, USA*
[2] *U.S. Geological Survey, Western Ecological Research Center, c/o Marine Science Institute, University of California, Santa Barbara, CA 93106, USA*

SUMMARY

Introduced marine species are a major environmental and economic problem. The rate of these biological invasions has substantially increased in recent years due to the globalization of the world's economies. The damage caused by invasive species is often a result of the higher densities and larger sizes they attain compared to where they are native. A prominent hypothesis explaining the success of introduced species is that they are relatively free of the effects of natural enemies. Most notably, they may encounter fewer parasites in their introduced range compared to their native range. Parasites are ubiquitous and pervasive in marine systems, yet their role in marine invasions is relatively unexplored. Although data on parasites of marine organisms exist, the extent to which parasites can mediate marine invasions, or the extent to which invasive parasites and pathogens are responsible for infecting or potentially decimating native marine species have not been examined. In this review, we present a theoretical framework to model invasion success and examine the evidence for a relationship between parasite presence and the success of introduced marine species. For this, we compare the prevalence and species richness of parasites in several introduced populations of marine species with populations where they are native. We also discuss the potential impacts of introduced marine parasites on native ecosystems.

Key words: Parasite, pathogen, disease, introduced species, biological invasion.

INTRODUCTION

Globalization of the world's economies has substantially increased the rate of biological invasions (Cohen & Carlton, 1998; Ewel *et al.* 1999) because most harmful exotic species arrive as hitchhikers on the vectors of international trade (Baskin, 1996). Advances in world-wide shipping and transportation have accelerated these processes, particularly in marine systems (Carlton, 1987). For example, ships regularly transit between biotic provinces and subsequently release up to hundreds of thousands of gallons of plankton-laden ballast water into new environments (Carlton, 1987). Carlton & Geller (1993) analyzed the ballast water of 159 Japanese cargo ships entering port in Coos Bay, Oregon, and found a minimum of 367 distinctly identifiable taxa. They estimated that, on any given day, ocean-going vessels transport over 3000 species. As international trade expands, so will the unintentional introduction of non-native species.

Introduced pests are a major threat to global biodiversity, ranked second only to habitat loss (Vitousek, 1990; Wilcove *et al.* 1998). The ecological and economic impacts of an introduced species are a direct result of its ecological success. Introduced species often attain unusually high population densities compared to both ecologically similar native species and to conspecific populations in their native region (for terrestrial examples see Elton, 1958; DeBach, 1974 – cottony cushion scale; Bird & Elgee, 1957 – European spruce sawfly; for aquatic and marine examples see Leech, 1992 – zebra mussel; Buttermore, Turner & Morrice, 1994 – northern Pacific seastar, Carlton *et al.* 1990 – Asian clam). In addition to increased population densities, introduced species often attain unusually large body sizes (Blaustein, Kuris & Alio, 1983; Crawley, 1987; Blossey & Notzhold, 1995; Torchin, Lafferty & Kuris, 2001). This suggests that introduced species are growing faster or surviving longer or both, compared to where they are native. What are the reasons for increased performance of introduced species relative to conspecifics in their native range? Is this correlated with invasion success? There are three main hypotheses for the success of introduced species relative to where they are native. The new habitat has (1) better environmental attributes, such as increased resources (Dobson, 1988), (2) fewer or poorer competitors (Crawley, 1986; Byers, 2000; Callaway & Aschehoug, 2000) and (3) a paucity of natural enemies, such as predators and parasites (Bird & Elgee, 1957; Elton, 1958; Baker & Stebbins, 1965; Huffaker, Messenger & DeBach, 1971; DeBach, 1974; Lawton & Brown, 1986; Dobson, 1988; Dobson & May, 1986; Lampo & Bayliss, 1996*a, b*; Meyer, 1996; Schoener & Spiller, 1995; Lafferty & Kuris, 1996; Torchin *et al.* 2001). Both (2) and (3) above may be characteristic of

* Corresponding author: Mark E. Torchin, Marine Science Institute, University of California, Santa Barbara, CA 93106, USA. Tel: 805-893-3998. Fax: 805-893-8062. E-mail: torchin@lifesci.ucsb.edu

disturbed environments (Lafferty, 1997; Naeem *et al.* 2000). Disturbance can interact positively or negatively with parasites (Lafferty, 1997; Dove, 1998) and, thus, parasites and introduced species may interact with environmental disturbances in a number of ways. Although these hypotheses are sometimes treated as alternatives (Blossey & Notzhold, 1995), we stress that they are not mutually exclusive and can act synergistically (for example, Settle & Wilson, 1990).

Both the physical environment and biotic interactions will determine the success of an introduction. For intentional introductions, humans may knowingly choose suitable environments where species are introduced to fill an apparently open ecological niche (e.g. trout in mountain streams). In contrast, for unintentional introductions, new regions should not, on average, be better than the native range with respect to resources, competitors and generalist predators. However, on average, those invasions that are successful should be in areas where all aspects of the environment are suitable and some are favourable. In particular, new regions should almost always contain fewer specific, co-evolved natural enemies compared to the native range. While plants and animals are being introduced at an increasing rate (Cohen & Carlton, 1998), their parasites may never reach the new region.

Many introduced species appear to lack most or all of their native, presumably coevolved, natural enemies. Founding populations of introduced vertebrates in terrestrial systems often carry fewer or limited subsets of the parasites found in native regions (Dobson, 1988; Freeland, 1993). Likewise, introduced plants that arrive as seeds often lack natural enemies (Elton, 1958). For marine invasions, larval stages introduced via ballast water lack parasites that can infect adult stages (Lafferty & Kuris, 1996). If a species does not bring its natural enemies with it, it should experience a release in its new geographic range. This release will be of a magnitude proportional to the ecological importance of the natural enemies left behind. By comparing native and introduced populations (Lampo & Bayliss, 1996 *a*, *b*; Calvo-Ugarteburu & McQuaid, 1998 *a*; Torchin *et al.* 2001) and examining interactions between introduced species and ecologically similar native species (Calvo-Ugarteburu & McQuaid, 1998 *b*; Byers, 2000; Callaway & Aschehoug, 2000; Torchin, Byers & Huspeni, unpublished observations), it is possible to evaluate these hypotheses. For example, in Australia, introduced cane toads achieve higher population densities compared to native South American populations. In contrast to the parasite-free introduced populations, native populations may be controlled by ticks which reduce frog biomass (Lampo & Bayliss, 1996 *a*, *b*). Although evidence for this exists for terrestrial species, there have been few such comparisons made

for marine systems (but see Calvo-Ugarteburu & McQuaid, 1998 *a*; Torchin *et al.* 2001).

We focus on parasites because they are ubiquitous and have the potential to affect host growth, reproduction, and survivorship. Both empirical and theoretical evidence suggest that parasites can reduce host density and potentially control host populations (Anderson & May, 1986; Scott, 1987; Kuris & Lafferty, 1992). Parasites can interact directly (metabolically reducing host growth, reproduction, and survivorship) or indirectly (interacting with predation or competition or both) to affect community structure and, thus, are likely mediators of invasion success. Although invasion success is often attributed to an escape from natural enemies, there are only a few studies where this has been systematically examined and the evidence from these is consistent with this hypothesis (Lampo & Bayliss, 1996 *a*, *b*; Torchin *et al.* 2001).

To assess the role of parasites in marine invasions, we review reasons why introduced species often leave their natural enemies behind. We define three types of invasion success and explore whether parasites can facilitate the success or failure of introduced marine species. For this, we use a theoretical framework developed to model invasion success. We then compare published studies of introduced species in both their native and introduced geographic ranges to examine the extent to which lack of parasites is a general phenomenon in marine invasions. We also review cases in which infected hosts invade new regions and explore the potential for introduced marine parasites and pathogens to impact native ecosystems. Finally, we discuss the importance of parasites as an indirect impact of invaders on food webs.

RELEASE FROM NATURAL ENEMIES

With Dogiel (1948), Bauer (1991), Kennedy (1994) and Dove (1998, 2000), we recognize that during the process of species invasions in a non-native region there will generally be a reduction in the parasite community and perhaps acquisition of native parasites. There are several reasons why introduced species may invade without their native parasites. Firstly, invasions often occur as a result of the introduction of relatively uninfected stages such as larvae or seeds. Although these stages sometimes harbour infectious agents and parasites, they are typically lost post-recruitment and do not affect other life-history stages (Polyanski, 1961; Rigby & Dufour, 1996; Cribb *et al.* 2000). In marine systems, invading species that arrived in ballast water, generally as larvae (Carlton, 1987), are rarely parasitized (Lafferty & Kuris, 1996). Secondly, even if adults are the source of the invasion, colonization by a low number of individuals reduces the probability of introducing parasitized hosts. Thirdly, even if parasites do invade, host population bottlenecks after introduction may break parasite transmission. A fundamental principle of epidemiology is that the spread of a directly-transmitted infectious agent through a population increases with the density of susceptible and infectious hosts (Anderson & May, 1986). Empirical comparative studies support the prediction that intensity and prevalence tend to increase with host population density (Anderson, 1982; Anderson & May, 1986; Arneberg *et al.* 1998). Epidemiological models (McKendrick, 1940; Kermack & McKendrick, 1927; Bailey, 1957) indicate that there is a host-threshold density below which a parasite cannot persist in a host population. Therefore, low-density populations should be less subject to infection by a host–specific infectious disease agent. Observations of the epidemiology of morbillivirus (e.g. measles) of humans (Black, 1966) and cattle (Aune & Schladweiler, 1992) support this prediction, as do the experimental studies of Stiven (1964, 1968) and the field manipulation of host density by Culver & Kuris (2000). This suggests that in a recently introduced host–parasite system, host density will often be too low for the parasite to establish a self-sustaining population. This will be particularly true for inefficiently transmitted (most microbial) diseases. Finally, parasites often have complex life-cycles, which necessitate two or more host species. If suitable hosts for all life cycle stages are not present, establishment will not occur. For example, *Schistosoma mansoni*, introduced with infected humans, established in areas of Latin America where an appropriate first intermediate host snail, *Biomphalaria glabrata*, was present (Rollinson & Southgate, 1987). Although *S. haematobium* was also probably introduced, no suitable first intermediate host snails were present, and, thus, *S. haematobium* never established. Similarly, avian malaria was repeatedly introduced to the Hawaiian Islands, but it was not until the introduction of its mosquito vector that transmission and eventual establishment occurred (Warner, 1969; Van Riper *et al.* 1986).

INVASION SUCCESS

Both the physical environment and biotic interactions will determine the success of an introduction. Using biotic interactions as well as life-history and demographic parameters of an invader, we developed a theoretical framework to model invasion success (Fig. 1). This model illustrates how resources, competition, predation and parasitism can affect invasion success and, perhaps more importantly, how they can interact to do so. We define three types of invasion success for which this model can be used: (1) absolute success (AS) in which the equilibrial

biomass of the invader is greater than zero, (2) relative conspecific success (RCS) in which the equilibrial biomass of the invader in its introduced range is greater than the biomass of the invader in its native range (controlling for area) and (3) relative interspecific success (RIS) in which the equilibrial biomass of the invader is greater than the biomass of native competitors in the introduced range (controlling for area). In this model, invasion success is quantified in terms of the biomass of the invader. Biomass is determined by both body size and population density. Growth and survival both positively affect body size while survival and natality both positively affect population density.

Direct effects are those in which a change in species A directly affects species B. Indirect effects are those, which are mediated through other species, for example if a change in species C affects species A, thereby affecting species B (Wootton, 1994). Direct effects are illustrated by arrows in Fig. 1 (and corresponding text). For example, invader growth, natality and survival directly increase with resources. Parasites can directly reduce invader growth and natality. Predators and some parasites directly reduce invader survival. Some parasites can directly negatively affect an invader's predators due to pathological effects of trophically-transmitted parasites on predators. An invader's predators will also directly positively affect some parasites by serving as hosts. Predators can directly reduce the abundance of some competitors. Some parasites can also directly reduce the abundance of an invader's competitors. Indirect effects are generally illustrated by the combination of two or more arrows (direct effects) operating in tandem. For example, resources are shared with competitors, so competitors indirectly decrease invader growth. Parasites and predators can also have an indirect positive effect on growth, survival and natality if they impact competitors, thereby releasing resources. Parasites can indirectly affect predation on the invader if they reduce predator density via pathology or increase predation efficiency by the effects of parasite-increased trophic transmission (Lafferty, 1999). Feedback in the model arises when the biomass of the invader becomes large enough to negatively impact resources and positively affect parasites and predators if the invader becomes a host or prey. Fig. 1 illustrates the potential importance and interconnectedness of resources, competition, predation and parasitism on invasion success. We point out that parasites, in particular, have the most linkages to life-history parameters either directly or through indirect routes via other biotic interactions. This feature contributes to their role in the stabilization of population dynamics, a role that is increasingly recognized (Price *et al.* 1986; Dobson, 1988; Freeland, 1993). Lacking parasites, introduced species may often perturb a system in unexpectedly powerful ways.

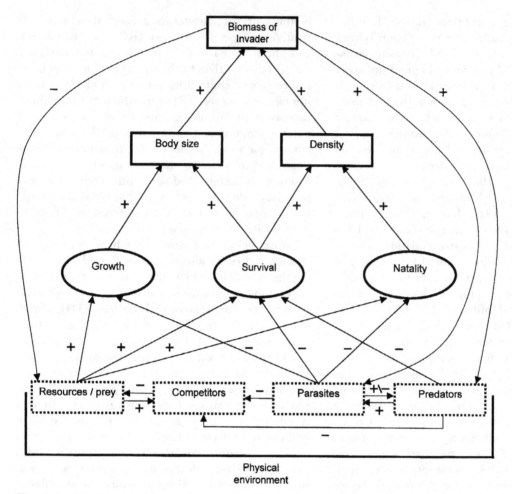

Fig. 1. In this theoretical framework, we consider invasion success to be proportional to invader biomass. This model represents a hierarchy of demographic parameters (solid boxes), life-history parameters (ovals) and biotic interactions (dashed boxes) of the invader. (+) indicates a positive effect, (−) indicates a negative effect and (+/−) indicates that the effect may be in either direction. Note that all biotic interactions are influenced by the physical environment which includes disturbances.

PARASITES IN NATIVE AND INTRODUCED POPULATIONS

To examine the extent to which a release from parasites is a general phenomenon in marine invasions, we compared parasitological studies of invasive species (Table 1). We limited our comparison to a selection of introduced species for which there were readily available studies providing ecological data in both the native and introduced ranges of the invader. We excluded species under culture from our comparison because these species are introduced intentionally and are often medically treated for parasites or have their parasites intentionally manipulated. Cultured species are also typically grown at unnaturally high densities which promotes disease transmission.

Table 1 shows that in native regions there are twice as many studies of parasites and about four times as many hosts examined for parasites from native populations compared to introduced populations. This is a potential bias to detect more

species of parasites in the native ranges. However, the large number of hosts examined in both regions makes it unlikely that rarefaction is the cause of detection of fewer species of parasites among the introduced populations. The species richness of parasites found in native regions (mean = 6) is on average three times greater than in introduced regions (mean = 2). The average prevalence of all parasites in the introduced range (30%) is over twice that in the native range (14%). Unlike species richness, prevalence is generally independent of sample size assuming that samples without infected individuals are included (Gregory & Blackburn, 1991). Other general patterns are that species presumably introduced via ballast water (*Hemigrapsus sanguineus*, *Mnemiopsis leidyi*, *Asterias amurensis*) tend to have the fewest parasites in the introduced range. The European green crab, *Carcinus maenas*, also has fewer parasites in regions where it was presumably introduced via ballast water (West Coast of North America, South Africa, Australia and Tasmania). On the East Coast of

Table 1. Comparison of parasites reported from introduced marine species in both their introduced and native ranges. *Invader* is the introduced host species. *Number of studies* is the number of useable references found, *Number of individuals* is the total number of parasite species and *Prevalence* is the average prevalence of all parasite species across all studies examined. n is data for the native range and i is data for the introduced range. Grand total (sum) is the sum of the values for all host species. Grand total (avg) is the average of values for all host species. * *Poecilia latipinna* is predominantly a brackish water fish

Invader	Parasite group	Number of studies		Number of individuals		Species richness		Prevalence		References
		n	i	n	i	n	i	n	i	
Cancer novaezelandiae (Pie-crust crab)	Trematoda	1	1	37	50	1	0	17	0	1
Carcinus maenas (Green crab)	Fecampida	1	0	534	—	2	—	—	—	2
	Trematoda	2	3	372	725	1	1	81	25	3, 4, 5, 6
	Cestoda	1	1	372	606	1	2	10	13	1, 6
	Nemertea	3	1	554	1285	1	1	57	26	6, 7, 8
	Acanthocephala	2	2	2311	725	1	1	23	2	4, 6, 9
	Nematoda	2	1	372	606	1	1	—	0	6, 10
	Copepoda	1	1	14	—	1	1	29	0	10, 11
	Rhizocephala	5	1	29437	1442	1	0	21	0	6, 13, 14, 15, 16
	Isopoda	3	1	2401	606	1	0	4	0	6, 13, 14, 17
Total (sum)‡						10	7	225	66	
Total (avg)						1	1	32	9	
Hemigrapsus sanguineus (Japanese shore crab)	Trematoda	1	2	181	1000	1	0	55	0	6, 18
	Nematoda	1	2	181	1000	0	1	0	0	6, 18
	Rhizocephala	2	2	181	1000	1	0	17	0	6, 18, 19
	Isopoda	1	2	181	1000	0	0	0	0	6, 18
	Nemertea	0	2	—	1000	0	0	—	0	6, 18
Total (sum)‡						2	1	72	0	
Total (avg)						0	0	18	0	
Batillaria attramentaria (Japanese mud snail)	Trematoda	3	1	1289	955	7	1	24	24	20, 21, 22, 23, 24†
Ilyanassa obsoleta (Atlantic mud snail)	Trematoda	7	1	23631	3852	11	5	37	31	25, 26, 27, 28, 29, 30, 31
Littorina littorea (Periwinkle snail)	Trematoda	8	3	7943	1169	12	3	37	8	31, 32, 33, 34, 35, 36, 37, 38, 39, 40
Littorina saxatilis (Periwinkle snail)	Trematoda	6	2	6657	2453	19	8	36	26	33, 36, 41, 42, 43, 44, 45, 46
Asterias amurensis (Pacific seastar)	Ciliophora	3	1	1028	2000	1	0	59	0	47, 48, 49, 50
	Copepoda	2	0	381	—	1	—	7	—	47, 48
Total (sum)‡						2	0	66	0	
Total (avg)						1	0	33	0	

Taxon								References
Mnemiopsis leidyi (Atlantic Ctenophore)								
Cnidaria	1	0	100	—	1	—	—	51
Trematoda	1	0	—	—	1	1	—	52
Nematoda	0	1	—	1	—	1	1	53
Amphipoda	1	0	—	1	1	—	1	54
Total (sum)‡					3	1	1	
Total (avg)					1	1	1	
*Poecilia latipinna** (Sailfin molly)								
Monogenea	1	0	—	—	1	—	—	55
Trematoda	2	1	60	40	15	100	85	55, 56
Nematoda	2	1	60	40	1	2	0	55, 56
Copepoda	2	1	60	40	1	28	0	55, 56
Isopoda	2	1	60	40	1	—	0	55, 56
Total (sum)‡					19	130	85	
Total (avg)					6	65	34	
Grand total (sum)‡	85	27	77863	21635		634	241	
Grand total (avg)	6	2	3244	941		30	13	

References: (1) Kuris & Gurney, 1997; (2) Kuris, Torchin & Lafferty, unpublished; (3) Stunkard, 1956; (4) Brattley et al. 1985; (5) Castilho & Barandela, 1990; (6) Torchin et al. 2001; (7) Comely & Ansell, 1989; (8) Torchin et al. 1996; (9) Thompson, 1985; (10) Plotz, 1982; (11) Gallien & Bloch, 1936; (12) Johnson, 1957; (13) Bourdon, 1963; (14) Rasmussen, 1973; (15) Minchin, 1997; (16) Mathieson et al. 1998; (17) Bourdon, 1964; (18) McDermott, 1998; (19) Yamaguchi et al. 1994; (20) Shimura & Ito, 1980; (21) Rybakov & Lukomskaya, 1988; (22) Harada & Suguri, 1989; (23) Torchin et al. unpublished; (24) McDermott, 1996†; (25) Grodhaus & Keh, 1958; (26) Stambaugh & McDermott, 1969; (27) Curtis & Hubbard, 1990; (28) Curtis, 1997; (29) Curtis & Tanner, 1999; (30) McCurdy, Boates & Forbes, 2000; (31) Pechenick, Fried & Simpkins, 2001; (32) Hoff, 1941; (33) James, 1968; (34) Robson & Williams, 1970; (35) Williams & Ellis, 1975; (36) Pohley, 1976; (37) Hughes & Answer, 1982; (38) Mathews, Montgomery & Hanna, 1985; (39) Laukner, 1987; (40) Evans, Irwin & Fitzpatrick, 1997; (41) James, 1965; (42) Threlfall & Goudie, 1977; (43) Irwin, 1983; (44) Newell, 1986; (45) Bustness & Galaktionov, 1999; (46) McCarthy, 2000; (47) Kuris, Lafferty and Grygier, 1996; (48) Goggin & Bauland, 1997; (49) Byrne et al. 1997; (50) Goggin, 1998; (51) Bumann & Puls, 1996; (52) Martorelli, 1996; (53) Gayerskaya & Mordvinova, 1994; (54) Cahoon, Tronzo & Howe, 1986; (55) Hoffman, 1999; (55) Torchin, unpublished data. †We did not include data from McDermott (1996) because repeated assessments of the parasite fauna of introduced populations of B. attramentaria are inconsistent with his findings (Torchin et al. unpublished). ‡We have included the summed prevalence of all parasite species for a given host (Total (sum)) as an indicator of the cumulative impact of parasites on a host species.

North America, where green crabs were introduced over 200 years ago, and not likely via ballast water, introduced populations are more frequently parasitized compared to other introduced populations (Torchin *et al.* 2001.). This may be due to the introduction of adult crabs and the longer residence time of crabs on the East Coast of North America, which could provide a greater opportunity for native parasites to colonize the invader. Further, the East Coast of North America is faunistically quite similar to the Atlantic Coast of Europe and the parasites recovered from the East Coast populations are either conspecific with or closely similar to related European species (Stunkard, 1956; Brattey *et al.* 1985; Thompson, 1985). Although the very extensive geographic range of introduced green crabs (4 regions, 3 continents) compared to its native region (Europe) increases the overall likelihood of introduced crabs being colonized by parasites from those regions, the introduced populations' prevalences are generally lower for the same types of parasites where the green crab is native. Certain types of parasites (parasitic castrators, parasitoids), although locally prevalent in Europe, have never been documented from introduced regions.

Populations of snails introduced as biological contaminants with shellfish (oysters) imported for aquaculture (*Batillaria attramentaria* and *Ilyanassa obsoleta*) or those intentionally introduced for food (*Littorina littorea* and *L. saxatillis*) typically harbour a subset of the parasite species present in their native range. This suggests that some infected snails were introduced and some of their native parasites became established in the introduced regions. This is a relatively frequent occurrence despite the complex, multihost life cycles of the trematodes that parasitize them. Host specificity, while typically high for the snail first intermediate hosts, is relatively broad for second intermediate hosts and definitive hosts.

Although the rate of marine invasions is increasing dramatically (Zibrowius, 1991; Furlani, 1996; Ruiz *et al.* 1997; Cohen & Carlton, 1998), there are relatively few studies of the parasites of these invaders. A cautious interpretation of the data presented in Table 1 suggests that introduced populations of marine species are less affected by parasites compared to populations in their native range. However, more studies on the parasites of introduced species, both from native and introduced populations, are necessary to better evaluate the role of parasites in the invasion success of exotic marine species. Our preliminary examination of these species suggests a more comprehensive study of randomly-selected introduced species across a broader range of taxa and different environments might reveal important general patterns pertaining to how host specificity, lifecycle stages, trophic categories and methods of introduction relate to the success of exotic species.

INTRODUCED PARASITES

If infected hosts invade a new locale and their parasites become established, these invasive parasites may impact native species if they can recruit to novel hosts. Table 2 illustrates parasite introductions in marine systems. The rhizocephalan barnacle, *Loxothylacus panopaei*, is native to the Gulf of Mexico and southern Florida. It was first discovered in Chesapeake Bay in 1964, presumably introduced with infected mud crabs associated with oysters transplanted from the Gulf of Mexico (Van Engel *et al.* 1965). It now parasitizes three crab species in its introduced range, including two which only appear to be infected within the introduced range (Hines, Alvarez & Reed, 1997). Another rhizocephalan barnacle, *Heterosaccus dollfusi*, followed its portunid host crab, *Charybdis longicollis*, from the Red Sea through the Suez Canal to the Mediterranean Sea. However, while *C. longicollis* invaded the Mediterranean before 1954 and is now well established, its parasite has only recently arrived (Galil & Lützen, 1995; Galil & Innocenti, 1999). This parasite has not been recovered from any species of native crab, nor from other introduced species of portunids, including *C. hellerii*, also native to the Red Sea. Other reports of rhizocephalans introduced with their hosts are anecdotal and lack confirmation (e.g. Boschma, 1972, Kinzelbach, 1965).

Parasitic copepods that infect shellfish have been widely introduced with the transport and culture of bivalves. *Mytilicola orientalis* and *Myicola ostrae* are both parasitic copepods of the Pacific oyster, *Crassostrea gigas*, in Asia, where they are native. *Mytilicola orientalis* has been accidentally introduced to Europe and the Pacific Coast of North America, while *Myicola* has only been reported from Europe (His, 1997; Stock, 1993; Holmes & Minchin, 1995; Minchin, 1996). They were likely introduced with infected oysters imported for culture. Both species infect native bivalves and *M. orientalis* is considered a serious pest (Holmes & Minchin, 1995). *Mytilicola intestinalis*, which was presumably introduced to northern Europe from blue mussels originating in the Mediterranean Sea, may have been transported in its host, *Mytillus galloprovincialis*, on the hulls of ships (Minchin, 1996). In the early 1950s an epidemic of *M. intestinalis* caused considerable damage to mussel fisheries and infections spread to other native bivalve species in the Netherlands (Stock, 1993).

Although exotic monogeneans have been reported more commonly from freshwater fish species than from marine fishes, we briefly review a few examples of monogeneans introduced with catadromous and anadromous fish species. These monogeneans are typically restricted to the freshwater portions of their hosts's life cycles. *Gyrodactylus salaris*, a monogenean of Atlantic salmon, has been introduced to,

Table 2. Introduced parasites. Asterisks indicate possible introductions

Parasite taxon	Parasite	Introduced region	Native region	Method of introduction	Impacts	References
Rhizocephala	*Loxothylacus panopaei*	ANA	GOM	Aquaculture	Natives	1, 2
	Heterosaccus dolfusi	MED	RED	Migrant	Unknown	3
Copepoda	*Mytilicola orientalis*	EUR, PNA	ASI	Aquaculture	Aquaculture/Natives	4, 5, 6, 7
	Mytilicola intestinalis	N. EUR	MED	Fouling	Fisheries	8, 9
	Myicola ostrae	EUR	ASI	Aquaculture	Aquaculture/Natives	8
Monogenae	*Gyrodactylus anguillae*	AUS, ASI, ANA	EUR?	Aquaculture	Aquaculture/Natives	10, 11
	Pseudodactylogyrus anguillae	EUR, ANA	ASI	Aquaculture	Aquaculture/Natives	12
	Pseudodactylogyrus bini	EUR, ANA	ASI	Aquaculture	Aquaculture/Natives	12
	Gyrodactylus salaries	NOR	—	Stocking	Aquaculture/Natives	13, 14
	*Neobenedenia melleni	JAP	HKG, HAN	Aquaculture	Aquaculture	15
	Nitzschia sturionis	ARA	CAS	Stocking	Natives	16, 17
Nematoda	*Anguillicola crassus*	EUR, ANA	ASI	Aquaculture	Aquaculture/Natives	18
Trematoda†	*Cercariae batillariae*	PNA	ASI	Aquaculture	Natives	19
Protozoa	*Haplosporidia*	PNA	JAP	Aquaculture	UnKnown	20
	*Haplosporidium nelsoni	ANA	ASI	Aquaculture	Aquaculture/Natives	21, 22
	*Bonamia ostreae	EUR	PNA	Aquaculture	Aquaculture/Natives	9, 23
	*Perkinsus marinus	N. ANA	S. ANA	Aquaculture	Aquaculture/Natives	24
	*Paramoeba invadens	N. ANA	Unknown	Unknown	Natives	25
Polychaeta	*Terebrasabella heterouncinata*	PNA	SAF	Aquaculture	Aquaculture/Natives	26, 27

Regions are: ANA Atlantic North America, MED Mediterranean, EUR Europe, PNA Pacific North America, AUS Australia, ASI Asia, NOR Norway, JAP Japan, ARA Aral Sea, GOM Gulf of Mexico, RED Red Sea, HKG Hong Kong, HAN Hainan, CAS Caspian Sea, SAF South Africa. References: (1) Van Engel *et al.* 1965; (2) Hines *et al.* 1997; (3) Galil & Lutzen, 1995; (4) His, 1977; (5) Stock, 1993; (6) Bernard, 1969; (7) Holmes & Minchin, 1995; (8) Stock, 1993; (9) Minchin, 1996; (10) Hayward *et al.* in press; (11) Ernst *et al.* 2000; (12) Hayward *et al.* 2001; (13) Johnsen & Jensen, 1991; (14) Hastein & Lindstad, 1991; (15) Ogawa *et al.* 1995; (16) Osmanov, 1971; (17) Zholdasova, 1997; (18) Barse & Secor, 1999; (19) Torchin, Byers & Huspeni, unpublished data; (20) Friedman, 1996; (21) Barber, 1997; (22) Andrews, 1980; (23) Chew, 1990; (24) Ford, 1996; (25) Scheibling & Hennigar, 1997; (26) Kuris & Culver, 1999; (27) Culver & Kuris, 2000. †Grodhaus & Keh (1958) report the invasive *Ilyanassa* (=*Nassarius*) *obsoletus* in San Francisco Bay, California to be infected with five larval trematode species (See Table 1). We suspect that they are likely introduced as well.

and now infects, wild salmon stocks in Norway. It causes heavy mortality in salmon parr (Johnsen & Jensen, 1991). It appears to have been introduced to Norway by stocking rivers with infected fish from Sweden (Johnsen & Jensen, 1991). *Gyrodactylus salaris* was geographically isolated in rivers because it could not survive in brackish or marine water. However, it was spread by the movement of infected fish among hatcheries (Johnsen & Jensen, 1988). Another monogenean, *Neobenedenia melleni* (= *girellae*), is a pest of marine fishes. Ogawa *et al.* (1995) suggest that this parasite was introduced to Japan with amberjack, *Seriola dumerili*, imported from Hainan and Hong Kong. This parasite has spread from the imported amberjack to cultured local marine fishes and causes mortality in heavily infected fish. Introduced monogeneans are pests in the eel trade. They often cause mortality and are capable of infecting native eel species (Hayward *et al.* 2001, in press). *Pseudodactylogyrus anguillae* and *P. bini* were introduced from Asia to North America and Europe with the importation of the Japanese eel, *Anguilla japonica* (Hayward *et al.* 2001). Both species now infect wild populations of the native North American eel, *Anguilla rostrata* (Cone & Marcogliese, 1995; Hayward *et al.* 2001) and wild populations of the native European eel, *Anguilla anguilla* (Gelnar *et al.* 1996; Hayward *et al.* 2001). Another probable invader, *Gyrodactylus anguillae*, occurs on four continents. It was likely introduced from Europe through the importation of infected European eels (Hayward *et al.* in press).

Another serious pest of eels is the swim bladder nematode, *Anguillicola crassus*. Native to Asia, *A. crassus* has been introduced to Europe and North America where it now infects native eels in both natural and cultured conditions (Barse & Secor, 1999). *Anguillicola crassus* can reach high prevalences in native eel populations and it can cause severe pathology in European and North American eels (Barse & Secor, 1999). Yet another parasite introduced with the importation of fish is the gill monogenean, *Nitzschia sturionis*. The first stocking of the stellate sturgeon from the Caspian to the Aral Sea introduced this monogenean. It caused massive mortality of the native ship sturgeon in the 1930s (Osmanov, 1971; Zholdasova, 1997).

The Japanese mud snail, *Batillaria attramentaria*, was accidentally introduced to the West Coast of North America with the importation of oysters for aquaculture (Bonnot, 1935). It now has a disjointed distribution: populations occur in Boundary Bay, British Columbia; Padilla Bay, Washington; Tomales Bay, California; Drakes Estero, California; Bolinas Lagoon, California and Elkhorn Slough, California (Byers, 1999). Although introduced populations of *B. attramentaria* are not infected with native trematodes, they are often infected, sometimes at high prevalences, with an exotic trematode

(Torchin, Byers & Huspeni, unpublished observations). This trematode, *Cercaria batillariae*, whose adult form has still not been described nor is its life cycle fully known (Shimura & Ito, 1980), likely parasitized some of the *B. attramentaria* associated with imported oysters. On the West Coast of North America, this trematode uses native fishes as second intermediate hosts and native birds as final hosts.

There are several marine protozoan parasites that are apparently introduced. Most are parasites of oysters, presumably introduced with infected aquaculture stocks. *Haplosporidium nelsoni*, agent of MSX disease, which caused massive mortalities of native oysters in Delaware and Chesapeake Bay may be introduced from Asia (Andrews, 1980; Barber, 1997). Undocumented introductions of the Japanese oyster, *Crassostrea gigas*, may have been the source of the invasion (Barber, 1997). Haplosporidian infections, that closely resemble *H. nelsoni*, were also found in *C. gigas* imported to California from Japan (Friedman, 1996). *Bonamia ostrae* which infects native European oysters may have been introduced from California (Chew, 1990). It was presumably introduced to France with a shipment of *Ostrea edulis*, cultured in California from broodstock originating in the Netherlands (Barber, 1997). It has since spread throughout Europe. The argument that it is introduced and not native to Europe stems from the fact that epizootics of *B. ostrae* were not recorded in Europe until 1979, soon after the introduction of oysters from California (Barber, 1997). There is also some evidence that the recent range extension of the oyster parasite, *Perkinsus marinus*, is a result of repeated introductions of infected native oysters to the northeastern USA (Ford, 1996). Recent outbreaks of the marine amoeba, *Paramoeba invadens*, have caused mass mortalities in the sea urchin, *Strongylocentrotus droebachiensis*, in the North Atlantic coast of Canada. *Paramoeba invadens* may be an exotic species, but further research is needed to determine its origin (Scheibling & Hennigar, 1997). The sabellid worm, *Terebrasabella heterouncinata* was accidentally imported from South Africa with infested abalone and became a major pest in abalone mariculture facilities in California in the early 1990s (Kuris & Culver, 1999). Although it does not derive any nutrition from its host, it infests the host's shell causing abnormal shell growth and heavily infested abalones do not grow (Kuris & Culver, 1999). This worm became established in nature near a mariculture facility where it infested abalone and other susceptible native gastropods. It has since been apparently eradicated by manually reducing the densities of potential host snails to levels too low for transmission to occur (Culver & Kuris, 2000).

Epizootics of previously undocumented parasites should not, *de facto*, be considered introduced species just because they are associated with the introduction of exotic hosts. If previously rare local infectious agents are eventually able to colonize introduced hosts that have already achieved a high abundance, this could fuel epizootics of such infectious agents. For putative parasite introductions associated with exotic hosts, examining hosts in the native range for parasites is the key to understanding the source of newly discovered parasites. Considering the number of introduced marine species (Zibrowius, 1991; Furlani, 1996; Ruiz *et al.* 1997; Cohen & Carlton, 1998), it is notable that there are relatively few reports of introduced marine parasites (and they are not included in the lists cited above). There are two probable explanations for this. Introduced parasites are understudied and it is rare that parasites accompany accidentally introduced hosts. We believe that both factors likely operate. An overwhelming majority of the exotic parasites in Table 2 have been introduced along with intentionally introduced hosts cultured by the aquaculture industry. The introduction of adult hosts from high host density to another region where hosts are kept at high densities is likely to facilitate the introduction and transmission of exotic parasites and diseases. Recently, recognition of this problem, with inspections and quarantine, has led to more cautious husbandry practices, and we predict that the role of aquaculture as a vector for unwanted introductions will decline. We also note that introduced parasites are more likely to be discovered in cultured species which are rigorously examined for diseases.

INDIRECT IMPACTS OF INVADERS ON NATIVE PARASITES AND HOSTS

Parasites may indirectly influence the impact of an introduced species in a novel environment. For example, when exotic species exclude native species, parasites specific to these native species will be excluded as well. If parasites with complex life cycles are excluded from the ecosystem there may be indirect effects on the suite of former hosts. As previously discussed, the Japanese mud snail *Batillaria attramentaria*, was introduced to the West Coast of North America. At the southern end of its introduced range, it overlaps with the California horn snail, *Cerithidea californica*. Both snails compete intensely for food (Whitlatch & Obrebski, 1980; Byers, 2000) and *B. attramentaria* appears to be competitively excluding *C. californica* (Carlton, 1975; McDermott, 1996; Byers, 2000; Byers & Goldwasser, 2001). *Cerithidea californica* serves as first intermediate host for at least 18 native trematode species throughout its range in California (Martin, 1972). Most of these trematodes are trophically-transmitted through the marsh community, infecting multiple hosts during their life cycle. If *C. californica* is excluded where it overlaps with *B. attramentaria*, its parasites will become locally extinct since none

are known to infect alternative first intermediate hosts, including *B. attramentaria*. Although the precise manifestations of these local extinctions on the marsh community remain unclear, the removal of several native trematode species due to local extinction of their first intermediate host will consequently eliminate infection by the trematodes in the second intermediate host molluscs, crustaceans and fishes. Because some trophically-transmitted parasites markedly increase the likelihood of predation of infected second intermediate hosts by final hosts (Holmes & Bethel, 1972; Lafferty & Morris, 1996; Lafferty, 1999), we predict that the loss of trematodes will reduce predation rates on these second intermediate hosts. This may then have a cascading effect, potentially altering the foraging strategies and abundances of the shore birds that serve as final hosts (Lafferty, 1992; Lafferty & Morris, 1996).

Another example of an invader with indirect effects in a native ecosystem is the introduced alga, *Caulerpa taxifolia*. In areas of the western Mediterranean Sea, *C. taxifolia* appears to be interfering with parasite transmission (Bartoli & Boudouresque, 1997). Trematodes which normally infect a native fish are absent in areas where *C. taxifolia* is present, yet they are prevalent in control sites without *C. taxifolia*. Bartoli & Boudouresque (1997) propose that the secondary metabolites produced by *C. taxifolia* may be responsible for the disappearance of these trematodes, rather than a reduction in the number of first intermediate hosts. It is also possible that the reduction in abundance and biodiversity of hosts other than molluscan first intermediate hosts could interrupt transmission of trematode life cycles in the *C. taxifolia* beds.

CONCLUSION

The pervasiveness of parasites calls for our increased attention to their role in the success and impacts of exotic species as well as their potential impacts as invaders. Our analysis of Table 1 supports the hypothesis that introduced species are typically released from specific natural enemies, namely, parasites. This calls for an additional comparison of the parasites of introduced species with those of sympatric and ecologically equivalent native species to further evaluate the role of parasites in invasion success, as well as, the impacts of the invader in the novel environment. While our focus is on the role of parasites in marine invasions, we recognize the importance of other biotic interactions and hope to illustrate the broader context in which parasites are embedded in these. As economic globalization continues, the unintentional introduction of exotic marine species will increase. Understanding pathways of accidental introductions will help reduce unwanted introductions. However, once established, understanding the reasons for the success of exotic species may help reduce the impacts of these invaders. The available evidence indicates that the release of introduced species from their native natural enemies, particularly parasites, is a significant contributor to their pest status.

ACKNOWLEDGEMENTS

This research was funded by a grant from the National Sea Grant College Program, National Oceanic and Atmospheric Administration (NOAA), U.S. Department of Commerce under grant number [NA06RG0142], project number [R/CZ-162] through the California Sea Grant College System, and in part by the California State Resources Agency. The views expressed herein are those of the authors and do not necessarily reflect the views of NOAA or any of its sub-agencies. The U.S. government is authorized to reproduce and distribute this paper for governmental purposes.

REFERENCES

ANDERSON, R. M. (1982). Epidemiology. In *Modern Parasitology*. (ed. Cox, F. E. G.), pp. 204–251. Oxford: Blackwell Scientific Publications.

ANDERSON, R. M. & MAY, R. M. (1986). The invasion, persistence and spread of infectious diseases within animal and plant communities. *Philosophical Transactions of the Royal Society of London* **314**, 533–570.

ANDREWS, J. D. (1980). A review of introductions of exotic oysters and biological planning for new importations. *Marine Fisheries Review* **42**, 1–11.

ARNEBERG, P., SKORPING, A., GRENFELL, B. & READ, A. F. (1998). Host densities as determinants of abundance in parasite communities. *Proceedings of the Royal Society of London B* **265**, 1283–1289.

AUNE, K. & SCHLADWEILER, P. (1992). *Wildlife Laboratory Annual Report*. Helena, Montana: Montana Department of Fisheries Wildlife and Parks.

BAILEY, N. T. J. (1957). *The Mathematical Theory of Epidemics*. New York: Hafner.

BAKER, H. G. & STEBBINS, G. L. (1965). *The Genetics of Colonizing Species*. New York: Academic Press.

BARBER, B. J. (1997). Impacts of bivalve introductions on marine ecosystems: a review. *Bulletin of the National Research Institute of Aquaculture* **Suppl. 3**, 141–153.

BARSE, A. M. & SECOR, D. H. (1999). An exotic nematode parasite of the American eel. *Fisheries* **24**, 6–10.

BARTOLI, P. & BOUDOURESQUE, C. F. (1997). Transmission failure of parasites (Digenea) in sites colonized by the recently introduced invasive alga *Caulerpa taxifolia*. *Marine Ecology Progress Series* **154**, 253–260.

BASKIN, Y. (1996). Curbing undesirable invaders. *BioScience* **46**, 732–736.

BAUER, O. N. (1991). Spread of parasites and diseases of aquatic organisms by acclimatization: a short review. *Journal of Fish Biology* **39**, 679–686.

BERNARD, F. R. (1969). Copepod *Mytilicola orientalis* in British Columbia bivalves. *Journal of the Fisheries Research Board of Canada* **26**, 190–191.

BIRD, F. T. & ELGEE, D. E. (1957). A virus disease and introduced parasites as factors controlling the European spruce sawfly, *Diprion hercyniae*, in central New Brunswick. *Canadian Entomologist* **89**, 371–378.

BLACK, F. L. (1966). Measles endemicity in insular populations: critical community size and its evolutionary implication. *Journal of Theoretical Biology* **11**, 207–211.

BLAUSTEIN, A. R., KURIS, A. M. & ALIÓ, J. J. (1983). Pest and parasite species-richness problems. *American Naturalist* **122**, 556–566.

BLOSSEY, B. & NOTZHOLD, R. (1995). Evolution of increased competitive ability in invasive nonindigenous plants: a hypothesis. *Journal of Ecology* **83**, 887–889.

BONNOT, P. (1935). The California oyster industry. *California Fish and Game* **21**, 65–80.

BOSCHMA, H. (1972). On the occurrence of *Carcinus maenas* (Linnaeus) and its parasite *Sacculina carcini* Thompson in Burma, with notes on the transport of crabs to new localities. *Zoologische Mededelingen* **47**, 145–155.

BOURDON, R. (1963). Epicarides et Rhizocéphales de Roscoff. *Cahiers de Biologie Marine* **4**, 415–434.

BOURDON, R. (1964). Epicarides et Rhizocéphales du Bassin D'Arcachon. *Procès-Verbaux de la Société Linnéenne de Bordeaux* **101**, 1–7.

BRATTEY, J., ELNER, R. W., UHAZY, L. S. & BAGNALL, A. E. (1985). Metazoan parasites and commensals of five crab (Brachyura) species from eastern Canada. *Canadian Journal of Zoology* **63**, 2224–2229.

BUMANN, D. & PULS, G. (1996). Infestation with larvae of the sea anemone *Edwardsia lineata* affects nutrition and growth of the ctenophore *Mnemiopsis leidyi*. *Parasitology* **113**, 123–128.

BUSTNES, J. O. & GALAKTIONOV, K. (1999). Anthropogenic influences on the infestation of intertidal gastropods by seabird trematode larvae on the southern Barents Sea Coast. *Marine Biology (Berlin)* **133**, 449–453.

BUTTERMORE, R. E., TURNER, E. & MORRICE, M. G. (1994). The introduced northern Pacific seastar, *Asterias amurensis* in Tasmania. *Memoirs of the Queensland Museum* **36**, 21–25.

BYERS, J. E. (1999). The distribution of an introduced mollusc and its role in the long-term demise of a native confamilial species. *Biological Invasions* **1**, 339–352.

BYERS, J. E. (2000). Competition between two estuarine snails: implications for invasions of exotic species. *Ecology* **81**, 1225–1239.

BYERS, J. E. & GOLDWASSER, L. (2001). Exposing the mechanism and timing of impact of nonindigenous species on native species. *Ecology* **82**, 1330–1343.

BYRNE, M., CERRA, A., NISHIGAKI, T. & HOSHI, M. (1997). Infestation of the testes of the Japanese sea star *Asterias amurensis* by the ciliate *Orchitophrya stellarum*: a caution against the use of the ciliate for biological control. *Diseases of Aquatic Organisms* **28**, 235–239.

CAHOON, L. B., TRONZO, C. R. & HOWE, J. C. (1986). Notes on the occurrence of *Hyperoche medusarum* (Kroyer) (Amphipoda, Hyperiidae) with Ctenopohora off North Carolina, USA. *Crustaceana* **51**, 95–97.

CALLAWAY, R. M. & ASCHEHOUG, E. T. (2000). Invasive

plants versus their new and old neighbors: a mechanism for exotic invasion. *Science* **290**, 521–523.

CALVO-UGARTEBURU, G. & McQUAID, C. D. (1998*a*). Parasitism and invasive species: effects of digenetic trematodes on mussels. *Marine Ecology Progress Series* **169**, 149–163.

CALVO-UGARTEBURU, G. & McQUAID, C. D. (1998*b*). Parasitism and introduced species: epidemiology of trematodes in the intertidal mussels *Perna perna* and *Mytilus galloprovincialis*. *Journal of Experimental Marine Biology and Ecology* **220**, 47–65.

CARLTON, J. T. (1975). Extinct and endangered populations of the endemic mud snail *Cerithidea californica* in Northern California. *Bulletin of the American Malacological Union* **41**, 65–66.

CARLTON, J. T. (1987). Patterns of transoceanic marine biological invasions in the Pacific Ocean. *Bulletin of Marine Science* **41**, 452–465.

CARLTON, J. T. & GELLAR, J. B. (1993). Ecological roulette: the global transport of nonindigenous marine organisms. *Science* **266**, 78–82.

CARLTON, J. T., THOMPSON, J. K., SCHEMEL, L. E. & NICHOLS, F. H. (1990). Remarkable invasion of San Francisco Bay (California, USA) by the Asian clam *Potamocorbula amurensis*. I. Introduction and dispersal. *Marine Ecology Progress Series* **66**, 81–94.

CASTILHO, F. & BARANDELA, T. (1990). Ultrastructural study on the spermatogenesis and spermatozoon of the metacercariae of *Microphallus primas* (Digenea), a parasite of *Carcinus maenas*. *Molecular Reproduction and Development* **25**, 140–146.

CHEW, K. K. (1990). Global bivalve shellfish introductions. *World Aquaculture* **21**, 9–22.

COHEN, A. N. & CARLTON, J. T. (1998). Accelerating invasion rate in a highly invaded estuary. *Science* **279**, 555–558.

COMELY, C. A. & ANSELL, A. D. (1989). The incidence of *Carcinonemertes carcinophila* (Kolliker) on some decapod crustaceans from the Scottish South Coast. *Ophelia* **30**, 225–233.

CONE, D. K. & MARCOGLIESE, D. J. (1995). *Pseudodactylogyrus anguillae* on *Anguilla rostrata* in Nova Scotia: an endemic or an introduction? *Journal of Fish Biology* **47**, 177–178.

CRAWLEY, M. J. (1986). The population biology of invaders. *Philosophical Transactions of the Royal Society of London B* **314**, 711–731.

CRAWLEY, M. J. (1987). What makes a community invasible? In *Colonization, Succession and Stability* (ed. Crawley, M. J. C., Gray, A. J. & Edwards, P. J.), pp. 482. Oxford: Blackwell Scientific Publications.

CRIBB, T. H., PICHELIN, S., DUFOUR, V., BRAY, R. A., CHAUVET, C., FALIEX, E., GALZIN, R., LO, C. M., LO-YAT, A., MORAND, S., RIGBY, M. C. & SASAL, P. (2000). Parasites of recruiting coral reef fish larvae in New Caledonia. *International Journal for Parasitology* **30**, 1445–1451.

CULVER, C. S. & KURIS, A. M. (2000). The apparent eradication of a locally established introduced marine pest. *Biological Invasions* **2**, 245–253.

CURTIS, L. A. (1997). *Ilyanassa obsoleta* (Gastropoda) as a host for trematodes in Delaware estuaries. *Journal of Parasitology* **83**, 793–803.

CURTIS, L. A. & HUBBARD, K. M. (1990). Trematode infections in a gastropod host misrepresented by observing shed cercariae. *Journal of Experimental Marine Biology and Ecology* **143**, 131–137.

CURTIS, L. A. & TANNER, N. L. (1999). Trematode accumulation by the estuarine gastropod *Ilyanassa obsoleta*. *Journal of Parasitology* **85**, 419–425.

DEBACH, P. (1974). *Biological Control by Natural Enemies*. Cambridge: Cambridge University Press.

DOBSON, A. P. (1988). Restoring island ecosystems: the potential of parasites to control introduced mammals. *Conservation Biology* **2**, 31–39.

DOBSON, A. P. & MAY, R. M. (1986). Patterns of invasions by pathogens and parasites. In *Ecology and Biological Invasions of North America and Hawaii* (ed. Mooney, H. A. & Drake, J. A.), Berlin: Springer-Verlag.

DOGIEL, V. A. (1948). Results and perspectives in parasitological research within the Leningrad State University. *Vestnik Leningradskogo Gosudarstvennogo Universiteta* **3**, 31–39.

DOVE, A. D. M. (1998). A silent tragedy: parasites and the exotic fishes of Australia. *Proceedings of the Royal Society of Queensland* **107**, 109–113.

DOVE, A. D. M. (2000). Richness patterns in the parasite communities of exotic poeciliid fishes. *Parasitology* **120**, 609–623.

ELTON, C. S. (1958). *The Ecology of Invasions by Animals and Plants*. London: Methuen and Company.

ERNST, I., FLETCHER, A. & HAYWARD, C. (2000). *Gyrodactylus anguillae* (Monogeneana: Gyrodactylidae) from anguillid eels (*Anguilla australis* and *Anguilla reinhardtii*) in Australia: a native or an exotic? *Journal of Parasitology* **86**, 1152–1156.

EVANS, D. W., IRWIN, S. W. B. & FITZPATRICK, S. M. (1997). Metacercarial encystment and *in vivo* cultivation of *Cercaria lebouri* Stunkard 1932 (Digenea: Notocotylidae) to adults identified as *Paramonostomum chabaudi* Van Strydonck 1965. *International Journal for Parasitology* **27**, 1299–1304.

EWEL, J. J., O'DOWD, D. J., BERGELSON, J., DAEHLER, C. C., D'ANTONIO, C. M., GOMEZ, L. D., GORDON, D. R., HOBBS, R. J., HOLT, A., HOPPER, K. R., HUGHES, C. E., LAHART, M., LEAKEY, R. R. B., LEE, W. G., LOOPE, L. L., LORENCE, D. H., LOUDA, S. M., LUGO, A. E., McEVOY, P. B., RICHARDSON, D. M. & VITOUSEK, P. M. (1999). Deliberate introductions of species: research needs. *BioScience* **49**, 619–630.

FORD, S. D. (1996). Range extension by the oyster parasite *Perkinsus marinus* into the Northeastern United States: response to climate change? *Journal of Shellfish Research* **15**, 45–56.

FREELAND, W. J. (1993). Parasites, pathogens and the impacts of introduced organisms on the balance of nature in Australia. In *Conservation biology in Australia and Oceania* (ed. Moritz, C. & Kikkawa, J.), pp. 171–180. Chipping Norton, New South Wales: Surrey Beatty & Sons.

FRIEDMAN, C. S. (1996). Haplosporidian infections of the Pacific oyster, *Crassostrea gigas* (Thunberg), in California and Japan. *Journal of Shellfish Research* **15**, 597–600.

FURLANI, D. M. (1996). *A Guide to the Introduced Marine Species in Australian Waters. Centre for Research in Introduced Marine Pests. Technical Report* **5**. Hobart, Australia.

GALIL, B. S. & INNOCENTI, G. (1999). Notes on the population structure of the portunid crab *Charybdis longicollis* Leene, parasitized by the rhizocephalan *Heterosaccus dollfusi* Boschma, off the Mediterranean coast of Israel. *Bulletin of Marine Science* **64**, 451–463.

GALIL, B. S. & LUTZEN, J. (1995). Biological observations on *Heterosaccus dollfusi* Boschma (Cirripedia: Rhizocephala), a parasite of *Charybdis longicollis* Leene (Decapoda: Brachyura), a Lessepsian migrant to the Mediterranean. *Journal of Crustacean Biology* **15**, 659–670.

GALLIEN, L. & BLOCH, F. (1936). Recherches sur *Lecithomyzon maenids* Bloch & Gallien, copépode parasite de la ponte de *Carcinus maenas* Pennant. *Bulletin Biologique de France et Belgique* **70**, 36–53.

GAYEVSKAYA, A. V. & MORDVINOVA, T. N. (1994). Occurrence of nematode larvae as parasites of the ctenophore *Mnemiopsis maccradyi* in the Black Sea. *Hydrobiological Journal* **30**, 108–110.

GELNAR, M., SCHOLZ, T., MATEJUSOVA, I. & KONECNY, R. (1996). Occurrence of *Pseudodactylogyrus anguillae* (Yin & Sproston, 1948) and *P. bini* (Kikuchi, 1929), parasites of eel, *Anguilla anguilla* L., in Austria (Monogeneana: Dactylogyridae). *Annalen des Naturhistorischen Museums in Wien Serie B* **98**, 1–4.

GOGGIN, C. L. (1998). Options for biological control of *Asterias amurensis*. *Proceedings of a meeting on the biology and management of the introduced seastar* Asterias amurensis *in Australian waters. Centre for Research on Introduced Marine Pests. Technical Report* **15**, 53–58.

GOGGIN, C. L. & BOULAND, C. (1997). The ciliate *Orchitophrya cf. stellarum* and other parasites and commensals of the Northern Pacific seastar *Asterias amurensis* from Japan. *International Journal for Parasitology* **27**, 1415–1418.

GREGORY, R. D. & BLACKBURN, T. M. (1991). Parasite prevalence and host sample size. *Parasitology Today* **7**, 316–318.

GRODHAUS, G. & KEH, B. (1958). The marine, dermatitis-producing cercaria of *Austrobilharzia variglandis* in California (Trematoda: Schistosomatidae). *Journal of Parasitology* **44**, 633–638.

HARADA, M. & SUGURI, S. (1989). Surveys on cercariae in brackish water snails in Kagawa Prefecture, Shikoku, Japan. *Japanese Journal of Parasitology* **38**, 388–391.

HASTEIN, T. & LINDSTAD, T. (1991). Diseases in wild and cultured salmon: possible interaction. *Aquaculture* **98**, 277–288.

HAYWARD, C. J., IWASHITA, M., CRANE, J. S. & OGAWA, K. (2001). First report of the invasive eel pest *Pseudodactylogyrus bini* in North America and in wild American eels. *Diseases of Aquatic Organisms* **44**, 53–60.

HAYWARD, C. J., IWASHITA, M., OGAWA, K. & ERNST, I. (in press). New evidence that *Gyrodactylus anguillae* (Monogeneana) is another invading pest of anguillid eels. *Biological Invasions*.

HINES, A. H., ALVAREZ, F. & REED, S. A. (1997). Introduced and native populations of a marine parasitic castrator: variation in prevalence of the rhizocephalan *Loxothylacus panopaei* in xanthid crabs. *Bulletin of Marine Science* **61**, 197–214.

HIS, E. (1977). Observations préliminaires sur la présence de *Mytilicola orientalis* Mori (1935) chez *Crassostrea gigas* Thunberg dans le Bassin D'Arcachon. *Bulletin de la Société Géologique de Normandie et des amis du muséum du Havre* **64**, 7–8.

HOFF, C. C. (1941). A case of correlation between infection of snail hosts with *Cryptocotyle lingua* and the habits of gulls. *Journal of Parasitology* **27**, 539.

HOFFMAN, G. L. (1999). *Parasites of North American Freshwater Fishes*. Ithaca, N.Y.: Comstock Pub. Associates.

HOLMES, J. C. & BETHEL, W. M. (1972). Modification of intermediate host behaviour by parasites. In *Behavioural Aspects of Parasite Transmission* (ed. Canning, E. U. & Wright, C. A.), New York: Academic Press.

HOLMES, J. M. C. & MINCHIN, D. (1995). Two exotic copepods imported into Ireland with Pacific oyster *Crassostrea gigas* (Thunberg). *Irish Naturalists Journal* **25**, 17–20.

HUFFAKER, C. B., MESSENGER, P. S. & DEBACH, P. (1971). The natural enemy component in natural control and the theory of biological control. In *Biological Control* (ed. Huffaker, C. B.), pp. 16–62. New York: Plenum Press.

HUGHES, R. N. & ANSWER, P. (1982). Growth, spawning and trematode infection of *Littorina littorea* (L.) from an exposed shore in North Wales. *Journal of Molluscan Studies* **2**, 321–336.

IRWIN, S. W. B. (1983). Incidence of trematode parasites in two populations of *Littorina saxatilis* (Olivi) from the North Shore of Belfast Lough. *Irish Naturalists Journal* **21**, 26–29.

JAMES, B. L. (1965). The effects of parasitism by larval Digenea on the digestive gland of intertidal prosobranch, *Littorina saxatilis* (Olivi) subsp. *tenebrosa* (Montagu). *Parasitology* **55**, 93–115.

JAMES, B. L. (1968). The distribution and keys of species in the family Littorinidae and of their Digenean parasites, in the region of Dale, Pembrokeshire. *Field Studies* **2**, 615–650.

JOHNSEN, B. O. & JENSEN, A. J. (1991). The *Gyrodactylus* story in Norway. *Aquaculture* **98**, 289–302.

JOHNSON, M. W. (1957). The copepod *Choniosphaera cancrorum* parasitizing a new host, the green crab *Carcinides maenas*. *Journal of Parasitology* **43**, 470–473.

KENNEDY, C. R. (1993). Introductions spread and colonization of new localities by fish helminth and crustacean parasites in the British Isles: a perspective and appraisal. *Journal of Fish Biology* **43**, 287–301.

KERMACK, W. O. & MCKENDRICK, A. G. (1927). Contributions to the mathematical theory of epidemics. *Proceedings of the Royal Society of London. Series A* **115**, 700–721.

KINZELBACH, R. (1965). Die blaue Schwimmkrabbe (*Callinectes sapidus*) ein Neuburger im Mittelmeer. *Natur und Museum* **95**, 293–296.

KURIS, A. M. & CULVER, C. S. (1999). An introduced sabellid polychaete pest infesting cultured abalones and its potential spread to other California gastropods. *Invertebrate Biology* **118**, 391–403.

KURIS, A. M. & GURNEY, R. (1997). Survey of Tasmanian Crabs for Parasites: A Progress Report. Proceedings of the first international workshop on the demography, impacts and management of the introduced populations of the European crab, *Carcinus maenas. Centre for Research on Introduced Marine Pests. Technical Report* **11**, 92–94.

KURIS, A. M. & LAFFERTY, K. D. (1992). Modelling crustacean fisheries: effects of parasites on management strategies. *Canadian Journal of Fisheries and Aquatic Sciences* **49**, 327–336.

KURIS, A. M., LAFFERTY, K. D. & GRYGIER, M. J. (1996). Detection and preliminary evaluation of natural enemies for possible biological control of the Northern Pacific seastar, *Asterias amurensis. Centre for Research on Introduced Marine Pests. Technical Report* **3**, 1–17.

LAFFERTY, K. D. (1992). Foraging on prey that are modified by parasites. *The American Naturalist* **140**, 854–867.

LAFFERTY, K. D. (1997). Environmental parasitology: what can parasites tell us about human impacts on the environment? In *Parasitology Today* **13**, 251–255.

LAFFERTY, K. D. (1999). The evolution of trophic transmission. *Parasitology Today* **15**, 111–115.

LAFFERTY, K. D. & KURIS, A. M. (1996). Biological control of marine pests. *Ecology* **77**, 1989–2000.

LAFFERTY, K. D. & MORRIS, A. M. (1996). Altered behavior of parasitized killifish increases susceptibility to predation by final bird hosts. *Ecology* **77**, 1390–1397.

LAMPO, M. & BAYLISS, P. (1996 *a*). The impact of ticks on *Bufo marinus* from native habitats. *Parasitology* **113**, 199–206.

LAMPO, M. & BAYLISS, P. (1996 *b*). Density estimates of cane toads from native populations based on mark-recapture data. *Wildlife Research* **23**, 305–315.

LAUCKNER, G. (1987). Ecological effects of larval trematode infestation on littoral marine invertebrate populations. *International Journal for Parasitology* **17**, 391–398.

LAWTON, J. H. & BROWN, K. C. (1986). The population and community ecology of invading insects. *Philosophical Transactions of the Royal Society of London B* **314**, 607–617.

LEECH, J. H. (1992). Impacts of the zebra mussel (*Dreissena polymorpha*) on water quality and fish spawning reefs in western Lake Erie. In *Zebra Mussels: Biology, Impacts, and Control* (ed. Nalepa, T. F. & Schlosser, D. S.), Florida: Lewis Publishers.

MARTIN, W. E. (1972). An annotated key to the cercariae that develop in the snail, *Cerithidea californica. Bulletin of the Southern California Academy of Sciences* **71**, 39–43.

MARTORELLI, S. R. (1996). First record of encysted metacercariae in hydrozoan jellyfishes and ctenophores of the Southern Atlantic. *Journal of Parasitology* **82**, 352–353.

MATHIESON, S., BERRY, A. J. & KENNEDY, S. (1998). The parasitic rhizocephalan barnacle *Sacculina carcini* in crabs of Forth Estuary, Scotland. *Journal of the Marine Biological Association of the United Kingdom* **78**, 665–667.

MATTHEWS, P. M., MONTGOMERY, W. I. & HANNA, R. E. B. (1985). Infestation of littorinids by larval Digenea around a small fishing port. *Parasitology* **90**, 277–287.

MCCARTHY, H. O., FITZPATRICK, S. & IRWIN, S. W. B. (2000). A transmissible trematode affects the direction and rhythm of movement in a marine gastropod. *Animal Behaviour* **59**, 1161–1166.

MCCURDY, D. G., BOATES, J. S. & FORBES, M. R. (2000). Spatial distribution of the intertidal snail *Ilyanassa obsoleta* in relation to parasitism by two species of trematodes. *Canadian Journal of Zoology* **78**, 1137–1143.

MCDERMOTT, J. J. (1998). The western Pacific brachyuran (*Hemigrapsus sanguineus*: Grapsidae), in its new habitat along the Atlantic Coast of the United States: a geographic distribution and ecology. *ICES Journal of Marine Science* **55**, 289–298.

MCDERMOTT, S. P. (1996). Parasites, density, and disturbance: factors influencing coexistence of *Cerithidea californica* and *Batillaria attrementaria*. M.S. Thesis, California State University, Fresno, California.

MCKENDRICK, A. (1940). The dynamics of crowd infections. *Edinburgh Medical Journal* (*N. Ser.*) **47**, 117–136.

MEYER, J. Y. (1996). Status of *Miconia calvescens* (Melastomataceae), a dominant invasive tree in the Society Islands (French Polynesia). *Pacific Science* **50**, 66–76.

MINCHIN, D. (1996). Management of the introduction and transfer of marine molluscs. *Aquatic Conservation* **6**, 229–244.

MINCHIN, D. (1997). The influence of the parasitic cirripede *Sacculina carcini* on its brachyuran host *Carcinus maenas* within its home range. *Proceedings of the First International Workshop on the Demography, Impacts and Management of the Introduced Populations of the European Crab*, Carcinus maenas. *Technical Report Number* **11**, 76–79.

NAEEM, S., KNOPS, J. M. H., TILMAN, D., HOWE, K. M., KENNEDY, T. & GALE, S. (2000). Plant diversity increases resistance to invasion in the absence of covarying extrinsic factors. *Oikos* **91**, 97–108.

NEWELL, C. R. (1986). The marine fauna and flora of the Isles of Scilly (United Kingdom): some marine digeneans from invertebrate hosts. *Journal of Natural History* **20**, 71–78.

OGAWA, K., BONDAD-REANTASO, M. G., FUKUDOME, M. & WAKABAYASHI, H. (1995). *Neobenedenia girellae* (Hargis, 1955) Yamaguti, 1963 (Monogeneana: Capsalidae) from cultured marine fishes of Japan. *Journal of Parasitology* **81**, 223–227.

OSMANOV, S. O. (1971). *Parasites of Fishes of Uzbekistan*. Tashkent, USSR: FAN.

PECHENIK, J. A., FRIED, B. & SIMPKINS, H. L. (2001). *Crepidula fornicata* is not a first intermediate host for trematodes: Who is? *Journal of Experimental Marine Biology and Ecology* **261**, 211–224.

PLOTZ, J. (1982). Uber den Lebenszyklus von *Paracuaria tridentata* and *Cosmocephalus obvelatus* (Nematoda, Acuariidae) von Seevogeln. *Seevogel* **1982**, 125–126.

POHLEY, W. J. (1976). Relationships among three species of *Littorina* and their larval digenea. *Marine Biology* **37**, 179–186.

POLYANSKI, Y. I. (1961). Ecology of parasites of marine fishes. In *Parasitology of Fishes* (ed. Dogiel, V. A., Petrushevski, G. K. & Polyanski, Y. I.), pp. 384. London: Oliver and Boyd Ltd.

PRICE, P. W., WESTOBY, M., RICE, B., ATSTATT, P. R., FRITZ, R. S., THOMPSON, J. N. & MOBLEY, K. (1986). Parasite mediation in ecological interactions. *Annual Review of Ecological and Systematics* **17**, 487–505.

RASMUSSEN, E. (1973). Systematics and the ecology of the Isefjord marine fauna (Denmark). *Ophelia* **11**, 142–165.

RIGBY, M. C. & DUFOUR, V. (1996). Parasites of coral reef fish recruits, *Epinephelus merra* (Serranidae), in French Polynesia. *Journal of Parasitology* **82**, 405–408.

ROBSON, E. M. & WILLIAMS, I. C. (1970). Relationships of some species of Digenea with the marine prosobranch *Littorina littorea* (L.) I. The occurrence of larval Digenea in *L. littorea* on the North Yorkshire Coast. *Journal of Helminthology* **44**, 153–168.

ROLLINSON, D. & SOUTHGATE, V. R. (1987). The genus *Schistosoma*: a taxonomic appraisal. In *The Biology of Schistosomes from Genes to Latrines* (ed. Rollinson, D. & Simpson, A. J. G.), pp. 472. London: Academic Press.

RUIZ, G. M., CARLTON, J. T., GROSHOLZ, E. D. & HINES, A. H. (1997). Global invasions of marine and estuarine habitats by non-indigenous species: mechanisms, extent, and consequences. *American Zoologist* **37**, 621–632.

RYBAKOV, A. V. & LUKOMSKAYA, O. G. (1988). On the life cycle of *Acanthoparyphium macracanthum* sp.n. (Trematoda, Echinostomatidae). *Parazitologiya* **22**, 224–229.

SCHEIBLING, R. E. & HENNIGAR, A. W. (1997). Recurrent outbreaks of disease in sea urchins *Strongylocentrotus droebachiensis* in Nova Scotia: evidence for a link with large-scale meteorologic and oceanographic events. *Marine Ecology Progress Series* **152**, 155–165.

SCHOENER, T. W. & SPILLER, D. A. (1995). Effect of predators and area on invasion: an experiment with island spiders. *Science* **267**, 1811–1813.

SCOTT, M. E. (1987). Regulation of mouse colony abundance by *Heligmosomoides polygyrus* (Nematoda). *Parasitology* **95**, 111–129.

SETTLE, W. H. & WILSON, L. T. (1990). Invasion by the variegated leafhopper and biotic interactions: parasitism, competition, and apparent competition. *Ecology* **71**, 1461–1470.

SHIMURA, S. & ITO, J. (1980). Two new species of marine cercariae from the Japanese intertidal gastropod, *Batillaria cumingii* (Crosse). *Japanese Journal of Parasitology* **29**, 369–375.

STAMBAUGH, J. E. & MCDERMOTT, J. J. (1969). The effects of trematode larvae on the locomotion of naturally infected *Nassarius obsoletus* (Gastropoda). *Proceedings of the Pennsylvania Academy of Science* **43**, 226–231.

STIVEN, A. E. (1964). Experimental studies on the host parasite system hydra and *Hydramoeba hydroxena* (Entz.). II. The components of a single epidemic. *Ecological Monographs* **34**, 119–142.

STIVEN, A. E. (1968). The components of a threshold in experimental epizootics of *Hydramoeba hydroxena* in populations of *Chlorohydra viridissima*. *Journal of Invertebrate Pathology* **11**, 348–357.

STOCK, J. H. (1993). Copepoda (Crustacea) associated with commercial and non-commercial Bivalvia in the East Scheldt, The Netherlands. *Bijdragen tot de Dierkunde* **63**, 61–64.

STUNKARD, H. W. (1956). Studies on the parasites of the green crab, *Carcinides maenas*. *Biological Bulletin* **111**, 295.

THOMPSON, A. B. (1985). Analysis of *Profilicollis botulus* (Acanthocephala: Echinorhynchidae) burdens in the shore crab, *Carcinus maenas*. *Journal of Animal Ecology* **54**, 595–604.

THRELFALL, W. & GOUDIE, R. I. (1977). Trematodes in the rough periwinkle, *Littorina saxatilis* (Olivi), from Newfoundland. *Proceedings of the Helminthological Society of Washington* **44**, 229–232.

TORCHIN, M. E., LAFFERTY, K. D. & KURIS, A. M. (1996). Infestation of an introduced host, the European green crab, *Carcinus maenas*, by a symbiotic nemertean egg predator, *Carcinonemertes epialti*. *Journal of Parasitology* **82**, 449–453.

TORCHIN, M. E., LAFFERTY, K. D. & KURIS, A. M. (2001). Release from parasites as natural enemies: increased performance of a globally introduced marine crab. *Biological Invasions*. **3** (4) in press.

VAN ENGEL, W. A., DILLON, W. A., ZWERNER, D. & ELDRIDGE, D. (1965). *Loxothylacus panopaei* (Cirripedia, Sacculinidae) an introduced parasite on the xanthid crab in Chesapeake Bay, U.S.A. *Crustaceana* **10**, 111–112.

VAN RIPER III, C., VAN RIPER, M. L., GOFF, M. L. & LAIRD, M. (1986). The epizootiology and ecological significance of malaria in Hawaiian land birds. *Ecological Monographs* **56**, 327–344.

VITOUSEK, P. M. (1990). Biological invasions and ecosystem processes: towards an integration of population biology and ecosystem studies. *Oikos* **57**, 7–13.

WARNER, R. E. (1969). The role of introduced diseases in the extinction of the endemic Hawaiian avifauna. *Condor* **70**, 101–120.

WHITLACH, R. B. & OBREBSKI, S. (1980). Feeding selectivity and coexistence in two deposit feeding gastropods. *Marine Biology* **58**, 219–225.

WILCOVE, D. S., ROTHSTEIN, D., DUBOW, J., PHILLIPS, A. & LOSOS, E. (1998). Quantifying threats to imperiled species in the United States. *Bioscience* **48**, 607–615.

WILLIAMS, I. C. & ELLIS, C. (1975). Movements of the common periwinkle, *Littorina littorea* (L.), on the Yorkshire Coast in winter and the influence of infection with larval digenea. *Journal of Experimental Marine Biology and Ecology* **17**, 47–58.

WOOTTON, J. T. (1994). The nature and consequences of indirect effects in ecological communities. *Annual Review of Ecology and Systematics* **25**, 443–466.

YAMAGUCHI, T, TOKUNAGA, S & ARATAKE, H. (1994). Contagious infection by the rhizocephalan parasite *Sacculina* sp. In the grapsid crab *Hemigrapsus sanguineus* (De Haan). *Crustacean Research* **23**, 89–101.

ZHOLDASOVA, I. (1997). Sturgeons and the Aral Sea ecological catastrophe. *Environmental Biology of Fishes* **48**, 373–380.

ZIBROWIUS, H. (1991). Ongoing modification of the Mediterranean marine fauna and flora by the establishment of exotic species. *Mesogee* **51**, 83–107.

Parasites as biological tags in population studies of marine organisms: an update

K. MACKENZIE

Department of Zoology, The University of Aberdeen, Tillydrone Avenue, Aberdeen AB24 2TZ, Scotland, UK

SUMMARY

This paper reviews the work published over the past decade on the use of parasites as biological tags in population studies of marine fish, mammals and invertebrates. Fish hosts are considered in taxonomic and ecological groups as follows: demersal, anadromous, small pelagic, large pelagic and elasmobranch. Most studies were carried out on demersal fish, particularly on members of the genera *Merluccius* (hake), *Sebastes* (rockfish) and on Atlantic cod *Gadus morhua* L., but Pacific salmonids and small pelagic fish of the genus *Trachurus* are also well-represented. A current multidisciplinary study of the population biology of horse mackerel *Trachurus trachurus* in European waters, which includes the use of parasites as tags, is described. Two studies recognize the potential for using parasites as tags for cetaceans but, in spite of the considerable potential for this approach in population studies of elasmobranchs, no original study has been carried out on this group for over ten years. Studies of parasites as tags for marine invertebrates have concentrated on squid. Recent trends in the use of parasites as biological tags for marine hosts are discussed.

Key words: Parasite tags, marine, fish, mammals, invertebrates.

INTRODUCTION

The first publication describing the use of a parasite as a biological tag in a population study of a purely marine organism appeared over 60 years ago (Herrington, Bearse & Firth, 1939). Since the 1960s, reviews of the subject by various authors have been published, with the emphasis on different host taxa and with some including freshwater as well as marine hosts. The most recent publication to examine the use of parasites as tags in population studies of marine fish in general is that of Mosquera, Gómez-Geistera & Pérez-Villar (2000), who showed how a simple mathematical model of macroparasitic infections could be successfully used for stock identification and to describe the migratory routes of marine fish. Another important recent review of the subject is that of Arthur (1997), while MacKenzie & Abaunza (1998) provided a guide to the procedures and methods employed in the use of parasites as tags for stock discrimination of marine fish.

The present paper reviews studies published since 1991 on the use of parasites as biological tags in population studies of marine fish, mammals and invertebrates. For a review of studies published before this date, see Williams, MacKenzie & McCarthy (1992). Fish hosts are considered under the sub-headings demersal, anadromous, small pelagic, large pelagic and elasmobranch.

MARINE FISH

Demersal fish

In general, the numbers of publications referring to the use of parasites as biological tags for different fish taxa reflect the commercial importance of the target fish, so it comes as no surprise to discover that over the past decade most publications on demersal fish have focused on members of the genera *Merluccius* (hake), *Sebastes* (rockfish) and on Atlantic cod, *Gadus morhua* L.

Five species in the genus *Merluccius* have been the subjects of biological tag studies. MacKenzie & Longshaw (1995) surveyed the protozoan and metazoan parasite faunas of *Merluccius hubbsi* and *M. australis* from 16 stations on the Falklands–Argentine Shelf and one off southern Chile. Ten parasite species were identified as potentially useful tags for stock identification and one adult digenean showed promise as a tag for following seasonal migrations. *Merluccius hubbsi* was also the subject of a study by Sardella & Timi (1999), who found significant differences between the parasite communities of hake from two zones of the Argentine Sea, indicating the potential value of parasites as tags for hake in this region. George-Nascimento & Arancibia (1994) studied the morphometrics and metazoan parasite faunas of *M. australis* from a series of inshore and offshore stations off the coast of southern Chile up to the Falkland Islands Shelf. Comparisons between areas suggested that there are at least four 'ecological' stocks in the study area, but that there is much interchange between them, so they may be treated as a single 'pure' stock. *Merluccius gayi* was the subject of another study by

Tel: +44 1224 272874. Fax: +44 1224 272396. E-mail: k.mackenzie@abdn.ac.uk

George-Nascimento (1996), who examined samples of this hake caught in three major fishing areas off central Chile for metazoan parasites. Multivariate analyses of the parasite data suggested the existence of at least two 'ecological' stocks in the study area. The author pointed out that although univariate analyses on selected parasite taxa from different samples may show no significant differences, multivariate analyses of entire parasite assemblages may reveal significant differences between parasite communities. Two other species of hake, *Merluccius capensis* and *M. paradoxus*, caught off the coast of Namibia, were examined for protozoan and metazoan parasites by Reimer (1993). Significant differences in the distributions of a myxozoan, a cestode postlarva and a larval nematode suggested the existence of distinct northern and southern stocks of both host species in the study area. Mattiucci *et al.* (2000) analysed the metazoan parasite communities of *Merluccius merluccius* from 13 stations in the Mediterranean and northeast Atlantic using a multivariate analysis and found significant differences between areas. Larvae of six different sibling species of the nematode genus *Anisakis* were recorded from these hake. Samples of another merluccid species, *Macruronus magellanicus*, from two fishing grounds in southern Chile, were examined for metazoan parasites by Oliva (2001). Evidence from univariate analysis of the infection data did not suggest the existence of separate stocks on the two grounds, but did suggest a possible migratory pattern from south to north. The most frequently used tag parasites in these studies were plerocercoids of the trypanorhynch cestode genus *Grillotia*, closely followed by anisakid nematode larvae, the adult digenean *Elytrophalloides oatesi* and the adult pseudophyllidean cestode *Clestobothrium crassiceps*. It should be noted, however, that protozoan and myxozoan parasites were not recorded in most studies.

Hemmingsen & MacKenzie (2001) reviewed the use of parasites as tags for Atlantic cod from 1961 to 1998. Six studies were reported in the last decade, three in the Barents Sea, two in the northwest Atlantic, and one in the Baltic Sea. Karasev (1994) assessed the literature on parasites of cod in the Barents Sea and identified potentially useful tags. In a subsequent study of the parasite faunas of different coastal populations of cod in the southern Barents Sea, Karasev (1998) found no evidence that any of the parasites recorded might be useful as tags. Larsen *et al.* (1997), on the other hand, successfully used two myxosporeans, an adult digenean and a parasitic copepod as tags in a study of the population structure of two types of cod in northern Norway. The two types were identified by otolith structure and evidence from this and the tag parasites suggested that the fjords contain local resident populations of Arcto-Norwegian cod and that only the coastal cod migrate between the fjords and offshore areas.

McClelland & Marcogliese (1994) used multivariate analyses of infection data on anisakid nematode larvae to distinguish between migrant and resident populations of cod in the Breton Shelf winter fishery off the Atlantic coast of Canada. Khan & Tuck (1995) used parasitological data from their own samples, together with data from earlier studies, to provide evidence of the existence of at least six stocks of cod, with some degree of mixing, in the Newfoundland–Labrador area. Sobecka & Kilian (2000) reported the preliminary results of a study of the parasite fauna of Baltic cod aimed at identifying potentially useful tags. The most frequently used tag parasite in all of these recent studies was the parasitic copepod *Lernaeocera branchialis*, but Jones & Taggart (1998) advised caution in the use of this parasite as a tag after analyses of data from artificial tagging experiments on cod and observations on the occurrence of the parasite showed evidence of differential survival of infected and uninfected cod.

Three large-scale studies of parasites as tags for rockfish of the genus *Sebastes* were reported, two from the Pacific coast of North America and one from the North Atlantic. Stanley, Lee & Whitaker (1992) surveyed the parasite fauna of yellowtail rockfish *Sebastes flavidus* from British Columbia to Central California. Only the gill monogenean *Microcotyle sebastis* was selected as a tag because it showed a clear trend of variation with latitude, with a progressive increase in prevalence from north to south, but the authors suggested that the parasitic copepod *Neobrachiella robusta* might also prove to be a useful tag given larger sample sizes. The authors concluded that the entire coastal population of rockfish could not be divided into discrete stocks, a conclusion consistent with results from other tagging methods. In Alaskan coastal waters, Moles, Heifitz & Love (1998) investigated the parasite faunas of two other rockfish species, *Sebastes borealis* and *S. aleutianus*. They selected the parasitic copepods *N. robusta* and *Trochopus trituba* and the acanthocephalan *Corynosoma* sp. as potentially useful tags. In particular, a major reduction in prevalence of all three parasites in *S. aleutianus* from the southeastern part of the study area suggested that the population there may constitute a separate stock. Bakay (1999) presented the preliminary results of a geographical analysis of the parasites infecting *Sebastes mentella* from 16 parts of the North Atlantic ranging from the Barents Sea to the Canadian coast. The analysis identified five areas distinguished by variations in the parasite communities.

Two studies were reported on walleye pollock *Theragra chalcogramma* in the northwest Pacific. Avdeev (1996) used infection data on two larval nematodes and two larval cestodes to follow migrations of immature fish in the Sea of Okhotsk. Data on the geographical distribution of infections and annual variations in infection led to the conclusion that immature pollock do not undergo extensive migrations but that a progressive increase in the population in the western part of the Sea of Okhotsk is due to immigration from eastern parts. The second study, by Avdeev & Avdeev (1998) focused on mature pollock during their pre-spawning period in the western part of the Bering Sea, the Comandor Islands and the east coast of Kamchatka. Mean abundances of 11 species of metazoan parasites, mostly long-lived larval forms, showed that the population around the Comandor Islands is a mixture of resident fish and migrants from the Bering Sea.

The plerocercoid of the trypanorhynch cestode *Callitetrarhynchus speciosus* was used by Cappo (1992) to follow migrations of Australian 'salmon' *Arripis truttaceus* off South Australia. Transmission of the cestode to salmon appeared to be largely restricted to one relatively small part of the study area, so migrants from that area could be identified by the presence of the parasite. In a larger study area off southern Australia, Sewell & Lester (1995) used multivariate analyses of infection data on nine taxa of metazoan parasites to identify two separate stocks of gemfish, *Rexea solandri*, and to suggest possible seasonal patterns of migration.

Three recent studies showed how parasites could be used as tags to identify populations of fish and to follow their movements over much smaller scales than had been attempted previously. Grutter (1998) showed that infections of a monogenean *Benedenia* sp. were significantly greater on samples of the coral reef fish *Hemigymnus melapterus* from a reef flat than on those from the reef slope – habitats only a few hundred metres apart. This suggested that populations of the host fish did not move between the two areas and that the parasite could be used for following small-scale movements of this species. Similarly, Cribb, Anderson & Dove (2000) described significant differences in infections of an acanthocephalan *Pomphorhynchus heronensis* between populations of another reef fish *Lutjanus carponotatus* separated by as little as 300 m. Sanmartín *et al.* (2000) compared the helminth parasite communities of conger eel, *Conger conger*, caught in two estuaries in Northwest Spain separated only by a narrow peninsula. Eight parasite species showed significant differences in levels of infection between the estuaries, suggesting the existence of separate populations of conger.

Two studies investigated the stock structure of Greenland halibut *Reinhardtius hippoglossoides* in the northwest Atlantic. Arthur & Albert (1993) examined halibut caught at eight stations in the Gulf of St. Lawrence and off the Newfoundland/Labrador coast. Five helminth parasites met all the selection criteria for use as biological tags. These criteria were: (1) that infections be of relatively long duration (years rather than months), (2) that no parasite reproduction occur on or in the host, and (3) that the

parasite be relatively abundant in at least one of the eight collections examined. By analysing prevalence and intensity data on the selected parasites the authors were able to separate stocks of halibut from the Gulf of St. Lawrence from those from two adjacent areas with an extremely high degree of accuracy. They concluded that this method provided fisheries management with a powerful tool to study the discreteness of stocks and to follow migrations of Greenland halibut in Canadian waters. Boje, Riget & Køie (1997) examined the metazoan parasite faunas of Greenland halibut from five stations around Greenland and one off Labrador and selected three helminths as stock discriminants. Analyses of the parasite data suggested that halibut populations off southwest Greenland and in the Denmark Strait were separate from those off the west coast of Greenland and off Labrador. Their conclusions with regard to halibut stock structure in the area supported those from previous investigations using other methods of stock discrimination.

New recruits to a seamount population of armorhead, *Pseudopentaceros wheeleri*, in the North Pacific were identified by Humphreys, Crossler & Rowland (1993) using a monogenean gill parasite, *Microcotyle macropharynx*. The parasite was absent from epipelagic pre-recruit fish but common on the seamount population. Identification of new recruits to the seamount was based on the absence of mature parasites. Comiskey & MacKenzie (2000) suggested the use of the acanthocephalans *Corynosoma* spp. to identify adult saithe, *Pollachius virens*, of Scottish coastal nursery origin in catches from offshore areas in the northern North Sea. The juvenile acanthocephalans were found in 19% of juvenile saithe from Scottish coastal nursery grounds, whereas saithe from Norwegian coastal nurseries were uninfected.

Small pelagic fish

Most of the recent biological tag studies on small pelagic fish have been on carangid species of the genus *Trachurus*, particularly the Pacific jack mackerel *Trachurus symmetricus murphyi*. A series of publications by a group of Chilean workers have led to the conclusion that there are two distinct stocks of jack mackerel along the Pacific coast of South America, one off the south-central coast of Chile and the other along the northern coast of Chile and the coast of Peru. George-Nascimento & Arancibia (1992) used a combination of parasite tags and host morphometrics to provide evidence that offshore mackerel migrated towards the coastal zone in winter in south-central Chile, but they detected no such pattern in populations further north. They found the most useful tag parasites to be larval nematodes of the genera *Anisakis* and *Hysterothylacium*, a juvenile acanthocephalan *Bolbosoma* sp., and an

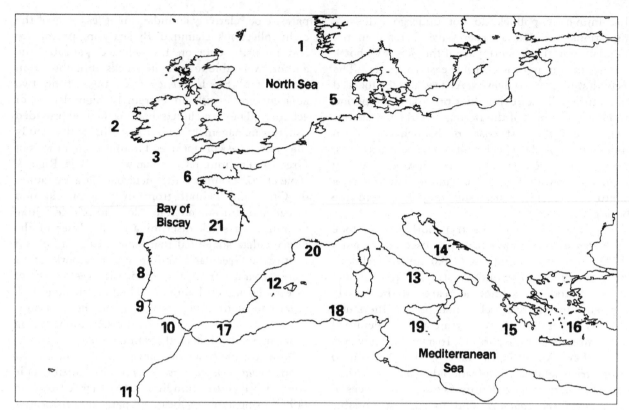

Fig. 1. Map of northeast Atlantic and Mediterranean study area of HOMSIR project on stock identification of
Trachurus trachurus, showing sampling stations.

isopod *Meinertia gaudichaudii*. Certain reproductive features of two other isopod species of the genus *Ceratothoa* showed significant differences between samples from jack mackerel caught off south-central and northern Chile, suggesting that the host populations in these regions belong to separate stocks (Aldana, Oyarzún & George-Nascimento, 1995). George-Nascimento (2000) then analysed the metazoan parasite communities of jack mackerel caught in these two fishing zones and found that his results supported the hypothesis of two stocks suggested by the earlier studies. Avdeev (1992) studied the geographical distributions of the two species of *Ceratothoa* used by Aldana *et al.* (1995) on *Trachurus* spp. across the Pacific Ocean and found evidence to support results from studies of host morphological characters that the oceanic population of *T. symmetricus murphyi* is distinct from that along the South American continental shelf.

Larvae of the nematode *Anisakis simplex* were used by Abaunza, Villamor & Pérez (1995) as tags for horse mackerel, *Trachurus trachurus*, caught to the north and northwest of Spain. They found the mean abundance of *A. simplex* to be significantly higher in Galician waters (northwest) than off the north coast, a result which casts doubts on the present definition of horse mackerel stocks in the area. The present author is a partner in the project HOMSIR: 'A Multidisciplinary Approach using Genetic Markers and Biological Tags in Horse Mackerel (*Trachurus*

trachurus) Stock Structure Analysis' (Fig. 1). The use of parasites as biological tags is one of several methods of stock identification employed in this study, and early data on the occurrence of long-lived anisakid nematode larvae of the genera *Anisakis* and *Hysterothylacium* show considerable promise. Horse mackerel caught in the North Sea are currently managed as a separate stock from those caught to the west of the British Isles. The mean abundances of the two nematode genera in horse mackerel show highly significant differences between these two areas (Fig. 2), a result which supports the current management structure. In the North Sea *Hysterothylacium* is the dominant genus, whereas *Anisakis* is dominant to the west of the British Isles. However, in the North Sea sample examined in 2000, three individual fish stood out starkly from the 50 other fish in the sample by virtue of the fact that they were each infected with large numbers (> 250) of *Anisakis*, whereas the mean abundance for the rest of the sample was only 5·0, with almost one-third of fish uninfected. Furthermore, the mean abundance of *Hysterothylacium* in these three anomalous fish was 53·3, as compared to 201·7 for the rest of the sample. This result suggests that some immigration of horse mackerel from the western stock to the North Sea stock does occur and provides a means of estimating the size of the immigrant component at any given time. Enzyme electrophoretic analyses of *Anisakis* samples from different parts of the study area have

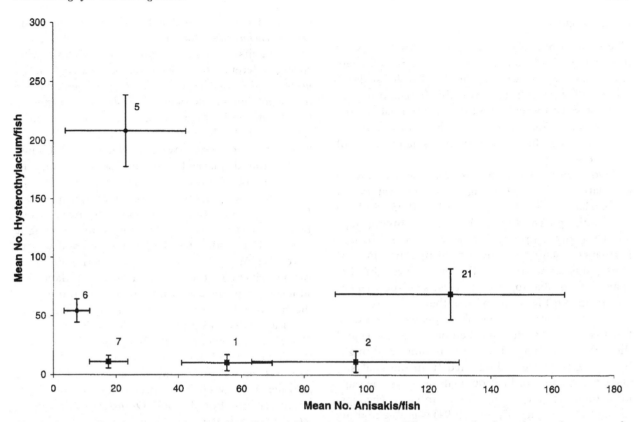

Fig. 2. Scatterplot of mean numbers of *Anisakis* and *Hysterothylacium* in different samples from the northern part of the HOMSIR study area (see Fig. 1). Horizontal and vertical lines are 95 % confidence limits. Note the isolated position of the North Sea sample No. 5.

so far revealed the presence of three different sibling species of the genus in horse mackerel, with markedly different geographical distributions (Dr S. Mattiucci, personal communication).

Somdal & Schram (1992) attempted to use eight species of ectoparasites (three monogeneans and five copepods) to identify mackerel *Scomber scombrus* originating from different nursery grounds in the Northwest Atlantic. They concluded that none were of value as tags using variations in their occurrence, but discussed the possible use of observed variations in the lengths of hamuli of the monogenean *Kuhnia scombri* as an indicator. A more successful study of stock identification of Japanese mackerel, *Scomber japonicus*, in the Northwest Pacific was carried out by Pozdnyakov & Vasilenko (1994), who collected data on helminth parasite infections of mackerel in the study area over a period of nine years. They found that mackerel populations in coastal and oceanic regions differed markedly in the composition of their parasite faunas. The parasite data supported other data on the distributions of different life history stages of mackerel by suggesting the existence of separate northern and southern populations. The parasite fauna of the northern population was dominated by parasites which complete their life cycles on the continental shelf, indicating no offshore feeding migration, whereas the parasite fauna of the

southern population included parasites characteristic of the oceanic region.

Another study based on long-term data is that of Moser & Hsieh (1992), who sampled Pacific herring, *Clupea harengus pallasi*, over a period of 10 years from three sites in northern California. The parasites selected as tags to distinguish between two spawning populations and to suggest that these populations remained separate outside the spawning season were anisakid nematode larvae, plerocercoids of the cestode *Lacistorhynchus dollfusi* and the adult digenean *Parahemiurus merus*.

Parasites enabled Yamaguchi & Honma (1992) to show the probable migration routes of Pacific saury, *Cololabis saira*, between the Sea of Okhotsk and parts of the western North Pacific. The parasites used were the copepods *Pennella* sp., *Caligus macarovi* and *Bomolochus* sp., the isopod *Irona melanostica*, and the acanthocephalan *Rhadinorhynchus* sp. Arthur & Albert (1996) investigated the parasite fauna of capelin, *Mallotus villosus*, in the St. Lawrence estuary and gulf and selected three parasites as potentially useful stock discriminators: the protozoan *Microsporidium* sp. and two larval anisakid nematodes. Nonparametric discriminant analyses of the infection data provided no evidence to support the concept of separate stocks of capelin within the study area.

Large pelagic fish

The main problem with using parasites as tags for this group of fish is that the hosts are in relatively scarce supply and are extremely valuable, so large numbers are not normally available for examination. Full use must therefore be made of each fish, so the most efficient approach is to analyse entire parasite communities rather than focus on a small number of parasite species.

Two biological tag studies were carried out on populations of large pelagic fish in waters off Queensland, Australia (Speare, 1994, 1995). In both studies the parasites were classified in three groups according to their probable longevity in the fish host: permanent, semi-permanent or temporary. Permanent parasites were those that had a probable life span of from one to several years, semi-permanent had probable life spans of less than one year, and temporary parasites were mobile species which were free to move between hosts. Temporary parasites and those that were free in the stomach lumen and might be regurgitated were eliminated from the analysis; classification and ordination analyses were then applied to infection data for the remaining two groups. Using this approach, Speare (1994) was able to draw conclusions from eight permanent and nine semi-permanent species of parasite regarding age-related migrations of black marlin, *Makaira indica*, along the coast of Queensland. He also applied the same methods to sailfish, *Istiophorus platypterus*, and found evidence of discrete subpopulations in northern and southern parts of his study area (Speare, 1995). Lester *et al.* (2001) also divided parasites into temporary and permanent groups in their impressive study of movement and stock structure of narrow-barred Spanish mackerel, *Scomberomorus commerson*. This study covered a huge area, with nine sampling sites to the west and north of Australia, including one in Indonesian waters. The aim of the study was to provide information that would help to provide a basis for fishery management agreements between Australia and Indonesia and between different Australian states. Multidimensional scaling and area cluster analysis were applied to seven temporary and seven permanent parasite species. The resultant data indicated extensive mixing of Australian west coast populations and more isolated populations along the north coast and in Indonesia. Temporary parasite data suggested different migratory patterns of male and female mackerel along the west coast of Australia. The temporary species used were all monogenean and copepod gill parasites; the permanent parasites were trypanorhynch metacestodes and anisakid nematode larvae embedded in fish tissue.

Anadromous fish

The use of freshwater parasites as tags to follow the oceanic distribution of Pacific salmon of the genus *Oncorhynchus* has a long history. This was reviewed by Margolis (1992) and Konovalov (1995), with special reference to sockeye salmon, *Oncorhynchus nerka*. In the past decade much use has been made of myxosporeans of the genus *Myxobolus*, parasites of the brain and spinal cord that infect juvenile salmonids in fresh water and serve as tags for mature fish on the high seas. Awakura, Nagasawa & Urawa (1995) found marked geographical variations in prevalence of *Myxobolus arcticus* and *M. neurobius* in adult masu salmon, *Oncorhynchus masou*, from seven Japanese rivers, and from maturing masu from Japanese coastal waters. Infections in sea-caught masu indicated that fish in Japanese coastal waters were a complex of mixed stocks originating from various rivers in the region. Prevalence data also supported a proposed autumn migration route from the Sea of Okhotsk to the Sea of Japan. Urawa & Nagasawa (1995) examined samples of five species of *Oncorhynchus* caught on the high seas of the North Pacific Ocean and the Bering Sea for infections with *M. arcticus*. They found marked differences in prevalence between host species and between different regions. For chinook, *Oncorhynchus tshawytscha*, in particular, prevalence was high in the northwest Pacific, but infected fish were rare in the northeast Pacific and in the Bering Sea. Urawa *et al.* (1998) reported that *M. arcticus* was present in 57–94 % of chinook originating from Asian rivers, but was rare in North America. Another species, *M. kisutch*, was found only in chinook originating from the Columbia River and vicinity, and so could serve as a marker for this population. The overall sample prevalences of *M. arcticus* in Asian and North American stocks were 67·7 % and 2·3 % respectively. Examinations of chinook caught on the high seas showed that fish of Asian origin were predominant west of 180° west and that up to 98 % of fish in the Bering Sea were of North American origin.

A related issue arising from the use of *M. arcticus* as a biological tag is its use in 'forensic parasitology'. A court case was described by Margolis (1993) in which a roadside fish vendor in British Columbia, Canada, was convicted of selling sockeye salmon that had been caught unlawfully. Part of the evidence leading to the conviction was based on the prevalence of *M. arcticus* in a sample of the suspect salmon. Prevalence had previously been shown to be markedly different in juvenile sockeye originating from each of the two nursery lakes within the river system. When adult sockeye returned on their spawning migration, the proportions native to the two lakes could be determined by the prevalence of *M. arcticus* in samples taken at different times from the lower reaches of the river. In this instance, examination of the suspect sample showed that the fish could only have been caught at a time when the commercial fishery had been closed.

Tseitlein & Afanasiev (1996) carried out isozyme analysis of *Anisakis* nematode larvae from humpback (pink) salmon, *Oncorhynchus gorbuscha*, caught at nine localities in Russian waters of the northwest Pacific from Sakhalin to Iturup Island. The results suggested the absence of segregated stocks and mixing of humpback populations in the study area.

Two reports reviewed and summarized information on the distribution and origins of Pacific salmonids caught on the high seas derived from the results of various studies, including the use of parasites. Burgner *et al.* (1992) described the use of two freshwater digeneans to determine the oceanic distribution of steelhead trout, *Oncorhynchus mykiss*, originating from rivers in the U.S. Pacific Northwest. The digeneans, adult *Plagioporus shawi* and metacercariae of *Nanophyetus salmincola*, occur in juvenile trout from this region only and serve as markers for this population in their adult marine phase. Myers *et al.* (1993) described the early results from the use of *Myxobolus* spp. as tags for chinook salmon.

Rulifson & Dadswell (1995) investigated the stock structure and migrations of striped bass, *Morone saxatilis*, in rivers and coastal waters of Atlantic Canada, by reviewing results from a variety of methods of stock discrimination, including parasites as tags. Data on parasite communities assisted in determining the range of ocean migration and site of origin of bass populations and contributed to the conclusion that some fish originating from rivers further south in the United States occurred in Canadian waters.

Elasmobranch fish

There appear to have been no published reports in the past decade of the use of parasites as tags in population studies of elasmobranch fish. This is disappointing considering the considerable potential that exists for the successful use of parasites in this field, as suggested by MacKenzie (1987) and Caira (1990).

MARINE MAMMALS

In the last decade two studies have been published dealing with parasites as tags in population studies of cetaceans. Balbuena & Raga (1994) examined specimens of the long-finned pilot whale, *Globicephala melas*, sampled from seven pods around the Faeroe Islands. They selected four of the most common helminth parasites as potentially useful tags and compared their mean abundances among pods. Significant differences between pods, particularly in the occurrence of the acanthocephalan *Bolbosoma capitatum*, supported previous pollutant and genetic studies in suggesting a degree of segregation, but the evidence was not considered conclusive enough to confirm the existence of separate stocks. Aznar *et al.*

(1995) compared the helminth parasite faunas of franciscanas, *Pontoporeia blainville*, from two localities off Argentina and one off Uruguay. They found considerable differences between parasite faunas in Argentinian and Uruguayan samples which suggested that the two populations may constitute separate stocks. The results must be treated with caution, however, because samples from the two regions were taken 15–18 years apart. The use of parasites as biological indicators for marine mammals was reviewed by Balbuena *et al.* (1995), who presented examples to illustrate both the value and limitations of the method, and observed that it has yet to reach its full potential for this host group.

MARINE INVERTEBRATES

Most studies carried out to date on parasites as tags for marine invertebrates relate to cephalopods, and particularly squid, as hosts. The past decade is no exception. Pascual & Hochberg (1996) discussed the use of parasites as tags in population studies of cephalopods and highlighted the problems involved with this particular host group. Foremost among these problems are the short life-spans and variable growth rates of the most heavily exploited cephalopod species, together with the fact that many species have protracted spawning seasons, so that multiple cohorts may be present in a population at any given time. Nevertheless, these authors showed that certain parasites can be used effectively to identify subpopulations within stocks of cephalopods. They particularly recommended the use of genetic studies of selected helminth larvae, particularly anisakid nematodes. Pascual *et al.* (1996) found that two species of short-finned squid caught off the northwest coast of Spain were infected with the larvae of only one sibling species of the nematode genus *Anisakis* – *A. simplex* B. From the known geographical distribution of this and another sibling species, *Anisakis typica*, with a more southern range, they concluded that the squid in their study area did not undertake long migrations, which would have taken them into the endemic area of *A. typica*. A study by Nagasawa, Mori & Okamura (1998) on parasites as tags for neon flying squid, *Ommastrephes bartrami*, covered a huge area of the North Pacific. Two cohorts of squid, winter-spring and autumn, were treated separately. Geographical differences in the prevalences and mean intensities of the larval nematode *Lappetascaris* sp. Type A, the tetraphyllidean metacestode *Phyllobothrium* sp. and the trypanorhynch metacestode *Tentacularia* sp., led to the conclusion that each cohort consists of two separate stocks. For González & Kroeck (2000), differences in parasite communities provided substantial additional evidence for the separate nature of the stock of shortfin squid, *Illex argentinus*, in the San Matías Gulf from stocks in neighbouring parts of the

Argentine continental shelf. Shukhgalter & Nigma-tullin (2001) surveyed the helminth faunas of jumbo squid, *Dosidicus gigas*, from four open ocean regions of the east Pacific. They found marked differences between Peruvian and east equatorial waters in the ontogenetic infection dynamics of several parasites, which supported the hypothesis of isolated populations in the two areas. Overall, anisakid nematode larvae have proved to be the most effective tags in population studies of squid, although cestode plerocercoids have also been used to good effect.

Timofeev (1997) suggested the use of the parasitic dinoflagellate *Ellobiopsis chattoni* to identify local populations of the calanoid copepod *Calanus finmarchicus* in the Norwegian Sea and Arctic Ocean. The author found highly significant differences in prevalence between samples from the Norwegian Sea and Spitzbergen.

CONCLUSIONS

The methods and criteria applied to studies of parasites as biological tags for populations of marine hosts have changed and evolved over the years. The main trends observed in studies carried out during the past decade or so are as follows.
(1) Multivariate statistical analyses of data on entire parasite communities are increasingly being used. Such studies are particularly useful and sometimes essential for studies of large valuable host species, when large samples are difficult or expensive to come by, but in recent years they have also been used effectively in studies of smaller, less valuable species, such as hake, *Merluccius* spp. It is time-consuming to count all parasites present in each individual host, but this problem can be minimized by taking subsamples of the more numerous parasite taxa, as in Lester *et al.* (2001). (2) Sibling species, for example of anisakid nematodes, identified in recent years, are increasingly being investigated as tools for stock discrimination of marine fish and cephalopods. The intensity of nematode infection is usually such that it would be impractical to identify each individual nematode to species, so the usual approach is to take subsamples and estimate proportions of the different sibling species present within any single host sample, as in the current HOMSIR project on European horse mackerel. (3) The term 'ecological stock' is being used with increasing frequency. This term is used to describe subpopulations which are distinguished by behavioural differences, but between which there is still a considerable amount of gene flow. One of the strengths of biological tagging is that parasites can often be used to identify such subpopulations where genetic studies may fail to do so. (4) More use is being made of the multidisciplinary approach, using a variety of methods including the use of biological tags, to investigate host population structure.

It is noticeable that many recent studies have concentrated on metazoan parasites and take no account of protozoans or myxozoans. This is unfortunate because these groups have provided useful tag species in the past. Another disappointing feature of recent biological tag history is the absence of any study on the population biology of an elasmobranch host. Elasmobranchs, with their distinctive parasite faunas including many highly host-specific species, would seem to be excellent subjects for such studies.

ACKNOWLEDGEMENTS

I am grateful to my partners in the HOMSIR project for their permission to use some unpublished results in this paper. Thanks are also due to Bob Kabata and Andrey Karasev for providing copies of some important North American and Russian publications.

REFERENCES

ABAUNZA, P., VILLAMOR, B. & PÉREZ, J. R. (1995). Infestation by larvae of *Anisakis simplex* (Nematoda: Ascaridata) in horse mackerel, *Trachurus trachurus*, and Atlantic mackerel, *Scomber scombrus*, in ICES Divisions VIIIb, VIIIc and IXa (N-NW of Spain). *Scientia Marina* **59**, 223–233.

ALDANA, M., OYARZÚN, J. & GEORGE-NASCIMENTO, M. (1995). Isopodos parasitos como indicadores poblacionales del jurel *Trachurus symmetricus murphyi* (Nichols, 1920) (Pisces: Carangidae) frente a las costas de Chile. *Biologia Pesquera* **24**, 23–32.

ARTHUR, J. R. (1997). Recent advances in the use of parasites as biological tags for marine fish. In *Diseases in Asian Aquaculture III* (ed. Flegel, T. W. & MacRae, I. H.), pp. 141–154. Manila, Fish Health Section, Asian Fisheries Society.

ARTHUR, J. R. & ALBERT, E. (1993). Use of parasites for separating stocks of Greenland halibut (*Reinhardtius hippoglossoides*) in the Canadian northwest Atlantic. *Canadian Journal of Fisheries and Aquatic Sciences* **50**, 2175–2181.

ARTHUR, J. R. & ALBERT, E. (1996). Parasites as potential biological tags for capelin (*Mallotus villosus*) in the St. Lawrence estuary and gulf. *Canadian Technical Report of Fisheries and Aquatic Sciences* **No. 2112**, 9 pp.

AVDEEV, G. V. (1996). Infestation by helminths and redistribution of immature walleye pollock *Teragra chalcogramma* in the Sea of Okhotsk. *Journal of Ichthyology* **36**, 665–673.

AVDEEV, G. V. & AVDEEV, V. V. (1998). Parasites as indicators of *Theragra chalcogramma* (Gadidae) populations from the Comandor Islands (In Russian). *Parazitologiya* **32**, 431–439.

AVDEEV, V. V. (1992). On the possible use of parasitic isopods as bioindicators of the migratory routes of horse mackerels in the Pacific Ocean. *Journal of Ichthyology* **32**, 14–21.

AWAKURA, T., NAGASAWA, K. & URAWA, S. (1995). Occurence of *Myxobolus arcticus* and *M. neurobius*

(Myxozoa: Myxosporea) in masu salmon (*Oncorhynchus masou*) from northern Japan. *Scientific Reports of the Hokkaido Salmon Hatchery* No. 49, 35–40.

AZNAR, F. J., RAGA, J. A., CARCUERA, J. & MANZÓN, F. (1995). Helminths as biological tags for franciscana (*Pontoporia blainvillei*) (Cetacea, Pontoporiidae) in Argentinian and Uruguayan waters. *Mammalia* 59, 427–435.

BAKAY, Y. (1999). Ecological and geographical analysis of *Sebastes mentella* parasitic fauna in the North Atlantic (Abstract). *Bulletin of the Scandinavian Society for Parasitology* 9, 28.

BALBUENA, J. A., AZNAR, F. J., FERNANDEZ, M. & RAGA, J. A. (1995). Parasites as indicators of social structure and stock identity of marine mammals. In *Whales, Seals, Fish and Man* (ed. Blix, A. S., Walloe, L. & Ulltang, O.), Developments in Marine Biology 4 pp. 133–140. Amsterdam, Elsevier.

BALBUENA, J. A. & RAGA, J. A. (1994). Intestinal helminths as indicators of segregation and social structure of pods of long-finned pilot whales (*Globicephala melas*) off the Faeroe Islands. *Canadian Journal of Zoology* 72, 443–448.

BOJE, J., RIGET, F. & KØIE, M. (1997). Helminth parasites as biological tags in population studies of Greenland halibut (*Reinhardtius hippoglossoides* (Walbaum)), in the north-west Atlantic. *ICES Journal of Marine Science* 54, 886–895.

BURGNER, R. L., LIGHT, J. T., MARGOLIS, L., OKAZAKI, T., TAUTZ, A. & ITO, S. (1992). Distribution and origins of steelhead trout (*Oncorhynchus mykiss*) in offshore waters of the North Pacific Ocean. *International North Pacific Fisheries Commission Bulletin* No. 51, 92 pp.

CAIRA, J. N. (1990). Metazoan parasites as indicators of elasmobranch biology. In *Elasmobranchs as Living Resources: Advances in the Biology, Ecology, Systematics, and the Status of the Fisheries.* (ed. Pratt, H. L. Jr., Gruber, S. H. & Taniuchi, T.), pp. 71–96. *NOAA Technical Report* 90.

CAPPO, M. (1992). Australian salmon. Parasites as biological tags. *Safish Magazine* 17, 7–10.

COMISKEY, P. & MACKENZIE, K. (2000). *Corynosoma* spp. may be useful biological tags for saithe in the northern North Sea. *Journal of Fish Biology* 57, 525–528.

CRIBB, T. H., ANDERSON, G. R. & DOVE, A. D. M. (2000). *Pomphorhynchus heronensis* and restricted movement of *Lutjanus carponotatus* on the Great Barrier Reef. *Journal of Helminthology* 74, 53–56.

GEORGE-NASCIMENTO, M. (1996). Populations and assemblages of parasites in hake, *Merluccius gayi*, from the southeastern Pacific Ocean: stock implications. *Journal of Fish Biology* 48, 557–568.

GEORGE-NASCIMENTO, M. (2000). Geographical variations in the jack mackerel *Trachurus symmetricus murphyi* populations in the southeastern Pacific Ocean as evidenced from the associated parasite communities. *Journal of Parasitology* 86, 929–932.

GEORGE-NASCIMENTO, M. & ARANCIBIA, H. (1992). Stocks ecológicos del jurel (*Trachurus symmetricus murphyi* Nichols) en tres zonas de pesca frente a Chile, detectados mediante comparación de su fauna parasitaria y morfometría. *Revista Chilena de Historia Natural* 65, 453–470.

GEORGE-NASCIMENTO, M. & ARANCIBIA, H. (1994). The parasite fauna and morphometry of the southern hake *Merluccius australis* (Hutton) as indicators of stock units. *Biologia Pesquera* 23, 31–47.

GONZÁLEZ, R. A. & KROECK, M. A. (2000). Enteric helminths of the shortfin squid *Illex argentinus* in San Matias Gulf (Argentina) as stock discriminants. *Acta Parasitologica* 45, 89–93.

GRUTTER, A. S. (1998). Habitat-related differences in the abundance of parasites from a coral reef fish: an indication of the movement patterns of *Hemigymnus melapterus*. *Journal of Fish Biology* 53, 49–57.

HEMMINGSEN, W. & MACKENZIE, K. (2001). The parasite fauna of the Atlantic cod, *Gadus morhua* L. *Advances in Marine Biology* 40, 1–80.

HERRINGTON, W. C., BEARSE, H. M. & FIRTH, F. E. (1939). Observations on the life history, occurrence and distribution of the redfish parasite *Sphyrion lumpi*. *United States Bureau of Fisheries Special Report* No. 5, 1–18.

HUMPHREYS, R. L. JR., CROSSLER, M. A. & ROWLAND, C. M. (1993). Use of a monogenean gill parasite and feasibility of condition indices for identifying new recruits to a seamount population of armorhead *Pseudopentaceros wheeleri* (Pentacerotidae). *Fishery Bulletin* 91, 455–463.

JONES, M. E. B. & TAGGART, C. T. (1998). Distribution of gill parasite (*Lernaeocera branchialis*) infection in Northwest Atlantic cod (*Gadus morhua*) and parasite-induced host mortality: inferences from tagging data. *Canadian Journal of Fisheries and Aquatic Sciences* 55, 364–375.

KARASEV, A. B. (1994). Use of parasites as biological tags when studying intraspecific structure of cod in the coastal areas of Russia and Norway (Abstract). *Bulletin of the Scandinavian Society for Parasitology* 4(2), 17.

KARASEV, A. B. (1998). On the use of parasites in studies of Arcto-Norwegian cod population structure (In Russian). In *Parasites and Diseases of Marine and Freshwater Fishes of the North Basin: Selected Papers* (ed. Karasev, A. B.), pp. 22–33. Murmansk, Russia, PINRO Press.

KHAN, R. A. & TUCK, C. (1995). Parasites as biological indicators of stocks of Atlantic cod (*Gadus morhua*) off Newfoundland, Canada. *Canadian Journal of Fisheries and Aquatic Sciences* 52 (Suppl. 1), 195–201.

KONOVALOV, S. M. (1995). Parasites as indicators of biological processes, with special reference to sockeye salmon (*Oncorhynchus nerka*). *Canadian Journal of Fisheries and Aquatic Sciences* 52 (Suppl. 1), 202–212.

LARSEN, G., HEMMINGSEN, W., MACKENZIE, K. & LYSNE, D. A. (1997). A population study of cod, *Gadus morhua* L. in northern Norway using otolith structure and parasite tags. *Fisheries Research* 32, 13–20.

LESTER, R. J. G., THOMPSON, C., MOSS, H. & BARKER, S. C. (2001). Movement and stock structure of narrow-banded Spanish mackerel as indicated by parasites. *Journal of Fish Biology* 59, 833–843.

MACKENZIE, K. (1987). Parasites as indicators of host populations. *International Journal for Parasitology* 17, 345–352.

MACKENZIE, K. & ABAUNZA, P. (1998). Parasites as biological tags for stock discrimination of marine fish: a guide to procedures and methods. *Fisheries Research* **38**, 45–56.

MACKENZIE, K. & LONGSHAW, M. (1995). Parasites of the hakes *Merluccius australis* and *M. hubbsi* in the waters around the Falkland Islands, southern Chile, and Argentina, with an assessment of their potential value as biological tags. *Canadian Journal of Fisheries and Aquatic Sciences* **52** (Suppl. 1), 213–224.

MARGOLIS, L. (1992). A brief history of Canadian research from 1955 to 1990 related to Pacific salmon (*Oncorhynchus* species) on the high seas. *International North Pacific Fisheries Commission Special Publication* No. 20, 79 pp.

MARGOLIS, L. (1993). A case of forensic parasitology. *Journal of Parasitology* **79**, 461–462.

MATTIUCCI, S., NASCETTI, G., TORINTI, E., RAMADORI, L., ABAUNZA, P. & PAGGI, L. (2000). Metazoan parasite communities of European hake (*Merluccius merluccius*) from Mediterranean and Atlantic waters: stock implications (Abstract). *Acta Parasitologica* **45**, 265.

MCCLELLAND, G. & MARCOGLIESE, D. J. (1994). Larval anisakine nematodes as biological indicators of cod (*Gadus morhua*) populations in the southern Gulf of St. Lawrence and on the Breton Shelf, Canada. *Bulletin of the Scandinavian Society for Parasitology* **4**, 97–116.

MOLES, A., HEIFITZ, J. & LOVE, D. C. (1998). Metazoan parasites as potential markers for selected Gulf of Alaska rockfishes. *Fishery Bulletin* **96**, 912–916.

MOSER, M. & HSIEH, J. (1992). Biological tags for stock separation in Pacific herring *Clupea harengus pallasi* in California. *Journal of Parasitology* **78**, 54–60.

MOSQUERA, J., GÓMEZ-GESTEIRA, M. & PÉREZ-VILLAR, V. (2000). Using parasites as biological tags of fish populations: a dynamical model. *Bulletin of Mathematical Biology* **62**, 87–99.

MYERS, K. W., HARRIS, C. K., ISHIDA, Y., MARGOLIS, L. & OGURA, M. (1993). Review of the Japanese landbased driftnet salmon fishery in the western North Pacific Ocean and the continent of origin of salmonids in this area. *International North Pacific Fisheries Commission Bulletin* No. 52, 86 pp.

NAGASAWA, K., MORI, J. & OKAMURA, H. (1998). Parasites as biological tags of stocks of neon flying squid (*Ommastrephes bartrami*) in the North Pacific Ocean. In *Contributed Papers to International Symposium on Large Pelagic Squids* (ed. Okutani, T.), pp. 49–64. Tokyo, Japan Marine Fisheries Research Center.

OLIVA, M. E. (2001). Metazoan parasites of *Macruronus magellanicus* from southern Chile as biological tags. *Journal of Fish Biology* **58**, 1617–1624.

PASCUAL, S., GONZALEZ, A., ARIAS, C. & GUERRA, A. (1996). Biotic relationships of *Illex coindetti* and *Todaropsis eblanae* (Cephalopoda, Ommastrephidae) in the northeast Atlantic: evidence from parasites. *Sarsia* **81**, 265–274.

PASCUAL, S. & HOCHBERG, F. G. (1996). Marine parasites as biological tags of cephalopod hosts. *Parasitology Today* **12**, 324–327.

POZDNYAKOV, S. E. & VASILENKO, A. V. (1994). Distribution, migration, and the helminth fauna of the Japanese mackerel, *Scomber japonicus*, in the Northwestern Pacific Ocean. *Journal of Ichthyology* **34**, 74–91.

REIMER, L. W. (1993). Parasites of *Merluccius capensis* and *M. paradoxus* from the coast of Namibia. *Applied Parasitology* **34**, 143–150.

RULIFSON, R. A. & DADSWELL, M. J. (1995). Life history and population characteristics of striped bass in Atlantic Canada. *Transactions of the American Fisheries Society* **124**, 477–507.

SANMARTÍN, M. I., ALVAREZ, M. F., PERIS, D., IGLESIAS, R. & LEIRO, J. (2000). Helminth parasite communities of the conger eel in the estuaries of Arousa and Muros (Galicia, north-west Spain). *Journal of Fish Biology* **57**, 1122–1133.

SARDELLA, N. H. & TIMI, J. T. (1999). Parasite communities of *Merluccius hubbsi* (hake) from two areas of the Argentine Sea (abstract). *Book of Abstracts. 5th International Symposium on Fish Parasites*, p. 129. Institute of Parasitology, Academy of Sciences of the Czech Republic, 9–13 August, 1999, ČeskéBudějovice, Czech Republic.

SEWELL, K. B. & LESTER, R. J. G. (1995). Stock composition and movement of gemfish, *Rexea solandri*, as indicated by parasites. *Canadian Journal of Fisheries and Aquatic Sciences* **52** (Suppl. 1), 225–232.

SHUKHGALTER, O. A. & NIGMATULLIN, C. M. (2001). Parasitic helminths of jumbo squid *Dosidicus gigas* (Cephalopoda: Ommastrephidae) in open waters of the central east Pacific. *Fisheries Research* **54**, 95–110.

SOBECKA, E. & KILIAN, K. (2000). Parasites of cod *Gadus morhua* Linnaeus, 1758 as biological tags (Abstract). *Acta Parasitologica* **45**, 266–267.

SOMDAL, O. & SCHRAM, T. A. (1992). Ectoparasites on Northeast Atlantic mackerel (*Scomber scombrus* L.) from Western and North Sea stocks. *Sarsia* **77**, 19–31.

SPEARE, P. (1994). Relationships among black marlin, *Makaira indica*, in eastern Australian coastal waters, inferred from parasites. *Australian Journal of Marine and Freshwater Research* **45**, 535–550.

SPEARE, P. (1995). Parasites as biological tags for sailfish *Istiophorus platypterus*, from east coast Australian waters. *Marine Ecology Progress Series* **118**, 43–50.

STANLEY, R. D., LEE, D. L. & WHITAKER, D. J. (1992). Parasites of yellowtail rockfish, *Sebastes flavidus* (Ayres, 1861) (Pisces: Teleostei), from the Pacific coast of North America as potential biological tags for stock identification. *Canadian Journal of Zoology* **70**, 1086–1096.

TIMOFEEV, S. F. (1997). Occurence of the parasitic dinoflagellate *Ellobiopsis chattoni* (Protozoa: Mastigophora) on the copepod *Calanus finmarchicus* (Crustacea: Copepoda) and the possibility of using the parasite as a biological tag of local populations (In Russian). *Parazitologiya* **31**, 334–340.

TSEITLEIN, D. G. & AFANASIEV, K. I. (1996) Isozyme analysis of *Anisakis simplex* (Ascaridida: Anisakidae) from humpback salmon of Sakhalin and Iturup Islands (In Russian). In *Voprosy Populyatsionnai Biologii Parazitiv*, pp. 122–130. Institute of Parasitology, Russian Academy of Sciences, Moscow.

URAWA, S. & NAGASAWA, K. (1995). Prevalence of *Myxobolus arcticus* (Myxozoa: Myxosporea) in five species of Pacific salmon in the North Pacific Ocean and Bering Sea. *Scientific Reports of the Hokkaido Salmon Hatchery* **No. 49**, 11–19.

URAWA, S., NAGASAWA, K., MARGOLIS, L. & MOLES, A. (1998). Stock identification of Chinook salmon (*Oncorhynchus tshawytscha*) in the North Pacific Ocean and Bering Sea by parasite tags. *North Pacific Anadromous Fish Commission Bulletin* **No. 1**, 199–204.

WILLIAMS, H. H., MACKENZIE, K. & MCCARTHY, A. M.

(1992). Parasites as biological indicators of the population biology, migrations, diet and phylogenetics of fish. *Reviews in Fish Biology and Fisheries* **2**, 144–176.

YAMAGUCHI, M. & HONMA, T. (1992). Parasitological study of the migration route of the Pacific saury, *Cololabis saira*, to the Okhotsk Sea (In Japanese). *Scientific Reports of the Hokkaido Fisheries Experimental Station* **39**, 35–44.

A review of the population biology and host–parasite interactions of the sea louse *Lepeophtheirus salmonis* (Copepoda: Caligidae)

O. TULLY[1] *and* D. T. NOLAN[2]

[1] *Department of Zoology, Trinity College Dublin, Dublin 2, Ireland*
[2] *Department of Animal Ecology and Ecophysiology, Faculty of Science, Toernooiveld 1, University of Nijmegen, 6525 ED Nijmegen, The Netherlands*

SUMMARY

Lepeophtheirus salmonis is a specific parasite of salmonids that occurs in the Atlantic and Pacific Oceans. When infestations are heavy fish mortality can occur although the factors that are responsible for causing epizootics, especially in wild salmonid populations are still largely unknown. Over the past 20 years this parasite has caused significant economic losses in farmed salmon production and possibly in wild salmonid populations locally. Understanding the connectivity between populations is crucial to an understanding of the epidemiology of infections and for management of infections in aquaculture. Data from genetics, pesticide resistance, larval dispersal models and spatial and temporal patterns of infestation in wild and farmed hosts suggests a spatially highly structured metapopulation the components of which have different levels of connectivity, probabilities of extinction and influence on the development of local infestations. The population structure is defined mainly by the dispersal dynamics of the planktonic stages and the behaviour of the host.

Until recently virtually nothing was known about the relationship between the parasite and the host, or how the host may influence lice at local or population level. Typically, impacts on the host have usually been reported in terms of pathological lesions caused by attachment and feeding of the adult stages, as well as localised mild epithelial responses to juvenile attachment. However many studies report pathology associated with severe infestation. Recent new studies on the host–parasite interactions of *L. salmonis* have shown that this parasite induces stress-related responses systemically in the host skin and gills and that the stress response and immune systems are modulated. In the second part of this review, these new studies are presented, together with results from other host–parasite model systems where data for caligid sea lice are missing. One of the most revealing methods reported recently is the application of a net confinement stressor to examine modulation of the stress response and immune system of the host fish. This approach has shown that although until now, infective stages of *L. salmonis* were not thought to affect the host, they do induce systematic effects in the host that result in a stress response and modulated immune system. Host–parasite interactions affecting these stress responses and the immune system may be key factors in facilitating epizootics by reducing the host's ability to reject the parasites, as well as reducing disease resistance under some environmental conditions. The host–parasite interaction therefore needs to be incorporated into any model of population structure and dynamics.

Key words: *Lepeophtheirus salmonis*, population structure, host–parasite interaction, stress, review.

INTRODUCTION

The salmon louse *Lepeophtheirus salmonis* Krøyer is a specific parasite of salmonid fishes. It occurs on all species of Pacific Salmon and on species of the genus *Salmo* in the Atlantic Ocean both in open ocean, coastal and estuarine locations (Kabata, 1979). It is an economically important parasite of farmed salmonids where it causes reduction in growth rate, mortality and increased production costs (Costello, 1993). It has also been implicated in incidences of mortality of wild Atlantic and Pacific salmon and sea trout (White, 1940; Johnson *et al.* 1996; Whelan & Poole, 1996).

Over the past 20 years much research has focused on this species because of its potential impact in salmonid aquaculture and on wild populations. Pike & Wadsworth (1999) have comprehensively review-ed the literature. The research is quite diverse from ultrastructural studies on the larval stages, to the mechanics of host location, distribution and behaviour of the free-living stages, the growth and rate of development of the parasitic stages, and the impact of the parasite on the host. A complete understanding of the population dynamics and population structure or the host–parasite interaction has not emerged however. Vital parameters of birth rates, mortality during the free living and parasitic stages, transmission rates, density dependent growth and mortality and the effects of the environment on these parameters are still largely lacking.

Quantitative studies on the distribution and behaviour of the free living stages has also proven difficult although the work of Costelloe *et al.* (1995, 1996, 1998, 1999) has significantly advanced this area of research. In this paper we review various aspects of the biology of *L. salmonis* and use it to interpret what the population structure of the

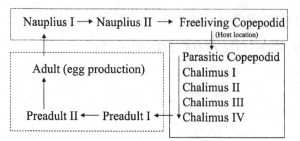

Fig. 1. The life cycle of *Lepeophtheirus salmonis*. The parasitic copepodid and chalimus stages cannot transfer between hosts. Preadults and adults can move freely over the surface of the host and transfer between hosts. The nauplius and free living copepodid are planktonic.

parasite is likely to be. Defining the population structure is vital to an understanding of the epidemiology of infestations. This may be determined mainly by the ecology and behaviour of the free-living stages and by the behaviour of the host population. The host–parasite interaction plays an important role however. This interaction controls vital rate processes of the parasite that affect its production. Similarly the impact of the parasite on the host fish can, during epizootics, cause significant host mortality and changes in behaviour that are also detrimental to the production of the parasite. As such, the host–parasite interaction significantly affects the development of infestations and therefore the potential for transfer of the parasite or gene flow between populations. With respect to the host-parasite interaction Pike & Wadsworth (1999) summarized the data on the pathological effects of sea lice on salmonids in terms of the mechanical disruption caused by parasite attachment and feeding activities. Aside from the pathology frequently reported in the disease state, the extent of the sea lice pathophysiology literature then was limited to the small number of published papers (namely Bjorn & Finstad, 1997; Dawson, 1998; Johnson & Albright, 1992*a*). Since this review many papers presenting data on various aspects of the host–parasite relationship between *L. salmonis* and its salmonid host have been published. This paper reviews this literature using examples of other host–parasite models where data for *L. salmonis* is lacking and identifies and presents promising new directions for further research.

THE LIFE CYCLE

The life cycle of *L. salmonis* and the morphology of the life history stages were described by Johnson & Albright (1991*a*) and by Schram (1993). The life cycle is similar to that of other caligid copepods (Kabata, 1979). Egg strings are produced and held by the adult female. A free-living planktonic nauplius hatches from the egg. Subsequent stages are nauplius II, the copepodid, 4 chalimus stages, 2

preadult stages and the adult male and female (Fig. 1). The copepodid is initially free living in the water column and finds and attaches to the host. Chalimus stages are fixed to the body surface of the host by a frontal filament whereas the pre-adults and adults can move freely over the surface of the host and can also transfer between hosts.

POPULATION STRUCTURE

A spatially explicit model

The sea louse life cycle has two distinct phases, the free-living planktonic phase and the parasitic phase. This cycle is analogous to any free-living marine organism with a planktonic stage and a specific post-settlement habitat requirement in the case where the juveniles and adults are sedentary. In such organisms the processes that control the spatial and temporal pattern of distribution of the larvae determine the population structure (essentially the level of connectivity or exchange that exists between patches of the sedentary adults). Where the connectivity is zero, larvae produced from a particular patch recruit only to that patch. At the other extreme, when connectivity is 1, then the population is open and all larvae produced in the population have equal opportunity to settle onto any habitat patch in the environment. Between those extremes a complex array of probabilities define the actual population structure. For each larva in the population there is a set of probabilities that predicts the patch where it is most likely to settle. For each patch, the recently settled larvae may have originated from any number of neighbouring patches. The patch of origin of each larva settling is determined by the processes that control the connectivity among patches. The level of connectivity between patches therefore dictates whether a habitat patch (in the case of sea lice this is a single fish) is open or closed to outside recruitment. In the case of sea lice it is unlikely to be either extreme; connectivity is neither 0 nor 1. Larvae originating from a fish have a high chance of recruiting to a fish in the same shoal but perhaps less chance to a fish in a separate shoal. How can the population structure be modelled in this case? The spatially explicit meta-population model described by Hanski (1999) for terrestrial populations may be an appropriate template for sea lice populations. In such a model (1) the probability that a population of sea lice becoming established in a habitat patch (a fish or a shoal of fish) is a function of the colonization rate or the rate of immigration from other patches and the extinction rate within the population: (2) Populations irrespective of size may be sources, where the intrinsic rate of natural increase (r) is > 0 or sinks (r < 0) for recruitment: (3) The rate at which populations become extinct is inversely related to patch size: (4) populations contribute to colonization depending on their size and distance from another population. This distance is set by the migration range of the host population and sea louse larvae.

These properties define the patch-specific long-term probability of occupancy or, in this case, the probability of an infestation being established and maintained. The number of patches and their area defines what Hanski calls the neighbourhood habitat area. Increasing this area increases the probability of habitat occupancy and the stability of the metapopulation. This may be described by a logistic function. In the case of *L. salmonis*, increases in the neighbourhood habitat area has occurred with the advent of salmonid farming. The stability of habitat patches has also increased for the parasite, as the host populations are now resident in coastal waters throughout the year where colonization rates may be higher because of lower distance between habitat patches. There is a high risk of extinction of the component populations given that the area of suitable habitat (the combined surface area of the fish in a shoal for instance) is extremely small and highly fragmented compared to the scale of dispersal and distribution of the larvae seeking that habitat. The habitat is moving in the case of wild fish stocks and sedentary in the case of farmed salmonids. In this case, the survival of the metapopulation is dependent on immigration and colonisation of new habitat patches (host populations). The advent of fish farming simply adds to the number and character of suitable habitat patches for the parasite. Farm populations are at high risk of extinction due to pest management programmes and fallowing of sites. Wild populations may be at risk due to low temperatures and rates of reproduction, scarcity of suitable hosts at certain times of year. The risk of population extinction may be a function of habitat area (the host population size) and that risk is possibly also dependent on the impact of the parasite on the host as a function of parasite density and the host parasite interaction generally (see below). The colonization pressure exerted by a population on neighbouring populations may decline exponentially with distance given the limited longevity of the larvae and the probable high mortality during the pelagic phase. There is a cut-off distance above which the connectivity between populations is not possible because of the limited larval longevity. The migratory behaviour of other salmonids increases the opportunity for immigration of the parasite into other populations.

Does the empirical evidence support the idea that *L. salmonis* is distributed as a metapopulation within a region? The region occupied by the metapopulation remains undefined but its boundaries can be found where the connectivity between populations tend to zero. Populations outside of a network of connected populations either exist as isolated populations or belong to a separate metapopulation.

Empirical evidence in support of a metapopulation

Genetic evidence. Allozyme data, randomly amplified polymorphic DNA (RAPDs), the 18S ribosomal RNA gene and microsatellite DNA methodologies have been used to detect population differentiation in *L. salmonis*. Allozyme data, which have the least discriminating power of the 3 methods, have revealed regional population differentiation over wide geographic scales. Isdal, Nylund & Naevdal (1997) found differences between northern and southern populations in Norway. Shinn *et al.* (2000) found differences between lice on the east and west coast of Scotland. The RAPDs method used by Todd *et al.* (1997) in Scotland found much finer scale differences. Differences were found between lice on farmed and wild salmon, between lice on farmed salmon in different bays and even putative individual 'farm markers' within lice populations infesting sea trout. This suggested a transfer of lice from farmed to wild fish. Microsatellite DNA methods (Shinn *et al.* 2000) revealed lower genetic variability in lice populations on farmed salmon compared to wild salmon suggesting a bottleneck to gene flow on farms. Nolan *et al.* (2000*e*) found different alleles in Norwegian, Scottish and Irish populations of lice.

Evidence from pesticide resistance. The development of pesticide resistance in sea lice populations has been demonstrated for hydrogen peroxide and dichlorvos. This resistance may have a genetic basis and as such reflects the degree of gene flow between populations that have had different levels of exposure to the pesticide. Treasurer *et al.* (2000) showed that there was definite resistance to hydrogen peroxide by evaluating the efficacy of controlled treatments on different farms. Jones *et al.* (1992) and Tully & McFadden (2000) demonstrated resistance to dichlorvos in Scottish and Irish populations respectively. In both studies, resistance varied between farms but in the Irish study, repeated sampling of the same farms also indicate inconsistent trends in resistance over time. This suggested periodic recruitment of lice into the farms with different resistance characteristics and a changing gene pool. The strong spatial differences in resistance to hydrogen peroxide found by Treasurer *et al.* (2000) suggest that there was a restricted recruitment of lice from sources outside of the farm.

Morphometrics and development rate data. Morphometric data have not been useful in identifying different populations. Morphological differences are mainly in response to changes in environmental conditions, in particular temperature during development. This phenotypic plasticity was clearly demonstrated by Nordhagen *et al.* (2000) who showed no morphological differences in progeny of lice from different sized parents.

Physiological data and especially the relationship between generation time and temperature have not been well studied although there are suggestions that the parameters of the temperature development rate function could be useful in discriminating populations. There are several reasons for this. Firstly, farmed lice are periodically or continuously exposed to pesticides. Under this regime, selection for faster development and early maturation should occur. Secondly, in lice populations generally, adaptation to different temperature regimes should occur. This capacity is present as demonstrated by cold acclimation in lice reared at very low temperature (Boxaspen & Næss, 2000) although as pointed out by Nordhagen, Heuch & Schram (2000), heritability of quantitative traits such as age at maturation may be quite low. There is also a reproductive cost associated with maturation at smaller size since fecundity and size are related. There is some evidence that differences in the temperature-development rate relationship exist. Generation times at 10 °C calculated for Pacific populations (Johnson & Albright, 1991*b*) and extrapolated to other temperatures by Tully (1992) are longer than those estimated by Heuch, Nordhagen & Schram (2000) for Norwegian populations. Development rate may also depend on the host species (Johnson & Albright, 1992*a*) and be influenced by the host parasite interaction (see below).

Distribution and behaviour of free-living stages. It has proven difficult to obtain quantitative data on the spatial and temporal distribution of the free-living stages. The studies by Costelloe *et al.* (1995, 1996, 1998, 1999) provide the most comprehensive data available in coastal waters. A number of studies on behaviour in response to various environmental stimuli have also been undertaken in the laboratory in order to predict larval depth distribution and behaviour in the field. No data are available on the distribution of the larvae in oceanic environments although the life cycle can be completed in oceanic conditions (Jacobsen & Gaard, 1997). Nauplii are found in the upper metres of the water column and migrate to and from surface waters on a tidal (Costelloe *et al.* 1996) or diel cycle (Heuch, Parsons & Boxaspen, 1995). Nauplii are in the surface waters prior to high tide and sink to lower depths on ebbing tides in environments where tides are pronounced. This behaviour, which can induce a net tidal stream transport towards the shore, is common in free-living planktonic species in coastal waters where retention in coastal waters or net transport towards the shore is important in recruitment. In non-tidal or weakly tidal environments the nauplii undertake a diel vertical migration coming to the surface during daylight hours and sinking to lower depths in darkness.

The horizontal distribution in coastal waters has been described by Costelloe *et al.* (1996). Larvae were found in high densities at salmon farms, in much lower densities in open water and were also abundant in estuarine areas close to river inflows and through which wild host populations migrated. Their occurrence here was periodic. A minimum salinity would be required for survival in these areas and their presence seems to be at least partially determined by rainfall and the volume of freshwater inflow.

The data of Costelloe *et al.* (1999) suggest the presence of two discrete populations of larvae in their study system, one originating from the resident farmed stock of salmon and a second which occurred in the estuary and which originated from migratory wild salmonids passing through the area. Densities of larvae decay rapidly with distance from the point source of the farm due to dilution in greater volumes of water (Anon, 1995). The distance that individual larvae can travel from the source, whether a farm or an individual wild fish, is determined by the rate of dispersion in the local hydrodynamic regime, the vertical migratory behaviour of the larvae taking advantage of selective tidal stream transport and the longevity of the larvae. Densities away from the point source will depend on the distance from the source, the original number hatched and the rate of natural mortality that occurs during the dispersion. Cumulative mortality will be a function of the rate and the time since hatching. These processes suggest that larval populations could travel considerable distances but that this distance is eventually limited by larval longevity at which point a sharp decline in density and transmission potential should occur. A model describing these processes has been formulated by Anon (1995) in order to predict the impact of infestations on wild sea trout from point sources of larval production from salmon farms.

The dilution and dispersion of larvae after hatching, considered from the point of view of the individual host, or the host population, suggest that larval density may generally be very low. Certainly in oceanic environments and in coastal open waters away from point sources of production this is likely to be the case. The distribution data of Costelloe *et al.* (1995), where high densities were recorded under specific conditions where host populations were aggregated, may represent unusual situations. Nevertheless, the high densities of larvae recorded in estuaries probably originated from wild salmon migrating upstream may be an important route for vertical transmission of the parasite to smolts migrating to sea.

Transmission can occur in open oceanic environments as indicated by the capture of infested hosts that have been in oceanic waters for a period longer than the generation time of the parasite at the prevailing temperature. Nevertheless, longevity of the female is in excess of 191 days during which time 11 egg strings may be produced at 7 °C (Nordhagen *et al.* 2000). Longevity could be significantly longer than this at the lower temperatures of approximately 3–4 °C experienced by salmon in the north Atlantic.

Larval development and rate of production. Larvae can be produced throughout the year over the majority of the species range. Egg development and hatching occur at temperatures as low as 2 °C and there is also evidence that cold adaptation occurs in winter populations (Boxaspen & Næss, 2000). The rate of development is temperature dependent, however, and at these lowest temperatures is extremely slow. The proportion of non-viable eggs is higher at low temperatures, smaller eggs with lower nutritional reserves are produced and larval survival may therefore be lower at low temperatures (Ritchie *et al.* 1993). The rate of larval production increases with temperature and a concentration of hatching occurs during this time because increasing temperatures has a synchronizing effect on development (Tully, 1992). In temperate latitudes, hatching and subsequent transmission levels may be higher, therefore, in spring and early summer. Reproductive investment per egg also increases at this time. Sampling for larvae is generally more successful in spring (Costelloe *et al.* 1995, 1996) supporting the view that higher levels of larval production and survival occur at this time. Successful transmission does occur in spring in wild salmonid smolts (Tully & Whelan, 1993) and corresponds, in Ireland, to the time of year when smolts are migrating to sea and adult salmon are returning to the coast.

Absolute daily production and re-transmission rates from wild and farmed populations are poorly known. Tully & Whelan (1993) estimated between 1 and 10 million larvae were produced daily from farmed salmon populations during spring on the west coast of Ireland and found a tentative relationship between this production and levels of parasite transmission to nearby wild populations during late spring. This relationship did not exist at other times of year, however, suggesting variable survival or transmission potential of larvae or differential susceptibility of the host to infestation over time.

The mechanism of host location. The responses of larvae to various environmental stimuli such as light and pressure ensure that they maintain themselves in surface waters where the host populations also occur. They display positive barokinesis and a positive phototactic response (Bron, Sommerville & Rae, 1993). The use of selective tidal stream transport in coastal tidal environments may also facilitate concentration of larval populations in estuarine areas

close to where host smolt populations will migrate. The location and attachment to the host require a separate set of responses by the copepodid.

Contact with the host is facilitated by a burst-swimming response to linear water accelerations, with a frequency of 3–12 Hz, which lasts for 1–3 s (Heuch & Karlsen, 1997). A fish moving within centimetres of the copepodid generates this type of acceleration. An interesting observation by Heuch & Karlsen (1997) is that this may originally have evolved as an escape response from fish predators in free-living copepods. Settlement occurs mainly on the fins (Tully *et al.* 1993 *b*) again supporting the idea that the copepodid responds to small-scale water movements generated in their proximity. Passive uptake of copepodids into the gill chamber with subsequent migration of the chalimus onto the fins and body surface has been suggested by Anon (1993) as a mechanism of transmission. This situation is likely to be an artefact of experimental attempts to infect fish with the copepodid and does not readily explain the lack of significant gill infestations on wild fish heavily infested with copepodids and chalimus stages (Tully *et al.* 1993 *b*).

The response to physical stimuli, such as small-scale water disturbances, does not explain the host specificity of the parasite as many fish species may generate similar disturbances. There is no chemotactic response prior to the attachment to the host (Bron *et al.* 1993). Chemosensory mechanisms may be important in determining if a copepodid attaches successfully to the host. There are chemoreceptors on the antennae. Settlement onto the host occurs by 'grappling' using the clawed antennae (Bron *et al.* 1993). If the fish is not a salmonid, settlement and moulting will not occur. It is unclear if the copepodid can successfully leave a non-salmonid host. It is more likely that it becomes enveloped in the host mucus and does not develop further although the precise details of this critical process are still unknown (see below).

Pre-adult and adult lice can live off the host for a number of days and can successfully transfer between hosts (Ritchie, 1997). This is certainly common in situations where host density is high, such as on farms and in experimental situations. Its importance in natural populations is unknown. These freely swimming life stages respond to overhead shadow and quickly transmit to fish in experimental situations.

The behaviour of the hosts. Although the processes controlling the distribution and dispersion of the free-living planktonic stages are important determinants of the population structure, the movement of host populations are equally important. Salmonids are anadromous fishes migrating from rivers bordering the north Atlantic and Pacific oceans. In Atlantic salmon concentrations of fish occur in surface waters both in oceanic feeding grounds and in coastal waters close to freshwater outflows where they may have to wait for suitable flow conditions before migrating upstream. Vertical transmission of the parasite may be facilitated by the relative timing of smolt migration and adult lice-infected salmon returning to the coast. The number of fish returning to the coast prior to or during the smolt migration as opposed to later in the year varies regionally however. Multi-sea winter fish return to the coast early in spring while single-sea winter fish (grilse) return to the coast later in summer. If the multi-sea winter stock is low it may represent a bottleneck to vertical transmission to smolts. The presence of hosts in coastal waters is intermittent in some areas where sea trout and Arctic charr are absent. Salmon farming, which now extends over a considerable range of the distribution of wild salmonids on coasts of the north Atlantic, adds a new dimension and opportunity for the parasite as resident host populations are available in coastal waters throughout the year. Transmission efficiency of the parasite is probably also enhanced by the concentration of hosts in small volumes of water which increases the encounter rate between host and parasite. There have been numerous documented instances of epizootics of the parasite under these conditions (Tully *et al.* 1993 *a*, *b*, 1999). Typically, smolt populations are infested with a range of parasitic stages whereas adult salmon returning from the ocean tend to be infested with adult lice. This may reflect an absence of transmission during the return migration from the ocean. Given the length of the oceanic phase however transmission must occur in the open ocean.

Evidence from epidemiological studies. Epidemiological studies of lice infestation on wild salmonid populations give some clues as to the scales of dispersal of lice from various sources of production and how the populations may be structured at local and regional scales. Tully *et al.* (1999) found that infestations of sea trout were significantly different at 3 spatial scales and over time within the same systems. Infestations were different in estuaries within the same bay, between bays within regions and between regions in Ireland. This indicated that the dispersal of the infective stage from the point of production, which may have been salmon farms or wild salmonid stocks, was limited or restricted in range, and probably also that the abundance of the hosts differed between systems. Groups of fish sampled in the same system on different dates also had different infestations suggesting temporally varying infestation pressure within the system. A number of reports from Ireland and the analysis in Tully *et al.* (1999) suggest limited dispersal capacity and a source-sink dynamic at local scales. In these examples, levels of infestation on wild sea trout appear to be related to the proximity of sources of

larval production from wild and farmed salmonids in coastal embayments. This effect is significant up to distances of 30 km and is a function of the rates of dispersal, the longevity of larvae and the movement of the host populations.

Differences in infestation over short spatial and temporal scales may be due to pulsed production of the larvae from local salmonid populations as suggested by Costelloe *et al.* (1995), or indicate aggregations of larvae in a heterogenous physical environment. The larval distribution data presented by Costelloe *et al.* (1995, 1996) indicate high variability in abundance in space and time.

Data from fish farms indicate site-specific patterns of transmission of the parasite. A high degree of self re-infection is often indicated on farms that are situated in areas with weak tidal flow. Bron *et al.* (1993) and Tully (1989) both found evidence of a clear succession of generations of the parasite at farm sites suggesting self re-infection. Jackson & Minchin (1993) show a variety of patterns depending on the location of the site. In the highly dynamic environments of offshore farm sites transmission rates were low. Mixed generations of fish on the farm lead to increased vertical transmission. The development of single-bay management strategies for sea lice control in Ireland (Jackson, 1997, 1998) is a recognition that cross-transmission of lice between farms separated by distances up to approximately 10 km is common. It is also apparent that transmission of the parasite to newly developed farm sites occurs gradually at a low level and probably from migratory wild salmonids in the locality (Jackson & Minchin, 1993). Logically, transmission, if it can occur between farms, may also occur from farms to wild salmonids and especially to populations that are resident in coastal waters such as sea trout as suggested by Tully & Whelan (1993).

CONCLUSIONS ON THE POPULATION STRUCTURE

The evidence from genetics, pesticide resistance, larval ecology, host behaviour and epidemiology all point to a high level of population differentiation or highly structured populations. The oceanic phase of salmonids, whereby salmon populations from large geographic regions may occupy the same feeding grounds, may result in a high level of gene exchange between lice originating from different coasts. Transmission in this environment may be quite low and inefficient however. On the coast, the epidemiological and genetic studies in particular point to population differences at local scales with connectivity varying as a function of distance between habitat patches (host populations) and where the scale is limited by the longevity of the larvae and migration of host fish.

Although the empirical data suggest the existence of metapopulations throughout the range of the species in the northeast Atlantic, the structure of those metapopulations remains vague because connectivity between habitat patches whether they are resident populations on farms or migratory wild fish has not been measured. The processes controlling larval dispersal are primarily physical but coupled to larval behaviour. There is no integrated biophysical study of dispersal in this species. Rates of mortality and dispersal, which determine how many larvae can reach a distant host population, have not been measured in different environments. Parameterisation of a spatially explicit metapopulation model remains elusive and estimates for these parameters will be difficult to obtain. Progress will require interdisciplinary research involving physical oceanographers, fish behavioural and plankton biologists. The demographics of the post-settlement stages are more easily studied. Reproductive output of the parasite is determined by rates of growth and mortality during development on the host. As described below, the host–parasite interaction (the quality of the habitat patch for the parasite) is a major determinant of this.

HOST–PARASITE INTERACTION

The interaction between the sea louse and its salmonid host is one of the most exciting areas of research, yet one that has been generally ignored until recently. Wikel, Ramachandra & Bergman (1994) defined a successful host–parasite relationship as a balance between limiting the parasite through host defences and the ability of the parasite to modulate, evade or restrict the host's responses. This clearly means that the host defences can influence the parasite population structure. Host–parasite interactions are generally characterized by highly complex processes (De Meeus *et al.* 1995; Gubbins & Gilligan, 1997; Wikel, 1999). Despite the economic importance of sea lice, comparatively little is known about the nature of their host–parasite relationship. Some limited studies in this area have been undertaken but it should be emphasized that results obtained from experimentally infected or diseased animals do not necessarily reflect the true nature of the host–parasite relationship. In many instances experimental infections result in parasite burdens that are more indicative of a disease state rather than the parasite burdens that occur when the host–parasite relationship is in balance. At high parasite burdens, approaching those that result in disease, the immunological response of the host may be reduced because of unconsidered factors such as stress (see below).

Pike & Wadsworth (1999) described and summarized the data on the pathological effects of sea lice on salmonids in terms of the mechanical disruption caused by parasite attachment and feeding activities.

Aside from the pathology frequently reported in the disease state, the extent of the sea lice pathophysiology literature then was limited to the small number of published papers (namely Bjorn & Finstad, 1997; Dawson *et al.* 1998; Johnson & Albright, 1992*a*). In the intervening years, several papers on various aspects of the host-parasite relationship between *L. salmonis* and its salmonid host have been published. There follows an overview of these data and, using examples of other host–parasite models where data for *L. salmonis* is lacking, promising new directions for further research are identified.

Stress and the host fish

A stressor is any stimulus (real or imagined) that disturbs or threatens to disturb the homeostasis of an animal. The stress response of fish is complex and involves a series of behavioural and physiological responses designed to be compensatory and/or adaptive which enable the animal to overcome the effects of the stressor. In this sense, the response to a stressor is designed to be beneficial for the fish and involves all levels of animal organisation. The role of the integrated host stress response in ectoparasite infestation is critical for understanding reduced parasite settlement and elimination of attached parasites.

The primary stress parameter measured and used to evaluate stress in fish is blood cortisol level. The primary stress response of fishes involves the massive release of catecholamines and corticosteroids and the secondary stress responses are the manifold immediate actions and effects of these hormones at all levels throughout the animal (Fig. 2). Increases in plasma glucose are the result of these hormones mobilising energy resources to fuel the stress responses. Although the immediate stress-related hyperglycaemia reported in many teleosts is mediated by the effects of catecholamines on glycogenolysis and glucose release from the liver, catecholamines also have effects on hydromineral balance and immune functions. The most disadvantageous side effect of catecholamines in relation to hydromineral balance is the increased permeability of tight junctions in the gills, which increases ion diffusion rates through the paracellular pathways of the branchial epithelium. Although the effects of catecholamines on the immune system of fishes have not been intensively studied, there is evidence for both stimulatory and inhibitory effects (left hand side of Fig. 2 and Wendelaar Bonga, 1997).

Cortisol mediates many of the stress-related changed in the skin of rainbow trout, including stimulation of mucus discharge and apoptosis, synthesis of vesicles in the cells of the upper epidermis, and epidermal infiltration by leukocytes

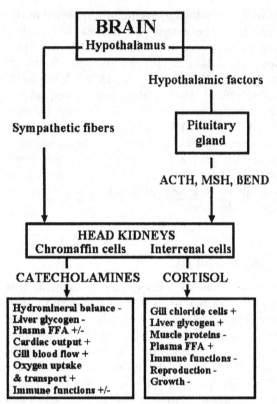

Fig. 2. The integrated stress response of fish involves two main axes illustrated schematically below. The activation of these axes by stimuli such as ectoparasite attachment, results in the liberation of catecholamines and glucocorticoids into the blood. Although the present understanding is far from complete, in general the effects of these axes on the physiological functions (including immune response) in the lower boxes are designated as positive/stimulatory (+) or negative/inhibitory (−). Abbreviations: ACTH, adrenocorticotropic hormone; MSH, melanocyte stimulating hormone; βEND, β-endorphin, FFA, free fatty acids.

(Iger *et al.* 1995). Further, cortisol is implicated in mediating the inhibitory effects of stressors on the immune response, thus decreasing disease resistance (Alford *et al.* 1994; Anderson, 1990; Ellis, 1981; Houghton & Matthews, 1990; Pickering & Pottinger, 1985, 1989). High cortisol levels in the blood are associated with reduced circulating lymphocyte levels and antibody production, lower mitogen-induced proliferation of these cells and inhibition of phagocytotic activity (Espelid *et al.* 1996; Kaattari & Tripp, 1987; Maule, Schreck & Kaattari, 1987; Stave & Roberson, 1985). Studies with Atlantic salmon and rainbow trout selected for high and low cortisol stress responses have demonstrated that a high cortisol response was correlated with reduced non-specific immunity and increased susceptibility to pathogens (Fevolden, Refstie & Gjerde, 1993; Fevolden & Roed, 1993; Fevolden, Roed & Gjerde, 1994).

Under aquaculture conditions, Nolan *et al.* (1999*b*) studied the effects of experimental infection

with relatively low numbers of *L. salmonis* on *S. salar* post-smolts. Without using a stressor, they showed that disrupted skin and gill epithelia over the whole body of lice-exposed fish were accompanied by minimal osmoregulatory disturbance (measured as serum Na and Cl) while the Na^+/K^+-ATPase activity in the gill was strongly elevated. This was associated with increased turnover in the chloride cell population (evidenced by increased levels of apoptosis and necrosis). Although the link with cortisol has been more difficult to demonstrate clearly, elevated cortisol has been associated with higher sea lice infestation levels (Bjorn & Finstad, 1997; Finstad *et al.* 2000). At lower parasite levels, elevation of cortisol may not occur for a prolonged period of time. In the longer term, chronic stress may lead to exhaustion of the corticotrophic cells in the interrenal gland. More data on the cortisol levels in fish infested with sea lice are needed.

A second approach to evaluate the effects of a stressful chronic treatment is the use of a second, acute stressor, typically net confinement or air exposure. The intensity of the response to a second stressor (as estimated with the above-mentioned stress parameters) reflects the condition of the chronically stressed fish. The condition of the chronically stressed fish is deduced from the intensity of the response to the second stressor, in comparison with the response of a non-chronically stressed control. Following from this, confinement stress has been used as a second stressor in a number of fish species to evaluate the effects of chronic stressors, including freshwater fish lice *Argulus* sp. and sea lice, *Lepeophtheirus salmonis* (Nolan *et al.* 2000*d*; Ruane *et al.* 1999, 2000). These experiments showed, under laboratory conditions, that infective stages and low ectoparasite levels modulate the acute stress response of the host fish. Such an approach, using stress, offers a promising new direction to reveal the more subtle modulatory effects of parasitic lice.

Direct effects of ectoparasitic lice on the host fish

Low levels of parasite infection generally do not induce a strong stress response in the host fish (see below and also Kabata, 1970). However, high levels of infection, other stressors or the development of a diseased state will result in immunosuppression of the host (Austin, 1999; Houghton & Matthews, 1990; Ross *et al.* 1996). This further exacerbates the situation, leading to circumstances that are deleterious to the host, as well as to the parasites that are dependent on it. The lack of effective immune responses during parasitic infections is well documented for a wide variety of taxa and is considered to be a hallmark of parasitic infections (Belley & Chadee, 1995). This is well known for the arthropods especially and these parasites produce numerous

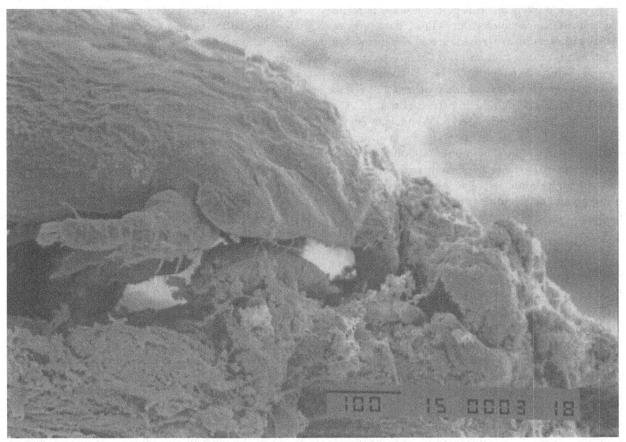

Fig. 3. Chalimus stage of the salmon louse, *Lepeophtheirus salmonis*, attached to the skin of the head of the Atlantic salmon, *Salmo salar*. The disruption of the skin observed is understood to be primarily mechanical in nature and caused by the penetration of the frontal filament. Beyond this, the effects of the juvenile stages of the parasite were very limited and until recently, this was the extent of the knowledge on the host–parasite relationship between these species. This is partially because localized effects were examined instead of more subtle systematic effects in the host (see text). Scale bar = 100 micrometres.

bioactive molecules in their saliva that play an important role in this phenomenon (Belley & Chadee, 1995; Nash *et al.* 1996; Roberts & Janovy, 1996; Wikel, 1999).

The fact that the inflammatory responses at the attachment and feeding sites are mild, or even absent, in many fish species (Fig. 3) supports the view that sea lice would also produce such substances (see Johnson & Albright, 1992*a*; Jones, Sommerville & Bron, 1990; Jonsdottir *et al.* 1992; MacKinnon, 1993). If so, it would be interesting to establish why there is no suppression of the localised inflammatory response reported in the coho salmon (Johnson & Albright, 1992*a, b*).

To date only a small amount of work on bioactive molecules produced by fish lice has been undertaken. From laboratory experiments, Ross *et al.* (2000) reported increased alkaline phosphatase and protease activity in the mucus of Atlantic salmon infected with *L. salmonis* and suggested that this parasite may produce these enzymes which are then secreted or excreted onto the host's body surface. Based on molecular weight, inhibition studies, affinity chromatography and Western blotting with an antibody

raised against Atlantic salmon trypsin, these proteases were shown to be a series of low molecular weight (18–25 kDa) trypsins produced by *L. salmonis* (Firth, Johnson & Ross, 1999). It is believed that the function of this trypsin is to aid in feeding activities and possibly to interfere with host defence mechanisms. O'Flaherty *et al.* (1999) examined mucus from wild *S. trutta* naturally infested with *L. salmonis* using SDS-PAGE electrophoresis and showed that there were conspicuous differences in the numbers of mucus protein bands between populations of *S. trutta*. More interestingly, they reported that all infected fish had intensely staining major bands between 14·2 and 24 kDa and postulated that a major band at 14·4 kDa could be lysozyme. It is also possible that some of these additional bands may be derived from the sea louse. Firth *et al.* (1999) reported that sea lice produced trypsin found in the mucus of infected Atlantic salmon had a molecular weight between 18–25 kDa.

Recently Bell, Bron & Sommerville (2000) demonstrated the presence of peroxidase-containing glands in nauplius, copepodid, chalimus, pre- and adult stages of both *L. salmonis* and *C. elongatus* (Bell *et al.*

2000) and it is likely that this peroxidase is a parasite-secreted bioactive molecule. These data are especially interesting as peroxidase is also an important component of the non-specific defence system of several fishes (Shephard, 1994; Wilhelm Filho, Giulivi & Boveris, 1993) and its synthesis and release into the mucus is high during stress (Brokken *et al.* 1998; Iger *et al.* 1995; Iger & Wendelaar Bonga, 1994). Sea lice-secreted peroxidase at the host epithelial barrier could be immunomodulatory if the enzyme destroys the cellular factors responsible for eliciting an immune response, or simply destructive if it damages cells and maintains the lesion open for feeding.

Indirect effects of ectoparasitic lice on the host fish

Several authors have looked at the effects of fish lice infections on host blood cortisol levels and glucose levels, as well as immune responses. Ruane *et al.* (1999, 2000) used net-confinement procedures to demonstrate *Argulus*-induced modulation of the cortisol and glucose stress response of the freshwater rainbow trout, *O. mykiss*. The effect of a low-level infection of rainbow trout with *A. foliaceus* on subsequent response to a stressful condition was examined after 21 days and no obvious signs of host immunomodulation were evident (Ruane *et al.* 1999). However when the parasitised fish were stressed by confinement, they showed a higher level of immunosuppression, as evidenced by decreased levels of serum lysozyme and alternative complement activity, and reduced oxygen radical production (respiratory burst), when compared with non-parasitised fish. Exposure to infective stages of *L. salmonis* resulted in a higher blood cortisol levels after four hours of net confinement, compared with control (unexposed) fish (Ruane *et al.* 2000), providing evidence for the first time that, although the response to the infective stages was believed to be limited to a localised cellular response, the consequences in relation to subsequent stress sensitivity were systemic, affecting the whole organism. This kind of approach is likely to produce the truest picture of the consequences of infection by even low numbers of juvenile *L. salmonis* on the host.

Infection of salmonids with the low numbers of the salmon louse, *L. salmonis*, has generally not resulted in significant increases in plasma cortisol levels (Bjorn & Finstad, 1997; Johnson & Albright, 1992*b*; Ross *et al.* 2000). However, with heavy infections, plasma cortisol levels often increase well beyond those levels known to cause immunosuppression (Mustafa & MacKinnon, 1999). Johnson & Albright (1992*a*) reported similar plasma cortisol values for *S. salar* in the laboratory before and after experimental infection with infective copepodids of *L. salmonis*, with values ranging from

10–77 ng · ml⁻¹. Bjorn & Finstad (1997) experimentally infected the sea trout *S. trutta* with *L. salmonis* in the laboratory and could not demonstrate differences in plasma cortisol between control and infected fish, a similar result to that reported by Ross *et al.* (2000). Dawson *et al.* (1998) reported significant differences between serum glucose in lice-infested wild seawater *S. trutta* sampled from different locations in the west of Ireland, but it cannot be excluded that the large variation in the values reported for this normally tightly regulated parameter resulted from the impact of the gill netting required to catch these fish before blood sampling. As the cortisol response in the blood of fishes begins rapidly after disturbance, experimental design and sampling protocols to establish basal or resting blood cortisol levels should take this into account (Nolan *et al.* 1999*a*; Pottinger, Moran & Cranwell, 1992; Quabius, Balm & Wendelaar Bonga, 1997). In the field, sampling true baseline cortisol values is problematic because of unavoidable catching stress, but mathematical extrapolation modelling methods may offer promising solutions (Poole, Nolan & Tully, 2000). It may be difficult to demonstrate significant differences in blood cortisol levels, but recently, a net-confinement procedure has been applied to demonstrate ectoparasite-induced modulation of the cortisol and glucose stress response of the rainbow trout *O. mykiss*. As mentioned above, exposure to infective stages of *L. salmonis* resulted in a higher blood cortisol levels after 4 h of net confinement, compared with control (unexposed) *O. mykiss* in sea water (Ruane *et al.* 2000), and similarly with freshwater *O. mykiss* infested for 21 days with adult *A. foliaceus* (Ruane *et al.* 1999). This new research direction shows that the effects of lice infestation can be studied and demonstrated, not in terms of differences in baseline parameter values between infested and uninfested fish, but rather in terms of the impact of the parasite on the stress response and the ability of infested fish to respond to and deal with a stressor. A fuller understanding of the immune-neuroendocrine interactions in fish would facilitate the development of novel ways to manipulate the fish immune system and improve disease resistance and fish health (Harris & Bird, 2000).

Effects of the host fish on ectoparasites

Since the early 1970s, it has been recognized that the host affects aspects of the biology of fish lice, but the mechanisms behind these effects are poorly understood. Host rejection of the cyclopoid copepods *Lernaea cyprinacea* and *L. polymorpha* has been reported in both naïve and previously infected fish (Shields & Goode, 1978; Woo & Shariff, 1990). This rejection is thought to be in part due to cellular responses. With respect to the salmon louse, *L. salmonis*, it has been demonstrated that for naïve hosts, salmon species differ in their susceptibility to infection (coho < chinook < Atlantic) (Johnson & Albright, 1992 *a*). Coho salmon, which were found to be the most resistant to sea lice establishment, mount localised strong tissue responses, leading to parasite loss. Naïve Atlantic salmon that were highly susceptible to infection failed to mount significant tissue responses to any of the developmental stages. Cortisol (hydrocortisone) implantation, which suppressed the inflammatory response and the development of epithelial hyperplasia, resulted in the inability of coho salmon to shed *L. salmonis* (Johnson & Albright 1992 *b*). Differences in the growth rate and the number of eggs carried by *L. salmonis* have been reported for different host species. *L. salmonis* grew faster and carried larger numbers of eggs on naïve Atlantic salmon when compared to naïve chinook salmon (Johnson, 1993). Host effects on egg production and viability have also been reported for *Ergasilus labracis* and *Lernaea cyprinacea* (Paperna & Zwerner, 1982; Woo & Shariff, 1990). Recently, as other species of sea lice become problematic in other countries where salmonids are farmed, differential host susceptibility is becoming apparent e.g. for *Caligus flexispina* in Chile (González, Carvajal & George-Nascimento, 2000). These species' differences offer tantalising clues for further research directions to find the key host responses that may underlie early parasite rejection. To date, no study has reported the host cellular immune response to experimental sea lice infection in resistant and susceptible salmonid species using functional assays, an approach that could identify the key response element(s).

The relative importance of innate and acquired immune responses in the relationship between fish lice and their hosts is not understood. In studies that have used naïve hosts, it appears that for some species, aspects of the innate immune response (inflammation and epithelial hyperplasia) are important in initial parasite infection (e.g. *L. salmonis* on coho salmon; Johnson & Albright 1992 *a*, *b*). Cellular responses have been attributed with causing changes in the distribution of *Lernaea piscinae* from the body to be cornea of the big head carp, *Aristichthys nobilis* (Shariff & Roberts, 1989). Rejection of the copepod, *Ergasilus labracis*, and a reduction in its egg sac production on striped bass, *Morone saxatilis*, has been attributed to be due to a well-developed tissue response (Paperna & Zwerner, 1982). In other species, such as naïve and naturally infected Atlantic salmon, only minor host tissue responses to the presence of the caligid copepods, *L. salmonis* and *C. elongatus*, have been reported at the point of attachment (Johnson & Albright 1992 *a*, *b*; Jones *et al.* 1990; Jonsdottir *et al.* 1992; MacKinnon, 1993) and this lack of extensive tissue response and

inflammation is thought to make Atlantic salmon highly susceptible to infection with *L. salmonis* (Johnson & Albright, 1992 *a*).

It has often been reported for other species of parasitic copepods, including those groups which penetrate and develop large hold-fast and feeding structures within the tissues of their host, that the tissue responses of the host are often only minor. The parasitic copepod, *Cardiodectes medusaeus*, parasitises the lanternfishes, *Diaphus theta* and *Tarletonbaenia crenularis*, and does this by penetrating the *bulbus arteriosis* of the heart and becoming enveloped in host tissue. Adult females extend their elongated body and egg sacs out to the external environment through the body wall of the host (Kabata, 1970; Sakuma *et al.* 1999) without eliciting any host response (Ho, J.-C., personal communication). Investigation of this host–parasite relationship may reveal common strategies between this copepodid and the caligid sea lice to evade host detection.

Few authors have looked at the antibody response of fish that are naturally infected with fish lice. Sakuma *et al.* (1999) and Thoney & Burreson (1988) were unable to detect a specific antibody response in *Leiostomus xanthurus* that were naturally infected with *Lernaea radiatus*. Grayson *et al.* (1991) reported a low-level antibody response in Atlantic salmon naturally infected with *L. salmonis*, but no antibody response in naturally infected rainbow trout. Woo & Shariff (1990) investigated the effects of previous exposure of the kissing gourami, *Helostoma temmincki*, to *Lernea cyprinicae* on subsequent susceptibility to infection. They reported that both naïve and previously infected fish rejected the copepods within 30 days after the appearance of the adult stages. At low challenge doses, adult copepods failed to develop on a proportion of previously infected fish; however at a higher challenge dose there was no difference between the previously infected and naïve hosts. Previously infected fish rejected adult copepods faster than naïve fish. Furthermore female copepods growing on previously infected hosts lost more of their egg sacs and produced fewer viable eggs than those growing on naïve hosts. It was proposed that copepod rejection in both groups was due to an acquired immune response, although the antibody response was not measured (Woo & Shariff 1990). With the exception of this study, there is little evidence for a protective specific immune response by fish against fish lice. Immunisation of naïve Atlantic salmon with partially purified extracts of *L. salmonis* did not significantly reduce the number of sea lice on the fish after challenge but did reduce the number of ovigerous females present and the number of eggs that they were carrying (Grayson *et al.* 1995). The relationship between the individual host responses and establishment and performance of *L. salmonis* is a very promising area for future research.

Disease susceptibility in relation to ectoparasites

The skin and gills of a fish are covered by complex epithelia comprised of living cells that are continuous over the body surface. They form the first barrier between the external and the internal environment and are protected with a chemically and functionally complex mucous coat that is discharged by specialised cells in the epidermis (Shephard, 1994). Several responses, such as increased apoptosis of branchial chloride cells and the pavement cells of the skin, are under control of the stress hormone cortisol, and will therefore also occur in areas not affected directly by skin parasites, such as in the gills. These stress parameters have been examined in skin and gills of *S. salar* post-smolts in relation to experimental infection with *L. salmonis* in a holistic approach to studying lice effects on these epithelia and to examine any potential link between the effects of sea lice and susceptibility to secondary infection (Nolan *et al.* 1999*b*). The direct tissue damage caused by parasite grazing, together with indirect epithelial effects and immunosuppression, offers opportunities for invasion by secondary pathogens, especially where the overall integrity of the epithelia is compromised. Infection with low numbers of pre-adult and adult sea louse, *L. salmonis*, induced strong stress-related changes in both skin and gill epithelia. These were indirect effects of the parasite, as they are reported from areas of the skin where the parasite had not been (Nolan *et al.* 1999*b*).

The effects of infection with the sea louse, *L. salmonis*, can then be divided into two distinct categories. The first is a direct effect of parasite attachment and feeding on the body surface, as reported by others (Johnson & Albright, 1992*a*, *b*; Jones *et al.* 1990; Jonsdottir *et al.* 1992; Wootten, Smith & Needham, 1982). The second is the effect of the integrated response on the integrity of the skin and gill epithelia, including increased levels of mucous cell discharge and increased apoptosis and necrosis in different cell types in the skin and gill epithelia (Nolan *et al.* 1999*b*). The osmoregulatory adjustments required to maintain hydromineral balance were observed in terms of increased gill Na^+/K^+-ATPase, as the lice-infected fish actively extruded the excessive sodium ions which entered the body compartment across these disrupted epithelia.

The stressor-induced disruption of the skin and gill epithelia of the fish caused by parasites (Nolan *et al.* 1999*b*, *c*, *d*) is significant in relation to disease resistance. The prolonged period of epithelial disruption leaves the fish open to invasion by opportunistic pathogens. If there is also immunosuppression, either because of elevated cortisol as a result of the stress response to the parasites (Houghton & Matthews, 1990; Nagae *et al.* 1994; Pickering & Pottinger, 1989), or by additional factors such as

toxic actions of organic pollutants (e.g. Dunier & Siwicki, 1993; Hart *et al.* 1998; Wong *et al.* 1992), the deleterious effects are compounded and disease resistance is lowered (e.g. Kakuta, 1977). Reduced levels of circulating lymphocytes in the blood of sea trout infested with *L. salmonis* have been reported (Bjorn & Finstad, 1997), while increased numbers of these and other leukocyte types have been reported outside of the blood vessels in the skin and gill epithelia of *L. salmonis*-infested Atlantic salmon (Nolan *et al.* 1999*b*). Extravasation of leukocytes from the blood vessels and their subsequent appearance in the peripheral tissues of stressed fish is part of the integrated stress response (Wendelaar Bonga, 1997) and is mediated, at least in part, by the stress hormone cortisol (Iger *et al.* 1995; Nolan *et al.* 1999*c*). Recently, enhanced susceptibility to the microsporidian, *Loma salmonae*, has been demonstrated in *S. salar* post-smolts during a primary infection with *L. salmonis* (Mustafa *et al.* 2000*b*) and this is associated with impaired macrophage function when the sea lice reach mobile pre- and adult stages (Mustafa *et al.* 2000*a*). Decreased numbers of mucus cells after sea lice infection indicated increased mucus discharge (Nolan *et al.* 1999*b*) but in the longer term, the effects of decreased numbers of mucus cells, with different mucopolysaccharide composition (Nolan *et al.* 2000*c*), may reduce the protective function of the normal mucus (Smith *et al.* 2000).

Another significant stressor-induced effect in the skin of salmonids involves changes in the numbers of vesicles in the upper cells of the epidermis. These are secretory vesicles that have been shown to contain endogenous peroxidase activity (Iger, Jenner & Wendelaar Bonga, 1994; Iger & Wendelaar Bonga, 1994) and their synthesis is enhanced in rainbow trout, *O. mykiss*, by cortisol administration (Iger *et al.* 1995). The contents of these vesicles are secreted out into the glycocalyx and appear also in the mucus of the fish. The numbers of these vesicles per filament cell are more than 10 fold higher in the mature sea trout smolt than in the rainbow trout and are influenced by stressors (Nolan *et al.* 2000*a*). Peroxidase is considered to be an antimicrobial component of the non-specific defence system of fish and has been demonstrated in the mucus and glycocalyx on the surface of the skin (Brokken *et al.* 1998; Iger *et al.* 1994; Iger & Wendelaar Bonga, 1994). The secreted skin peroxidase is a biochemically distinct isoform from the peroxidase of the blood (Brokken *et al.* 1998). However, although the significance of the enhanced secretion of peroxidase during stress is unknown at present, the stimulation of vesicle synthesis by cortisol is associated with reduced establishment by the ectoparasite *A. foliaceus* on *O. mykiss* (Nolan *et al.* 2000*d*). An interesting related finding is that specialised peroxidase-containing glands also occur over the body of the caligid sea lice, *L. salmonis* and *C. elongatus* (Bell *et al.* 2000), possibly indicating a counter-production by these ectoparasites. As these glands were associated with the mouth tube (as well as other locations), and were found mainly in pre-adult and adult stages, they probably function in the feeding process. Previously, proteases from *L. salmonis* and *C. elongatus* have been reported, including serine proteases (Ellis, Masson & Munro, 1990), and *C. elongatus*, which has a wider range of host species, contained greater numbers and types of these enzymes. A role for these enzymes in the host-parasite interaction, or in the increased susceptibility to secondary infections has yet to be demonstrated. Recently, the antibacterial properties of *O. mykiss* mucus have been demonstrated (Smith *et al.* 2000) and this kind of approach may prove useful in studying the host and the parasite derived enzymes and their effects on epithelial cells and mucous properties.

NEW PERSPECTIVES ON HOST–PARASITE INTERACTIONS

Further research into the host–parasite relationship of the economically important parasites such as the caligid sea louse, *L. salmonis*, is necessary in order to fully understand the role of the interaction on the production of the parasite, how the host can influence the parasite development and production, and as a result, for identifying new factors which may prove the key to enhancing host rejection of the parasite (Scholz, 1999). Such knowledge will also provide a scientific basis for addressing the parasite-related problems that are obstructing the development of aquaculture in other countries (Dadzie, 1992; Hecht & Endemann, 1998; Rodgers & Furones, 1998; Roth, Richards & Sommerville, 1993). The potential for cell culture methods to contribute to studying the host–parasite interaction of fish lice has been recently explored in Nolan & Johnson (2000). Although there is little host response to the frontal filament (MacKinnon, 1993; Pike, MacKenzie & Rowland, 1993), the recent localisation and ontogeny of the filament producing cell groups in *L. salmonis* demonstrated by Gonzalez-Alanis *et al.* (2001) paves the way for investigating and identifying the stage in the parasite life cycle where these parasite-secreted compounds are produced. Although cell cultures from caligid lice have not been reported to date, recent advances with crustacean (and other invertebrate) tissues and cells appear very promising (Mothersill & Austin, 2000), and existing cell culture methods for fish cells are reasonably well advanced (Babich & Borenfreund, 1991; Magwood & George, 1996; Neumann, Fagan & Belosevich, 1995; Nolan *et al.* 2002).

CONCLUSIONS

Significant advances have been made in under-

standing critical issues of sea lice biology and the host parasite interaction over the past 20 years. Given the level of research on this host–parasite system it may be a valuable model and help advance understanding of other marine host parasite relationships. Significant gaps remain in the understanding of the population structure although the framework proposed in this paper may be a useful focus for study. In marine parasites with direct life cycles and free-living planktonic stages, a holistic and multi-disciplinary approach is required for a complete development of the model of population structure. The review of the host–parasite interaction presented here also provides ample evidence that the important parameters of this interaction must also be incorporated into such a model because the host has an effect on the development and production of the parasite on the one hand and the parasite can also cause significant host mortality.

ACKNOWLEDGEMENTS

The authors are grateful to Dr S. C. Johnson for input and helpful discussion on host–parasite interactions for this paper. Fig. 3 is credited to M. Paulissen from his undergraduate study in Biology at University of Nijmegen, The Netherlands.

REFERENCES

ALFORD III, P. B., TOMASSO, J. R., BODINE, A. B. & KENDALL, C. (1994). Apoptotic death of peripheral leukocytes in channel catfish: effect of confinement-induced stress. *Journal of Aquatic Animal Health* **6**, 64–69.

ANDERSON, D. P. (1990). Immunological indicators: effects of environmental stress on immune protection and disease outbreaks. *American Fisheries Society Symposium* **8**, 38–50.

ANON (1993). Report of the Sea Trout Working Group. Department of the Marine and Natural Resources, Dublin, Ireland.

ANON (1995). Report of the Sea Trout Working Group. Department of the Marine and Natural Resources, Dublin, Ireland.

AUSTIN, B. (1999). The effects of pollution on fish health. *Journal of Applied Microbiology* **85**, 234S–242S.

BABICH, H. & BORENFREUND, E. (1991). Cytotoxicity and genotoxicity assays with cultured fish cells: a review. *Toxicology In Vitro* **5**, 91–100.

BELL, S., BRON, J. E. & SOMMERVILLE, C. (2000). The distribution of exocrine glands in *Lepeophtheirus salmonis* and *Caligus elongatus* (Copepoda: Caligidae). *Contributions to Zoology* **69**, 9–20.

BELLEY, A. & CHADEE, K. (1995). Eicosanoid production by parasites: from pathogenesis to immunomodulation. *Parasitology Today* **11**, 327–334.

BJORN, P. A. & FINSTAD, B. (1997). The physiological effects of salmon lice infection on sea trout post smolts. *Nordic Journal of Freshwater Research* **73**, 60–72.

BOXASPEN, K. & NÆSS, T. (2000). Development of eggs and the planktonic stages of salmon lice (*Lepeophtheirus salmonis*) at low temperatures. *Contributions to Zoology* **69**, 51–56.

BROKKEN, L. J. S., VERBOST, P. M., ATSMA, W. & WENDELAAR BONGA, S. E. (1998). Isolation, partial characterization and localization of integumental peroxidase, a stress-related enzyme in the skin of a teleostean fish (*Cyprinus carpio* L.). *Fish Physiology and Biochemistry* **18**, 331–342.

BRON, J. E., SOMMERVILLE, C. & RAE, G. H. (1993). Aspects of the behaviour of copepodid larvae of the salmon louse *Lepeophtheirus salmonis* (Krøyer, 1837). In *Pathogens of Wild and Farmed Fish: Sea Lice* (ed. Boxshall, G. A. & Defaye, D.), pp. 125–142. Chichester, Ellis Horwood Ltd.

COSTELLO, M. J. (1993). Review of methods to control sea lice (Caligidae: Crustacea) infestations on salmon (*Salmo salar*) farms. In *Pathogens of Wild and Farmed Fish: Sea Lice* (ed. Boxshall, G. A. & Defaye, D.), pp. 219–252. Chichester, Ellis Horwood Limited.

COSTELLOE, M., COSTELLOE, J., COGHLAN, N., O'DONOGHOE, G. & O'CONNOR, B. (1998). Distribution of the larval stages of *Lepeophtheirus salmonis* in three bays on the west coast of Ireland. *ICES Journal of Marine Science* **55**, 181–187.

COSTELLOE, M., COSTELLOE, J., O'DONOGHOE, G., COGHLAN, N. & O'CONNOR, B. (1999). A review of field studies on the sea louse, *Lepeophtheirus salmonis* Krøyer on the west coast of Ireland. *Bulletin of the European Association of Fish Pathologists* **19**, 260–264.

COSTELLOE, J., COSTELLOE, M. & ROCHE, N. (1995). Variation in sea lice infestation on Atlantic salmon smolts in Killary Harbour, West Coast of Ireland. *Aquaculture International* **3**, 379–393.

COSTELLOE, M., COSTELLOE, J. & ROCHE, N. (1996). Planktonic dispersion of larval salmon lice, *Lepeophtheirus salmonis*, associated with cultured salmon, *Salmo salar*, in western Ireland. *Journal of the Marine Biological Association of the United Kingdom* **76**, 141–149.

DADZIE, S. (1992). An overview of aquaculture in Eastern Africa. *Hydrobiologia* **232**, 99–110.

DAWSON, L. H. J. (1998). The physiological effects of salmon lice (*Lepeophtheirus salmonis*) infections on returning post-smolt sea trout (*Salmo trutta* L.) in western Ireland. *ICES Journal of Marine Science* **55**, 193–200.

DAWSON, L. H. J., PIKE, A. W., HOULIHAN, D. F. & Mc VICAR, A. H. (1998). Effects of salmon lice *Lepeophtheirus salmonis* on sea trout *Salmo trutta* at different times after seawater transfer. *Diseases of Aquatic Organisms* **33**, 179–186.

DE MEEUS, T., MORAND, S., MAGNAN, N., CHI, T. D. & RENAUD, F. (1995). Comparative host-parasite relationship of two copepod species ectoparasitic on three fish species. *Acta Oecologica* **16**, 361–374.

DUNIER, M. & SIWICKI, A. K. (1993). Effects of pesticides and other organic pollutants in the aquatic environment on immunity of fish: a review. *Fish and Shellfish Immunology* **3**, 423–438.

ELLIS, A. E. (1981). Stress and the modulation of defence mechanisms in fish. In *Stress and Fish* (ed. Pickering, A. D.), pp. 147–169. London, Academic Press.

ELLIS, A. E., MASSON, N. & MUNRO, A. L. S. (1990). A comparison of proteases extracted from *Caligus elongatus* (Nordmann, 1832) and *Lepeophtheirus salmonis* (Krøyer, 1838). *Journal of Fish Diseases* **13**, 163–165.

ESPELID, S., LOKKEN, G. B., STEIRO, K. & BOGWALD, J. (1996). Effects of cortisol and stress on the immune system in Atlantic salmon (*Salmo salar* L.). *Fish and Shellfish Immunology* **6**, 95–110.

FEVOLDEN, S. E., REFSTIE, T. & GJERDE, B. (1993). Genetic and phenotypic parameters for cortisol and glucose stress response in Atlantic salmon and rainbow trout. *Aquaculture* **118**, 205–216.

FEVOLDEN, S. E. & ROED, K. H. (1993). Cortisol and immune characteristics in rainbow trout (*Oncorhynchus mykiss*) selected from high or low tolerance to stress. *Journal of Fish Biology* **43**, 919–930.

FEVOLDEN, S. E., ROED, K. H. & GJERDE, B. (1994). Genetic components of post-stress cortisol and lysozyme activity in Atlantic salmon; correlations to disease resistance. *Fish and Shellfish Immunology* **4**, 507–519.

FINSTAD, B., BJORN, P. A., GRIMNES, A. & HVIDSTEN, N. A. (2000). Laboratory and field investigations of salmon lice *Lepeophtheirus salmonis* (Krøyer) infestation on Atlantic salmon (*Salmo salar* L.) post-smolts. *Aquaculture Research* **31**, 795–803.

FIRTH, K. J., JOHNSON, S. C. & ROSS, N. W. (1999). Characterization of proteases in the skin mucus of Atlantic salmon (*Salmo salar*) infected with the salmon louse *Lepeophtheirus salmonis* and in *L. salmonis* whole body homogenates. *Journal of Parasitology* **86**, 1199–1205.

GONZALEZ, L., CARVAJAL, J. & GEORGE-NASCIMENTO, M. (2000). Differential infectivity of *Caligus flexispina* (Copepoda, Caligidae) in three farmed salmonids in Chile. *Aquaculture* **183**, 13–23.

GONZALEZ-ALANIS, P., WRIGHT, G. M., JOHNSON, S. C. & BURKA, J. F. (2001). Frontal filament morphogenesis in the salmon louse *Lepeophtheirus salmonis*. *Journal of Parasitology* **87**, 561–574.

GRAYSON, T. H., JENKINS, P. G., WRATHMELL, A. B. & HARRIS, J. E. (1991). Serum responses to the salmon louse, *Lepeophtheirus salmonis* (Krøyer, 1838), in naturally infected salmonids and immunised rainbow trout, *Oncorhynchus mykiss* (Walbaum), and rabbits. *Fish and Shellfish Immunology* **1**, 141–155.

GRAYSON, T. H., JOHN, R. J., WADSWORTH, S., GREAVES, K., COX, D., ROPER, J., WRATHMELL, A. B., GILPIN, M. L. & HARRIS, J. E. (1995). Immunization of Atlantic salmon against louse: identification of antigens and effects on louse fecundity. *Journal of Fish Biology* **47** (Supplement A), 85–94.

GUBBINS, S. & GILLIGAN, C. A. (1997). Persistence of host–parasite interactions in a disturbed environment. *Journal of Theoretical Biology* **188**, 241–258.

HANSKI, I. (1999). *Metapopulation Ecology*. Oxford Series in Ecology and Evolution. Oxford, Oxford University Press.

HARRIS, J. & BIRD, D. J. (2000). Modulation of the fish immune system by hormones. *Veterinary Immunology and Immunopathology* **77**, 163–176.

HART, L. J., SMITH, S. A., SMITH, B. J., ROBERTSON, J., BESTEMAN, E. G. & HOLLADAY, S. D. (1998). Subacute immunotoxic effects of the polycyclic aromatic hydrocarbon 7,12-dimethylbenzanthracene (DMBA) on spleen and pronephros leukocytic cell counts and phagocytic cell activity in tilapia (*Oreochromis niloticus*). *Aquatic Toxicology* **41**, 17–29.

HECHT, T. & ENDEMANN, F. (1998). The impact of parasites, infections and diseases on the development of aquaculture in sub-Saharan Africa. *Journal of Applied Ichthyology* **14**, 213–221.

HEUCH, P. A. & KARLSEN, H. E. (1997). Detection of infrasonic oscillations by copepodids of *Lepeophtheirus salmonis* (Copepoda: Caligidae). *Journal of Plankton Research* **19**, 735–747.

HEUCH, P. A., NORDHAGEN, J. R. & SCHRAM, T. A. (2000). Egg production in the salmon louse (*Lepeophtheirus salmonis* (Krøyer)) in relation to origin and water temperature. *Aquaculture Research* **31**, 805–814.

HEUCH, P. A., PARSONS, A. & BOXASPEN, K. (1995). Diel vertical migration: a possible host-finding mechanism in salmon louse (*Lepeophtheirus salmonis*) copepodids? *Canadian Journal of Fisheries and Aquatic Sciences* **52**, 681–689.

HOUGHTON, G. & MATTHEWS, R. A. (1990). Immunosuppression in juvenile carp, *Cyprinus carpio* L.: the effects of the corticosteroids triamcinolone acetonide and hydrocortisone 21-hemisuccinate (cortisol) on acquired immunity and the humoral antibody response to *Ichthyophthirius multifiliis*. *Journal of Fish Diseases* **13**, 269–280.

IGER, Y., BALM, P. H. M., JENNER, H. A. & WENDELAAR BONGA, S. E. (1995). Cortisol induces stress-related changes in the skin of rainbow trout (*Oncorhynchus mykiss*). *General and Comparative Endocrinology* **97**, 188–198.

IGER, Y., JENNER, H. A. & WENDELAAR BONGA, S. E. (1994). Cellular responses in the skin of the trout (*Oncorhynchus mykiss*) exposed to temperature elevation. *Journal of Fish Biology* **44**, 921–935.

IGER, Y. & WENDELAAR BONGA, S. E. (1994*d*). Cellular responses of the skin of carp (*Cyprinus carpio*) exposed to acidified water. *Cell and Tissue Research* **275**, 481–492.

ISDAL, E., NYLUND, A. & NAEVDAL, G. (1997). Genetic differences among salmon lice (*Lepeophtheirus salmonis*) from six Norwegian coastal sites: evidence from allozymes. *Bulletin of the European Association of Fish Pathologists* **17**, 17–22.

JACOBSEN, J. & GAARD, E. (1997). Open-ocean infestation by salmon lice (*Lepeophtheirus salmonis*): comparison with wild and escaped farmed Atlantic Salmon (*Salmo salar* L.). *ICES Journal of Marine Science* **54**, 1113–1119.

JACKSON, D. & MINCHIN, D. (1993). Lice infestation of farmed salmon in Ireland. In *Pathogens of Wild and Farmed Fish: Sea Lice* (ed. Boxshall, G. A. & Defaye, D.), pp. 188–201. Chichester, Ellis Horwood.

JACKSON, D. (1997). Single Bay Management. *Aquaculture Newsletter* **24**, 4–5.

JACKSON, D. (1998). Developments in sea lice management in Irish salmon farming. *Caligus* **4**, 2–3.

JOHNSON, S. C. (1993). A comparison of development and growth rates of *Lepeophtheirus salmonis* (Copepoda: Caligidae) on naive Atlantic salmon (*Salmo salar*) and

chinook (*Oncorhynchus tshawytscha*) salmon. In *Pathogens of Wild and Farmed Fish: Sea Lice* (ed. Boxshall, G. A. & Defaye, D.), pp. 68–80. Chichester, Ellis Horwood Limited.

JOHNSON, S. C. & ALBRIGHT, L. J. (1991*a*). The developmental stages of *Lepeophtheirus salmonis* (Krøyer, 1837) (Copepoda: Caligidae). *Canadian Journal of Zoology* **69**, 929–950.

JOHNSON, S. C. & ALBRIGHT, L. J. (1991*b*). Development, growth, and survival of *Lepeophtheirus salmonis* (Copepoda, Caligidae) under laboratory conditions. *Journal of the Marine Biological Association of the United Kingdom* **71**, 425–436.

JOHNSON, S. C. & ALBRIGHT, L. J. (1992*a*). Comparative susceptibility and histopathology of the response of naive Atlantic, chinook and coho salmon to experimental infection with *Lepeophtheirus salmonis* (Copepoda: Caligidae). *Diseases of Aquatic Organisms* **14**, 179–193.

JOHNSON, S. C. & ALBRIGHT, L. J. (1992*b*). Effects of cortisol implants on the susceptibility and the histopathology of the responses of naive coho salmon *Oncorhynchus kisutch* to experimental infection with *Lepeophtheirus salmonis* (Copepoda: Caligidae). *Diseases of Aquatic Organisms* **14**, 195–205.

JOHNSON, S. C., BLAYLOCK, R. B., ELPHICK, J. & HYATT, K. (1996). Disease induced by the sea louse (*Lepeophtheirus salmonis*) (Copepoda: Caligidae) in wild sockeye salmon (*Oncorhynchus nerka*) stocks of Alberni Inlet, British Columbia. *Canadian Journal of Fisheries and Aquatic Sciences* **53**, 2888–2897.

JONES, M. W., SOMMERVILLE, C. & BRON, J. (1990). The histopathology associated with the juvenile stages of *Lepeophtheirus salmonis* on the Atlantic salmon, *Salmo salar* L. *Journal of Fish Diseases* **13**, 303–310.

JONES, M. W., SOMMERVILLE, C. & WOOTTEN, R. (1992). Reduced sensitivity of the salmon louse, *Lepeophtheirus salmonis*, to the organophosphate dichlorvos. *Journal of Fish Diseases* **15**, 197–202.

JONSDOTTIR, H., BRON, J. E., WOOTTEN, R. & TURNBULL, J. F. (1992). The histopathology associated with the pre-adult and adult stages of *Lepeophtheirus salmonis* on the Atlantic salmon, *Salmo salar* L. *Journal of Fish Diseases* **15**, 521–527.

KAATTARI, S. L. & TRIPP, R. A. (1987). Cellular mechanisms of glucocorticoid immunosuppression in salmon. *Journal of Fish Biology* **31** (Supplement A), 129–132.

KABATA, Z. (1970). *Diseases of Fishes. Book 1: Crustacea as Enemies of Fishes*. Jersey City, N.J., T.F.H. Publications.

KABATA, I. (1979). *Parasitic Copepoda of British Fishes*. London, Ray Society.

KAKUTA, Z. (1997). Effect of sewage on blood parameters and the resistance against bacterial infection of goldfish, *Carassius auratus*. *Environmental Toxicology and Water Quality* **12**, 43–51.

MacKINNON, B. M. (1993). Host response of Atlantic salmon (*Salmo salar*) to infection by sea lice (*Caligus elongatus*). *Canadian Journal of Fisheries and Aquatic Sciences* **50**, 789–792.

MAGWOOD, S. & GEORGE, S. (1996). *In vitro* alternatives to whole animal testing. Comparative cytotoxicity studies of divalent metals in established cell lines derived from tropical and temperate water fish species in a neutral red assay. *Marine Environmental Research* **42**, 37–40.

MAULE, A. G., SCHRECK, C. B. & KAATTARI, S. L. (1987). Changes in the immune system of coho salmon (*Oncorhynchus kisutch*) during parr-to-smolt transformation and after implantation of cortisol. *Canadian Journal of Fisheries and Aquatic Sciences* **44**, 161–166.

MOTHERSILL, C. & AUSTIN, B. (2000). *Aquatic Invertebrate Cell Culture*. Chichester, Praxis Publishing U.K.

MUSTAFA, A. & MacKINNON, B. M. (1999). Atlantic salmon, *Salmo salar* L., and Artic char, *Salvelinus alpinus* (L.): comparative correlation between iodine-iodide levels, plasma cortisol levels, and infection intensity with the sea louse *Caligus elongatus*. *Canadian Journal of Zoology* **77**, 1092–1101.

MUSTAFA, A., MacWILLIAMS, C., FERNANDEZ, N., MATCHETT, K., CONBOY, G. A. & BURKA, J. F. (2000a). Effects of sea lice (*Lepeophtheirus salmonis* Krøyer, 1837) infestation on macrophage functions in Atlantic salmon (*Salmo salar* L.). *Fish and Shellfish Immunology* **10**, 47–59.

MUSTAFA, A., SPEARA, D. J., DALEY, J., CONBOY, G. A. & BURKA, J. F. (2000b). Enhanced susceptibility of seawater cultured rainbow trout, *Oncorhynchus mykiss* (Walbaum), to the microsporidian *Loma salmonae* during a primary infection with the sea louse, *Lepeophtheirus salmonis*. *Journal of Fish Diseases* **23**, 337–341.

NAGAE, M., FUDA, H., URA, K., KAWAMURA, H., ADACHI, S., HARA, A. & YAMAUCHI, K. (1994). The effect of cortisol administration on blood plasma immunoglobulin M (IgM) concentrations in masu salmon (*Oncorhynchus masou*). *Fish Physiology and Biochemistry* **13**, 41–48.

NASH, A. D., EGAN, P. J., KIMPTON, W., ELHAY, M. J. & BOWLES, V. M. (1996). Local cell traffic and cytokine production associated with ectoparasitic infection. *Veterinary Immunology and Immunopathology* **54**, 269–279.

NEUMANN, N. F., FAGAN, D. & BELOSEVIC, M. (1995). Macrophage activating factor(s) secreted by mitogen stimulated goldfish kidney leukocytes synergize with bacterial lipopolysaccharide to induce nitric oxide production in teleost macrophages. *Developmental and Comparative Immunology* **19**, 473–482.

NOLAN, D. T., HADDERINGH, R. H., SPANINGS, F. A. T., JENNER, H. A. & WENDELAAR BONGA, S. E. (2000a). Effects of short-term acute temperature elevation on sea trout smolts (*Salmo trutta* L.) in tap water and in Rhine water: effects on skin and gill epithelia, hydromineral balance and gill specific Na⁺/K⁺-ATPase activity. *Canadian Journal of Fisheries and Aquatic Sciences* **57**, 708–718.

NOLAN, D. T. & JOHNSON, S. C. (2000). Interaction between crustacean ectoparasites and their hosts: a tissue culture perspective. In *Aquatic Invertebrate Cell Culture* (ed. Mothersill, C. & Austin, B.), pp. 135–164. Chichester, Praxis Publishing U.K.

NOLAN, D. T., NABBEN, I., JIE, L. & WENDELAAR BONGA, S. E. (2002). Primary cell culture of rainbow trout skin explants: growth, cell composition, proliferation and apoptosis. *In Vitro Cellular and Developmental Biology (Animal)* (in press).

NOLAN, D. T., OP'T VELD, R. L. J. M., BALM, P. H. M. & WENDELAAR BONGA, S. E. (1999a). Ambient salinity modulates the stress response of the tilapia, *Oreochromis mossambicus*, (Peters) to net confinement. *Aquaculture* **177**, 297–309.

NOLAN, D. T., REILLY, P. & WENDELAAR BONGA, S. E. (1999b). Infection with low numbers of the sea louse *Lepeophtheirus salmonis* (Krøyer) induces stress-related effects in post-smolt Atlantic salmon (*Salmo salar* L.). *Canadian Journal of Fisheries and Aquatic Sciences* **56**, 947–959.

NOLAN, D. T., RUANE, N. M., VAN DER HEIJDEN, Y., QUABIUS, E. S., COSTELLOE, J. & WENDELAAR BONGA, S. E. (2000c). Juvenile *Lepeophtheirus salmonis* (Krøyer) affect the skin and gills of rainbow trout *Oncorhynchus mykiss* (Walbaum) and the host response to a handling procedure. *Aquaculture Research* **31**, 823–833.

NOLAN, D. T., VAN DER SALM, A. L. & WENDELAAR BONGA, S. E. (2000d). The host–parasite relationship between the rainbow trout (*Oncorhynchus mykiss*) and the ectoparasite *Argulus foliaceus* (Crustacea: Branchiura): Epithelial mucous cell response, cortisol and factors which may influence parasite establishment. *Contributions to Zoology* **69**, 57–63.

NOLAN, D. T., VAN DER SALM, A. L. & WENDELAAR BONGA, S. E. (1999c). *In vitro* effects of short-term cortisol exposure on proliferation and apoptosis in the skin epidermis of rainbow trout (*Oncorhynchus mykiss* Walbaum). In *Recent Developments in Comparative Endocrinology and Neurobiology* (ed. Roubos, E. Wendelaar Bonga, S. E. Vaudry, H. & De Loof, A.), pp. 161–162. Maastricht, Shaker Publishers.

NOLAN, D. V., MARTIN, S. A. M., KELLY, Y., GLENNON, K., PALMER, R., SMITH, T., McCORMACK, G. P. & POWELL, R. (2000e). Development of microsatellite PCR typing methodology for the sea louse *Lepeophtheirus salmonis* (Køyer). *Aquaculture Research* **31**, 815–822.

NORDHAGEN, J. R., HEUCH, P. A. & SCHRAM, T. A. (2000). Size as indicator of origin of salmon lice *Lepeophtheirus salmonis* (Copepoda: Caligidae). *Contributions to Zoology* **69**, 99–108.

O'FLAHERTY, G., RYAN, R., MAC EVILLY, U., POOLE, W. R., TULLY, O. & NOLAN, D. T. (1999). Biochemical responses in relation to sea lice (*Lepeophtheirus salmonis* Krøyer) infestation in wild sea trout (*Salmo trutta*, L.) post smolts. Oral presentation at the 4th International Conference on Sea Lice, Trinity College Dublin, 28–30 June 1999.

PAPERNA, I. & ZWERNER, D. E. (1982). Host–parasite relationship of *Ergasilus labracis* Krøyer (Cyclopidea, Ergasilidae) and the striped bass, *Morone saxatilis* (Walbaum) from the lower Chesapeake Bay. *Annales De Parasitologie (Paris)* **57**, 393–405.

PICKERING, A. D. & POTTINGER, T. G. (1985). Cortisol can increase the susceptibility of brown trout, *Salmo trutta* L., to disease without reducing the white blood cell count. *Journal of Fish Biology* **27**, 611–619.

PICKERING, A. D. & POTTINGER, T. G. (1989). Stress responses and disease resistance in salmonid fish: effects of chronic elevation of plasma cortisol. *Fish Physiology and Biochemistry* **7**, 253–258.

PIKE, A. W., MacKENZIE, K. & ROWLAND, A. (1993). Ultrastructure of the frontal filament in chalimus larvae of *Caligus elongatus* and *Lepeophtheirus salmonis* from Atlantic salmon, *Salmo salar*. In *Pathogens of Wild and Farmed Fish : Sea Lice* (ed. Boxshall, G. A. & Defaye, D.), pp. 99–113. Chichester, Ellis Horwood Limited.

PIKE, A. W. & WADSWORTH, S. L. (1999). Sea lice on salmonids : their biology and control. *Advances in Parasitology* **44**, 233–337.

POOLE, W. R., NOLAN, D. T. & TULLY, O. (2000). Modelling the effects of capture and sea lice *Lepeophtheirus salmonis* (Krøyer) infestation on the cortisol stress response in trout. *Aquaculture Research* **31**, 835–841.

POTTINGER, T. G., MORAN, T. A. & CRANWELL, P. A. (1992). The biliary accumulation of corticosteroids in rainbow trout, *Onchorhynchus mykiss*, during acute and chronic stress. *Fish Physiology and Biochemistry* **10**, 55–66.

QUABIUS, E. S., BALM, P. H. M. & WENDELAAR BONGA, S. E. (1997). Interrenal stress responsiveness of tilapia (*Oreochromis mossambicus*) is impaired by dietary exposure to PCR 126. *General and Comparative Endocrinology* **108**, 472–482.

RITCHIE, G. (1997). The host transfer ability of *Lepeophtheirus salmonis* (Copepoda : Caligidae) from farmed Atlantic salmon, *Salmo salar* L. *Journal of Fish Diseases* **20**, 153–157.

RITCHIE, G., MORDUE (LUNTZ), A. J., PIKE, A. W. & RAE, G. H. (1993). The reproductive output of *Lepeophtheirus salmonis* adult females in relation to seasonal variability of temperature and photoperiod. In *Pathogens of Wild and Farmed Fish : Sea Lice* (ed. Boxshall, G. A. & Defaye, D.), pp. 153–165. Chichester, Ellis Horwood Limited.

ROBERTS, L. S. & JANOVY JR, J. (1996). Basic principles and concepts II : immunology and pathology. In *Gerald D. Schmidt & Larry S. Roberts' Foundations of Parasitology*. pp. 21–34. Dubuque, IA, Wm. C. Brown Publishers.

RODGERS, C. J. & FURONES, M. D. (1998). Disease problems in cultured marine fish in the Mediterranean. *Fish Pathology* **33**, 157–164.

ROSS, N. W., FIRTH, K. J., WANG, A. P., BURKA, J. F. & JOHNSON, S. C. (2000). Changes in hydrolytic enzyme activities of naive Atlantic salmon *Salmo salar* skin mucus due to infection with the salmon louse *Lepeophtheirus salmonis* and cortisol implantation. *Diseases of Aquatic Organisms* **41**, 43–51.

ROSS, P. S., DE SWART, R. L., VAN LOVEREN, H., OSTERHAUS, A. D. M. E. & VOS, J. G. (1996). The immunotoxicity of environmental contaminants to marine wildlife : a review. *Annual Review of Fish Diseases* **6**, 151–165.

ROTH, M., RICHARDS, R. H. & SOMMERVILLE, C. (1993). Current practices in the chemotherapeutic control of sea lice infestations in aquaculture : a review. *Journal of Fish Diseases* **16**, 1–26.

RUANE, N. M., NOLAN, D. T., ROTLLANT, J., COSTELLOE, J. & WENDELAAR BONGA, S. E. (2000). Experimental exposure of rainbow trout *Oncorhynchus mykiss* (Walbaum) to the infective stages of the sea louse *Lepeophtheirus salmonis* (Krøyer) influences the physiological response to an acute stressor. *Fish and Shellfish Immunology* **10**, 451–463.

RUANE, N. M., NOLAN, D. T., ROTLLANT, J., TORT, L., BALM, P. H. M. & WENDELAAR BONGA, S. E. (1999). Modulation of the response of rainbow trout (*Oncorhynchus mykiss*

Walbaum) to confinement, by an ectoparasitic (*Argulus foliaceus* L.) infestation and cortisol feeding. *Fish Physiology and Biochemistry* **20**, 43–51.

SAKUMA, K. M., RALSTON, S., LANARZ, W. H. & EMBURY, M. (1999). Effects of the parasitic copepod *Cardiodectes medusaeus* on the lanternfishes *Diaphus theta* and *Tarletonbeania crenularis* off central California. *Environmental Biology of Fishes* **55**, 423–430.

SCHOLZ, T. (1999). Parasites in cultured and feral fish. *Veterinary Parasitology* **84**, 317–335.

SCHRAM, T. A. (1993). Supplementary descriptions of the developmental stages of *Lepeophtheirus salmonis* (Krøyer, 1837) (Copepoda : Caligidae). In *Pathogens of Wild and Farmed Fish : Sea Lice* (ed. Boxshall, G. A. & Defaye, D.), pp. 30–47. Chichester, Ellis Horwood Limited.

SHARIFF, M. & ROBERTS, R. (1989). The experimental histopathology of *Lernaea polymorpha* Wu, 1938 infection in naive *Aristichthys nobilis* (Richardson) and a comparison with the lesion on naturally infected clinically resistant fish. *Journal of Fish Diseases* **12**, 405–414.

SHEPHARD, K. L. (1994). Functions for fish mucus. *Reviews in Fish Biology and Fisheries* **4**, 401–429.

SHIELDS, R. J. & GOODE, R. P. (1978). Host rejection of *Lernaea cyprinacea* L. (Copepoda). *Crustaceana* **35**, 301–307.

SHINN, A. P., BANKS, B. A., TANGE, N., BRON, A. P., SOMMERVILLE, C., AOKI, T. & WOOTTEN, R. (2000). Utility of 18S rDNA and ITS sequences as population markers for *Lepeophtheirus salmonis* (Copepoda : Caligidae) parasitising Atlantic salmon (*Salmo salar*) in Scotland. *Contributions to Zoology* **69**, 79–87.

SMITH, V. J., FERNANDES, J. M. O., JONES, S. J., KEMP, G. D. & TATNER, M. F. (2000). Antibacterial proteins in rainbow trout, *Oncorhynchus mykiss*. *Fish and Shellfish Immunology* **10**, 243–260.

STAVE, J. W. & ROBERSON, B. S. (1985). Hydrocortisone suppresses the chemiluminescent response of striped bass phagocytes. *Developmental and Comparative Immunology* **9**, 77–84.

THONEY, D. A. & BURRESON, E. M. (1988). Lack of a specific humoral response in *Leiostomus xanthurus* (Pisces : Sciaenidae) to parasitic copepods and monogeans. *Journal of Parasitology* **74**, 191–194.

TODD, C. D., WALKER, A. M., WOLFF, K., NORTHCOTT, S. J., WALKER, A. F., RITCHIE, M. G., HOSKINS, R., ABBOTT, R. J. & HAZON, N. (1997). Genetic differentiation of populations of the copepod sea louse *Lepeophtheirus salmonis* (Krøyer) ectoparasitic on wild and farmed salmonids around the coasts of Scotland : evidence from RAPD markers. *Journal of Experimental Marine Biology and Ecology* **210**, 251–274.

TREASURER, J. M., WADSWORTH, S. & GRANT, A. (2000). Resistance of sea lice, *Lepeophtheirus salmonis* (Krøyer), to hydrogen peroxide on farmed Atlantic salmon, *Salmo salar* L. *Aquaculture Research* **11**, 855–860.

TULLY, O. (1989). The succession of generations and growth of the caligid copepods *Caligus elongatus* and *Lepeophtheirus salmonis* parasitising farmed Atlantic salmon smolts (*Salmo salar*). *Journal of the Marine*

Biological Association of the United Kingdom **69**, 279–287.

TULLY, O. (1992). Predicting infestation parameters and impacts of caligid copepods in wild and cultured fish populations. *Invertebrate Reproduction and Development* **22**, 91–102.

TULLY, O., GARGAN, P., POOLE, W. R. & WHELAN, K. F. (1999). Spatial and temporal variation in the infestation of sea trout (*Salmo trutta* L.) by the caligid copepod *Lepeophtheirus salmonis* (Krøyer) in relation to sources of infection in Ireland. *Parasitology* **119**, 41–51.

TULLY, O. & MCFADDEN, Y. (2000). Variation in sensitivity of sea lice (*Lepeophtheirus salmonis* (Krøyer)) to dichlorvos on Irish salmon farms in 1991–1992. *Aquaculture Research* **31**, 849–854.

TULLY, O., POOLE, W. R. & WHELAN, K. F. (1993a). Infestation parameters for *Lepeophtheirus salmonis* (Krøyer) (Copepoda: Caligidae) parasitic on sea trout, *Salmo trutta* L. off the west coast of Ireland during 1990 and 1991. *Aquaculture and Fisheries Management* **24**, 545–555.

TULLY, O., POOLE, W. R., WHELAN, K. F. & MERIGOUX, S. (1993b). Parameters and possible causes of epizooties of *Lepeophtheirus salmonis* (Krøyer) infesting sea trout (*Salmo trutta* L.) off the west coast of Ireland. In *Pathogens of Wild and Farmed Fish: Sea Lice* (ed. Boxshall, G. A. & Defaye, D.), pp. 202–213. Chichester, Ellis Horwood Limited.

TULLY, O. & WHELAN, K. F. (1993). Production of nauplii of *Lepeophtheirus salmonis* (Krøyer) (Copepoda: Caligidae) from farmed and wild salmon and its relation to the infestation of wild sea trout (*Salmo trutta* L.) off the west coast of Ireland in 1991. *Fisheries Research* **17**, 187–200.

WENDELAAR BONGA, S. E. (1997). The stress response in fish. *Physiological Reviews* **77**, 591–625.

WHELAN, K. F. & POOLE, W. R. (1996). The sea trout stock collapse, 1989–1992. In *The Conservation of Aquatic Systems* (ed. Reynolds, J. D.). Dublin, Royal Irish Academy.

WHITE, H. C. (1940). "Sealice" (*Lepeophtheirus*) and death of salmon. *Journal of the Fisheries Research Board of Canada* **5**, 172–175.

WIKEL, S. K. (1999). Modulation of the host immune system by ectoparasitic arthropods. *BioScience* **49**, 311–320.

WIKEL, S. K., RAMACHANDRA, R. N. & BERGMAN, D. K. (1994). Tick-induced modulation of the host immune response. *International Journal for Parasitology* **24**, 59–66.

WILHELM FILHO, D., GIULIVI, C. & BOVERIS, A. (1993). Antioxidant defences in marine fish: 1. Teleosts. *Comparative Biochemistry and Physiology* **106**, 409–413.

WONG, S., FOURNIER, M., CODERRE, D., BANSKA, W. & KRZYSTYNIAK, K. (1992). Environmental immunotoxicology. In *Animal Markers as Pollutioncology Indicators*. (ed. Peakall, D.), pp. 167–189. London, Chapman & Hall.

WOO, P. T. K. & SHARIFF, M. (1990). *Lernaea cyprinacea* L. (Copepoda: Caligidea) in *Helostoma temminicki* Cuvier & Valenciennes: the dynamics of resistance in recovered and naive fish. *Journal of Fish Diseases* **13**, 485–493.

WOOTTEN, R., SMITH, J. W. & NEEDHAM, E. A. (1982). Aspects of the biology of the parasite copepods *Lepeophtheirus salmonis* and *Caligus elongatus* on farmed salmonids, and their treatment. *Proceedings of the Royal Society of Edinburgh* **81B**, 185–197.

The trouble with sealworms (*Pseudoterranova decipiens* species complex, Nematoda): a review

G. McCLELLAND

Department of Fisheries and Oceans, Gulf Fisheries Centre, P.O. Box 5030, Moncton, New Brunswick, Canada E1C 9B6

SUMMARY

Sealworms or codworms, larvae of ascaridoid nematodes belonging to the *Pseudoterranova decipiens* species complex, infect the flesh of numerous species of marine and euryhaline fish, and have proven a chronic and costly cosmetic problem for seafood processors. Moreover, the parasite may cause abdominal discomfort in humans when consumed in raw, undercooked or lightly marinated fish. In this review, the phylogeny, life cycle and distributions of sealworms are discussed along with biotic and abiotic factors which may influence distributions of these parasites in their intermediate and final hosts. Also considered here are efforts to control the problem through commercial fishing practices, fish processing technology, and the reduction of infection parameters in marine fish populations by biological means. Ironically, concern over sealworm problem has subsided in some fisheries in recent years, not as a result of falling infection parameters in fish stocks or innovations in processing technology, but as a consequence of declines in abundance and size of groundfish.

Key words: Sealworms, *Pseudoterranova* spp., phylogeny, life cycle, distribution, controls.

INTRODUCTION

Sealworms or codworms, larval ascaridoid nematodes belonging to the *Pseudoterranova decipiens* species complex (Paggi *et al*. 2000), which infect the flesh of marine and euryhaline fish, have proven a chronic and costly cosmetic problem for fish processors. Detection and removal of the parasites from the flesh of Atlantic cod (*Gadus morhua*) and other demersal species, and the resultant downgrading and discard of product have been estimated to cost processors in Atlantic Canada a total of $26·6 (Malouf, 1986) to $50 million (Aryee & Poehlman, 1991) annually. Bublitz & Choudhury (1992) conclude that labour-intensive 'candling' of fillets for sealworm and other parasites accounts for approximately half of production costs for Pacific cod (*G. macrocephalus*) from the Bering Sea and Gulf of Alaska, and may cause delays which promote microbial growth and enzymatic degradation of product. Sealworm infestations are also a chronic problem in British, Norwegian and Icelandic cod fisheries (Desportes & McClelland, 2001), and a cause for some concern in the Chilean (Carvajal & Cattan, 1985) and Argentinean hake (*Merluccius* spp.) fisheries (Herreras *et al*. 2000).

From a public health standpoint, larval sealworms can infect humans when consumed in raw, undercooked, or lightly marinated fish, and are capable of causing clinical signs of anisakiasis, such as nausea, severe epigastric pain, vomiting and other abdominal discomforts (Margolis, 1977). The majority of 'archetypal' cases of sealworm anisakiasis, involving penetration of alimentary tract and associated organs

Tel: +1(506)851-6218. Fax: +1(506)851-2079. E-mail: McClellandG@dfo-mpo.gc.ca

and severe pathology (Smith, 1999) are reported from Japan, where seafood is a major component of the diet and traditionally served raw (Oshima, 1987). Of > 12000 cases of clinical anisakiasis documented from Japan, however, only 335 are attributed to *Pseudoterranova* sp. infection, with *Anisakis* sp. infection being diagnosed in the remainder (Ishikura *et al*. 1993). Cases of 'extragastrointestinal' or 'oropharangeal' sealworm anisakiasis are even rarer (Amin *et al*. 2000). The great majority of cases diagnosed in Europe and the Americas can be classified as 'transient luminal' (Smith, 1999). The latter are asymptomatic, but for a tingling sensation in the throat or mild abdominal discomfort, and the nematodes are expelled by coughing, vomiting or defaecation.

Detailed accounts of pathology, diagnoses and treatment of sealworm anisakiasis are found in numerous case histories and reviews (e.g. Bier *et al*. 1987; Bouree, Paugam & Petithory, 1995; Fair, 2000). In addition to human cases, Smith (1999) reviews pathology associated with sealworm infections in natural fish and pinniped hosts, and in nonspecific ('accidental' and experimental) mammalian hosts. However, since the last comprehensive review of the sealworm problem (Hafsteinsson & Rizvi, 1987) there have been significant developments in sealworm phylogeny and taxonomy (Mattiucci *et al*., 1998, Paggi *et al*., 2000). Moreover, two workshops (Bowen, 1990, Desportes & McClelland, 2001) and a review article (Burt, 1994) have addressed the life cycle and distributions of the parasites in the North Atlantic, and efforts to mitigate the sealworm problem through seafood processing technology and biological controls. The latter issues are updated and reviewed here, and biotic and abiotic

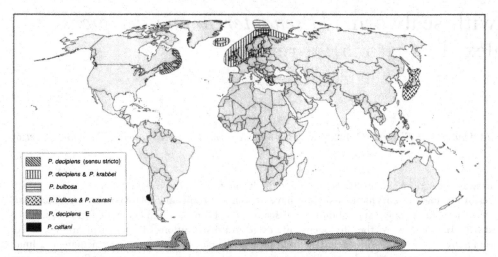

Fig. 1. Known distributions of sealworm species *Pseudoterranova krabbei*, *P. decipiens* (*sensu stricto*), *P. bulbosa*, *P. azarasi*, *P. decipiens* E, and *P. cattani* (Paggi *et al.* 1991, 2000; Brattey & Stenson, 1993; Brattey & Davidson, 1996; Bullini *et al.* 1997; Mattiucci *et al.* 1998; George-Nascimento & Llanos, 1995; George-Nascimento & Urrutia, 2000). As the specific geographic origins of specimens of *P. decipiens* E were not reported, the range shown is arbitrary.

influences on the spatial and temporal distributions of sealworms in their intermediate and final hosts are discussed.

TAXONOMY

After a confusing taxonomic history in which sealworm was placed variously in the genera *Porrocaecum*, *Terranova* and *Phocanema* (Hafsteinsson & Rizvi, 1987). Gibson (1983) established precedence for *Pseudoterranova*, a genus including *P. decipiens* from pinnipeds, *P. kogiae* from pygmy sperm whale (*Kogia breviceps*) and *P. ceticola* from the dwarf sperm whale (*Kogia simus*). The parasite is commonly referred to as codworm or sealworm, although, as Margolis (1977) points out, the latter term is more appropriate given that the definitive hosts of the nematode are seals.

With the confusion regarding the generic status resolved, sealworms were, for a short time, considered to belong to a single cosmopolitan species, *Pseudoterranova decipiens*. Multilocus electrophoresis of enzyme systems of larval and adult nematodes (Paggi *et al.* 1991), however, revealed that *Pseudoterranova* from North Atlantic fish and seals consist of three structurally similar but reproductively isolated sibling species. In the Northeast Atlantic, *P. decipiens* A and B are sympatric in waters off northern Europe and Iceland, while sibling species C is confined to the Barents Sea. Sibling species A is lacking in the Northwest Atlantic, but sibling species B occurs off eastern Canada, from northern Labrador to the Gulf of Maine and is sympatric with species C off Labrador and northern Newfoundland (Brattey & Stenson, 1993; Brattey & Davidson, 1996). A fourth sibling, *P. decipiens* D is sympatric with species C in Japanese waters (Mattiucci *et al.* 1998), and a fifth, *P. decipiens* E is found in the Antarctic (Bullini *et al.* 1997). Citing structural and morpho-

metric differences in the caudal regions of adult males, Paggi *et al.* (2000) proposed *P. krabbei* and *P. decipiens* (*sensu stricto*) as specific designations for species A and B respectively, while Mattiucci *et al.* (1998) adopted *P. bulbosa* and *P. azarasi* for species C and D. Finally, on the basis of similar electrophoretic and morphometric analyses, a sixth sealworm species, *P. cattani*, was described from sea lions and fish in Chilean waters (George-Nascimento & Llanos, 1995; George-Nascimento & Urrutia, 2000).

As evident from Fig. 1, surveys employing allozyme and morphometric analyses provide rather modest information on the geographical distributions of six members of the *P. decipiens* species complex. Results of future surveys may extend the ranges of some or all of these species and, perhaps, reveal additional species. Even in previously surveyed areas, where nematodes were identified from relatively few, widely distributed fish and seal hosts, some species may have escaped detection. For example, isoelectric focusing of soluble proteins from whole larvae (Appleton & Burt, 1991) reveal two variants of *Pseudoterranova* in fish from the southern Gulf of St Lawrence and the Lower Bay of Fundy, areas of eastern Canada where, according to Paggi *et al.* (2000), only *P. decipiens* (*sensu stricto*) occurs. A subsequent study of sealworms from fish and seals in the Lower Bay of Fundy (Quarrar, 1999) describes two variants of the parasite differing in regard to early life cycle, final host and morphometry of the lips of fourth stage larvae (L4) and adults.

LIFE CYCLE

Laboratory and field studies of life cycles and distributions of sealworms have been conducted primarily in countries having commercial fishing interests in the North Atlantic and adjacent waters

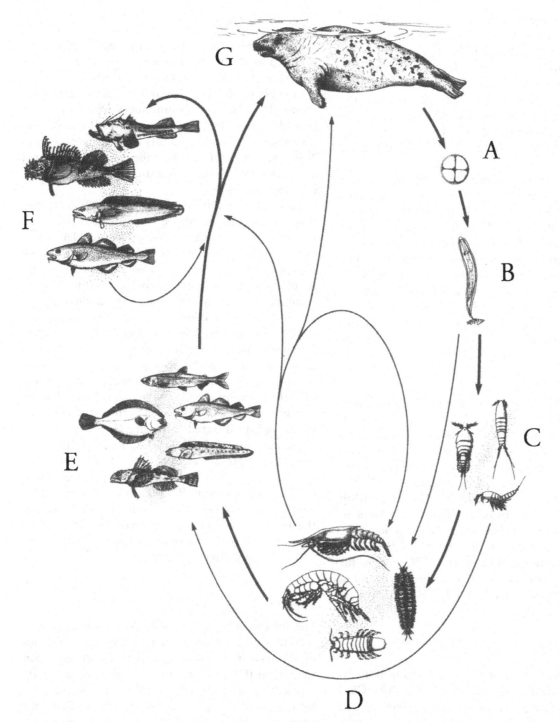

Fig. 2. Life cycle of *P. decipiens* (*sensu stricto*) in eastern Canada: A, egg; B, free-living ensheathed L2 (L3?); C, copepod hosts; D, macroinvertebrate hosts; E, primary fish hosts; F, secondary fish hosts; G, seal host (from McClelland, Misra & Martell, 1990).

(Bowen, 1990; Desportes & McClelland, 2001). The life cycle described here (Fig. 2) is derived largely from laboratory experiments and surveys involving sealworm and sealworm hosts from the southern Gulf of St Lawrence and the Breton and Scotian shelves in eastern Canada. *P. decipiens* (*sensu stricto*) is the only sealworm species identified from these areas (Paggi *et al.* 1991), and the life cycle may be unique to this species.

Ova

P. decipiens ova are 0·04–0·05 mm in diameter and contain embryos at the 2 to 16 cell stage when passed with the faeces of the seal host (McClelland *et al.* 1990). The ova are negatively buoyant, with a free settling rate of $\sim 1 \times 10^{-4}$ m sec^{-1} (McConnell, Marcogliese & Stacey, 1997) and presumably descend to the seabed before hatching. Mean hatching time has an inverse curvilinear (hyperbolic) relationship to temperature (McClelland, 1982) and varies from 7 d at 22 °C to 125 d at 1·7 °C (McClelland *et al.* 1990). Embryos do not develop beyond the tadpole stage at 0 °C, and ova held at this temperature for a year do not hatch when subsequently incubated at higher temperatures (Measures, 1996). Ova also fail to complete development and hatch at temperatures approaching 25 °C (McClelland, 1982).

There is some uncertainty regarding the number of moults performed in the egg. TEM analyses (Measures & Hong, 1995) indicate that there is only one moult, with cuticle of the first-stage larva (L1) being retained as a sheath by the second-stage larvae (L2) which emerges from the egg. On the other hand, artificial eclosion of eggs prior to hatch (Køie, Berland & Burt, 1995) reveals a delicate L1 cuticle, which evidently remains in the shell, while the third-stage larva (L3) emerges from the egg ensheathed in a sturdier L2 cuticle.

Ensheathed larvae

Newly hatched L2 (L3) larvae are 0·200–0·215 mm in length and retain the L1 (L2) cuticle as a sheath (McClelland *et al.* 1990). The larvae adhere to the substrate by their tails, alternately arching and extending their bodies to produce a flicking motion. They are extremely active at temperatures >10 °C, but sluggish to immobile at < 5 °C. The relationship of post-hatch longevity of the free-living larva to temperature is again described by hyperbolic curve. Larvae held at temperatures ⩾ 20 °C survive 24–48 h, while those maintained at 0–5 °C persist for 90–140 d and remain infective to copepods for a maximum of 111 d (McClelland, 1982; Measures, 1996).

Meiofaunal hosts

While records of natural *P. decipiens* infections in copepods are lacking, experimental evidence (McClelland, 1982) indicates that newly hatched, ensheathed larvae readily infect adult and fifth copepodite larvae of various marine benthic and epibenthic copepods, belonging to the Harpacticoida and Cyclopoida. Ensheathed larvae are also infective to temporary meiofauna such as larval gammaridean amphipods. The nematodes exsheathe in the crustacean gut and penetrate to the haemocoel. Within the brief life span of their copepod host, which varies from 3–7 days post-exposure (PE) at 15 °C to 20–35 d at 5 °C, sealworm larvae grow to 0·30–0·50 mm in length. While larvae in copepod haemocoels are not infective to fish, they are capable of infecting various macro-invertebrates (below) which are not susceptible to infection on direct exposure to ensheathed larvae (McClelland, 1995).

Macroinvertebrate hosts

Larval sealworm in the 1–9 mm length range have been reported from 13 crustacean and one polychaete species in the North Atlantic and adjacent waters (Marcogliese, 2001 *a*). Natural hosts include a sea mouse (Polychaeta), a gammaridean amphipod and a caprellid from the White Sea, Russia, a shrimp (Decapoda) from the Barents Sea, an isopod from Norway, a mysid from the Elbe estuary, Germany, and three mysid and five gammaridean species from eastern Canada.

Laboratory studies (McClelland, 1990) show that juvenile gammarideans are susceptible to infection by exposure to freshly hatched ensheathed larvae but transmission is enhanced with the participation of copepod carrier hosts. Moreover, macroinvertebrates (polychaetes, nudibranchs, mysids, mature amphipods, isopods, cumaceans and decapods), not susceptible to infection by ensheathed larvae, are readily infected by larger exsheathed larvae from copepods. After penetrating to the haemocoel of an amphipod, larval sealworms grow at an exponential rate until they reach 2–3 mm in length. Subsequent growth is asymptotic. There is no evidence that the parasite moults in the macroinvertebrate host. Larvae in lightly infected (intensity $\{I\} = 1$–2) amphipods, held for 84 d PE at 15 °C reach 6–10 mm in length, similar in size to L3s found in small benthophagous fish and L3 and fourth stage larvae (L4s) in seal stomachs. They have also developed the anatomical characteristics of infective L3s, such as the intestinal caecum, and primordial L4 lips and genitalia. Growth is slower at lower temperatures and in heavily infected hosts.

Fish hosts

Despite evidence that larval *P. decipiens* may become infective to seals in an invertebrate host, fish are clearly essential sealworm hosts in that they participate in the temporal and spatial dispersal of the larvae, thereby increasing the likelihood of ingestion by definitive hosts (McClelland, 1995). They also support significant larval growth, thus improving the parasite's ability to establish itself and survive to maturity in the gastrointestinal tract of the final host.

Helminth parasites of cold-ocean fishes often have low host specificities (Holmes, 1990) and larval sealworms, which infect marine and euryhaline fishes in polar and temperate waters, appear not to be an exception in this regard. *Pseudoterranova* sp. larvae are reported from > 75 species belonging to 29 families, 10 orders and three classes of fish in the North Atlantic (McClelland, Misra & Martell, 1990; Desportes & McClelland, 2001). As relatively few larvae have been subjected to allozyme analyses, the species status of the parasite is uncertain in the majority of these hosts. Paggi *et al.* (1991, 2000)

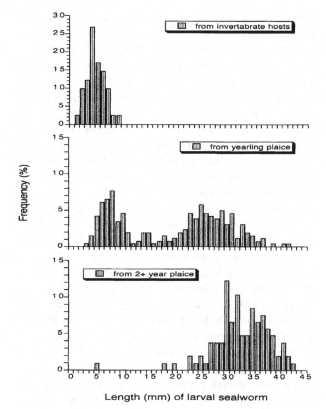

Fig. 3. Length frequency distributions of larval sealworms *Pseudoterranova* sp(p). in invertebrate hosts from the North Atlantic, and in juvenile plaice *Hippoglossoides platessoides* sampled on the central Scotian Shelf in spring: lengths of nematodes from invertebrate hosts are reported in documents reviewed by Marcogliese (2001 a).

identify *P. krabbei* in Norwegian and Faroe Island gadids (*Gadus morhua*, *Melanogrammus aeglefinus* and *Pollachius virens*) and Scottish flatfishes (*Psetta maxima* and *Hippoglossoides platessoides*), *P. decipiens* (*sensu stricto*) in Icelandic and Norwegian gadids (*Brosme brosme* and *G. morhua*) and Scottish dab (*H. platessoides*) and *P. bulbosa* (Mattiucci *et al.* 1998) in *H. platessoides* from the Barents Sea. In the Northwest Atlantic, *P. decipiens* (*sensu stricto*) is reported in gadids (*Boreogadus saida*, *G. morhua*, *G. ogac* and *P. virens*), cottids (*Hemitripterus americanus* and *Myoxocephalus scorpius*) and pleuronectids (*Hippoglossus hippoglossus* and *Reinhardtius hippoglossoides*) ranging from southern Labrador to the Gulf of Maine and *P. bulbosa* in flatfishes (*H. platessoides* and *R. hippoglossoides*) from northern Labrador (Paggi *et al.* 1991, 2000; Brattey & Davidson, 1996).

In Japanese waters, fish hosts of *P. bulbosa* remain unknown, while larval *P. azarasi* is identified in only one fish species, *Gadus macrocephalus* (Mattiucci *et al.* 1998). The endemic Antarctic species *P. decipiens* E (Bullini *et al.* 1997) has yet to be identified from Antarctic fish, although larval sealworms are found in dozens of fish species in the area (Palm, Andersen & Klöser, 1994; Palm, 1999). George-Nascimento & Urrutia (2000) conclude that larvae described from

Chilean fish species (*Merluccius gayi*, *Genypterus maculatus*, *Paralichthys microps* and *Cilus gilberti*) (George- Nascimento & Llanos, 1995) belong to *P. cattani*.

Primary fish hosts. The primary fish hosts of larval sealworms are benthic consumers which acquire the parasite directly from invertebrate hosts (McClelland & Martell, 2001 a). The parasite occurs in at least 35 species of benthophagous fish in Norwegian, Icelandic and eastern Canadian fisheries (Desportes & McClelland, 2001). These hosts include juveniles of larger commercially-exploited species as well as unexploited species which, even when fully grown, are too small to be caught in commercial gear.

Laboratory experiments (McClelland, 1995) reveal that larval *P. decipiens*, transmitted serially through copepod and amphipod hosts, infect a broad spectrum of fish species including those (e.g. *Pollachius virens* and *Pleuronectus americanus*) which, as a consequence of natural ecological and behavioural barriers (below), are seldom infected in the wild. Larval sealworms as small as 1·4 mm, but usually > 2 mm in length, are infective to fish. Larvae transmitted at 15 °C penetrate the gut wall and migrate to the somatic musculature within 3–170 hours post-infection (PE). The migration rate is directly related to the size of the nematode and temperature, and inversely related to size of the fish host. Larvae from the gut wall, viscera and musculature were 1·4–6·2 mm in length at 24 hours PE. They subsequently grew at a linear rate reaching 27·3 (23·6–30·7), 10·1 (6·0–21·8) and 12·4 (9·5–14·4) mm in length in rainbow smelt (*Osmerus mordax*), mummichog (*Fundulus heteroclitus*) and winter flounder (*Pleuronectus americanus*), respectively, by 56 days PE. As evident below, sealworm growth rates are considerably slower at temperatures favoured by fish hosts in their natural environment. Distribution of larval sealworms between the body cavity and musculature of experimental fish hosts at 7–56 d varied with host species, but the majority of viable nematodes (67–100 %) occupied the musculature.

In recent surveys employing pepsin-HCl digestion and microscopy (McClelland, 2000; McClelland & Martell, 2001 a), *P. decipiens* larvae similar in size (3–9 mm in length) and anatomy to those infecting invertebrate hosts have been recovered from small benthic consumers on the Scotian shelf (Fig. 3). The latter hosts include juveniles of commercially important species (*G. morhua*, *M. aeglefinus*, *Hippoglossoides platessoides* and *Hippoglossus hippoglossus*) and juvenile and adults of small unexploited species (*Enchelyopus cimbrius*, *Triglops murrayi* and *Aspidophoroides monopterygius*). The parasites are apparently acquired in pulses (Fig. 3) reflecting seasonal variation in the availability of infected prey (McClelland, 2000). Infected mysids (*Mysis mixta*) have

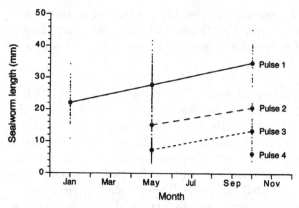

Fig. 4. Growth of larval sealworms *Pseudoterranova decipiens* (*sensu stricto*) in 0 group and yearling plaice from the central Scotian Shelf as indicated by the relationships of mean nematode length in length frequency pulses, and time of year.

been found in stomachs of juvenile Canadian plaice (long rough dab) (*H. platessoides*) (Martell & McClelland, 1995), which exploit *M. mixta* primarily in winter (Martell & McClelland, 1994). Larval sealworm reach the asymptotic length range (25–60 mm) after 1–2 y in plaice from the central Scotian Shelf (Fig. 4). Based on records from naturally infected captive plaice (McClelland, 2000), *P. decipiens* may survive 6 y and possibly much longer in a fish host.

Secondary fish hosts. In light of experimental evidence of fish to fish transmission (Burt *et al.* 1990; McClelland, 1995; Jensen, 1997) and the heavy infections found in large [> 50 cm in total length (TL)] demersal piscivores, it is apparent that larval sealworms may pass through one or more fish hosts which acquire the parasite by preying on smaller fish (McClelland *et al.* 1990). Secondary fish hosts in the North Atlantic include monkfish (*Lophius americanus*), Atlantic cod (*Gadus morhua*), cusk (torsk) (*Brosme brome*) and sea raven (*Hemitripterus americanus*), species which may be considered primary hosts as juveniles but become increasingly piscivorous as they mature (McClelland & Martell, 2001 *a*). Large demersal piscivores generally host the greatest numbers of sealworm larvae in terms of parasite prevalence (P) (percentage of hosts infected in a population or sample), abundance (A) (mean worm count in a host population or sample) and intensity (I) (nos. in individual hosts). However, while infections in the latter hosts may be a serious cosmetic nuisance for fish processors, they probably represent a cul-de-sac in the sealworm life cycle. Sealworms seem to lose vigour when subjected to serial transmissions through fish hosts (Burt *et al.* 1990) and rather than re-establishing themselves in successive fish hosts, may simply be lost from the system. Moreover, as evident from the large numbers of necrotic, encapsulated worms occupying the body cavity and hypaxial musculature of piscivorous

groundfish, many of the sealworm larvae acquired by secondary fish hosts may succumb to the host tissue response (McClelland, 1995). Lastly, demersal fish consumed by seals are smaller (< 35 cm TL) than those exploited commercially (Bowen, Lawson & Beck, 1993; Bowen & Harrison, 1996). Hence, there seems to be little likelihood that sealworms, which survive transmission to large demersal fish, will subsequently be ingested by the definitive hosts.

Apparently, distributions of larval sealworms in fish host tissues vary, not only with host species and size (age), but also with species of *Pseudoterranova*. While usually confined to the fillets of smaller (younger) demersal fishes in eastern Canadian waters, larvae of *P. decipiens* (*sensu stricto*) become increasingly prevalent in the body cavities and napes (hypaxial musculature surrounding the body cavity) of larger (older) groundfish (McClelland, Misra & Martell, 1990). In a survey of large gadids from the central Scotian Shelf, McClelland & Martell (2001 *a*) report that, of the sealworm larvae infecting cod > 70 cm TL, 47 % occupied the napes and 22 %, the body cavity. In cusk (> 50 cm TL) and white hake (*Urophycis tenuis*) (> 60 cm TL), 70 % of larval sealworms occurred in the napes. In Northeast Atlantic cod fisheries, where *P. decipiens* is sympatric with *P. krabbei* (Paggi *et al.* 2000), larval sealworms occupying the musculature are evenly distributed between the fillets and napes of smaller (< 50 cm TL) fish, but accumulate primarily in the napes of larger fish (Young, 1972; Platt, 1975; Wootten & Waddell, 1977). On the other hand, *P. bulbosa* occurs almost exclusively on the liver of plaice (dab) (*H. platessoides*) from the Barents Sea (Bristow & Berland, 1992). Similarly, larval sealworms (*P. decipiens* E ?) in Antarctic fish are found on the surface of the liver and elsewhere in the body cavity, but not in the flesh (Palm, 1999).

SEAL HOSTS

Since allozyme analyses have been conducted on only a few hundred mature nematodes from widely scattered sources, seal host spectra and geographical ranges of the various sealworm species are largely unknown. A survey of sealworms from North Atlantic seals (Paggi *et al.* 1991, 2001) has shown that the sympatric species, *P. krabbei* and *P. decipiens* (*sensu stricto*), may occur in either harbour (*Phoca vitulina*) or grey seals (*Halichoerus grypus*) in the Northeast Atlantic, but the former nematode species is more common in grey seals, and the latter in harbour seals. While *P. krabbei* has not been found in the Northwest Atlantic, *P. decipiens* infects harbour, grey and hooded seals (*Cystophora cristata*) in eastern Canada. Evidently, bearded seals (*Erignathus barbatus*) are hosts of *P. bulbosa* from the Barents Sea, Labrador, northeastern Newfoundland and Japan, while *P. azarasi* reaches maturity in Stellar's sea lion (*Eumatopius jubatus*), bearded seal, ribbon seal (*Phoca fasciata*) and northern fur seal (*Callorhinus ursinus*) in Japanese waters (Bristow & Berland, 1992; Brattey & Stenson, 1993; Mattiucci *et al.* 1998). Chilean sealworm, *P. cattani*, and Antarctic sealworm, *P. decipiens* E, are respectively described from single seal hosts, the South American seal lion (*Otario byronia*) (George-Nascimento & Urrutia, 2000) and weddell seals (*Leptonychotes weddelli*) (Bullini *et al.* 1997).

Results of experimental transmissions where juvenile harbour and grey seals received single doses of larvae (n = 100–1000) from Scotian Shelf cod, indicate that *P. decipiens* completes the moult to the (assumed) L4 within 2–5 d PE, and the final moult (assumed fourth moult or M4), between 5–15 d PE (McClelland, 1980 a, b). Maturity of both male and female nematodes is reached in 15–25 d but may be delayed (> 30 d) in heavily infected (I > 500) seals. Ova are first passed in host faeces by 16–30 d PE, and the patency period is 15–45 d in duration. Sealworms grew larger and more fecund in grey seals. By the sixth week PE, adult female and male nematodes in harbour seals were 61 (41–76) and 54 (46–61) mm in length respectively, while respective lengths of female and male sealworms from infections of similar duration in grey seals were 82 (70–104) and 64 (54–73) mm.

Adult and some larval sealworms are found among stomach contents in seals which have fed recently, but between meals, larvae and adults are attached, singly or in clusters, to the stomach wall (primarily in the fundus) with their anterior extremities embedded in host tissues (McClelland, 1980 c). L3s and L4s penetrate the fundic wall to the submucosa, while adults and immature nematodes, which have completed the final moult, are superficially embedded in the mucosa. The heads of larvae and adults are anchored to surrounding host tissues by a cap of hyaline-like substance, which is presumably secreted by the nematodes. Clusters of *P. decipiens* are seldom observed in the stomachs of free-living seals, although clusters of a related species, *Contracaecum osculatum*, are not uncommon.

DISTRIBUTIONS

In invertebrate hosts

Although a broad spectrum of benthic invertebrates (crustaceans, polychaetes, and nudibranchs) has proven susceptible to infection with *P. decipiens* (*sensu stricto*) in the laboratory (McClelland, 1990), little is known about the distribution of the parasite in natural invertebrate hosts. A review of host records from eastern Canada and the White Sea, Russia (Marcogliese, 2001 a) shows that sealworm infection parameters in invertebrates are generally quite low, even in those hosts which appear to play

an important role in the transmission of the parasite. Sealworm abundance varies from 0·0002 to 0·0046 in amphipods and mysids from Sable Island on the central Scotian Shelf, and inshore waters of Nova Scotia and the Lower Bay of Fundy. Respective abundances of 0·0075 and 0·0001 are reported for the gammarid, *Marinogammarus obtusatus*, and the polychaete, *Lepidonotus squamatus* from the White Sea. In surveys conducted by Marcogliese and colleagues (Marcogliese, 2001 a), worms were recovered by mass digestion of invertebrates in a modified Baermann apparatus and, hence, prevalence and intensities in individual hosts were not determined. Given the low abundances of the parasite, however, one might conclude that the maximum intensity of infection was one, and that prevalence varied from 0·02 to 0·75 %. Evidently, sealworm prevalence in mysids (*M. mixta*) consumed by plaice in the Sable Island area was somewhat greater (P = 4·6 %) (Martell & McClelland, 1995).

In fish hosts

As phylogenetically diverse species of marine and euryhaline fishes seem equally susceptible to *P. decipiens* infection in the laboratory, light infection or absence of infection in natural populations of certain fish species is probably attributable to ecological, behavioural and physiological (e.g. host response) barriers to the transmission of the parasite (McClelland, 1995). Worm count frequencies are positively skewed, being best described by a Poisson lognormal distribution for length specific data, and a negative binomial distribution for age specific data (Myers & Brattey, 1990). While the parasite is most prevalent and abundant in large demersal piscivores (secondary fish hosts), it occurs in greatest density in small benthic consumers (primary fish hosts) (Desportes & McClelland, 2001). In a recent survey of demersal fish from the central Scotian Shelf off Nova Scotia Canada, sealworm abundance was greatest (A = 151) in mature sea raven and the most heavily infected individual fish (I = 721) was a 10 kg cod (McClelland & Martell, 2001 a). Sealworm densities, however, were greatest (D = 665–2465 nematodes kg^{-1} body weight) in fourbeard rockling (*Enchelyopus cimbrius*), mailed sculpin (*Triglops murrayi*), alligatorfish (*Aspidophoroides monopterygius*) and juvenile plaice, fish ranging from two to 80 g in weight. As a rule, sealworm prevalence and abundance generally increase with host age (length, weight) within a given host population and this is especially true in species which progress from a benthic to a piscivorous diet as they mature. Sealworm abundance is usually more closely related to host length than to host age (Platt, 1975; des Clers, 1989).

Larval sealworms (*P. decipiens* and *P. krabbei*) are found in inshore populations of Atlantic cod and other demersal species throughout northern Europe

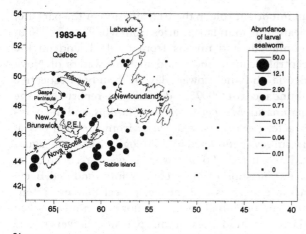

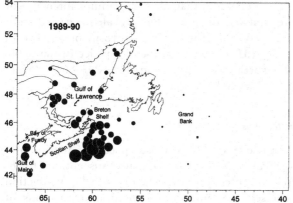

Fig. 5. Distributions of larval sealworms
Pseudoterranova decipiens (*sensu stricto*) in plaice
Hippoglossoides platessoides in eastern Canada in 1983–84
and 1989–90 (McClelland, Misra & Martell, 2000).

(Jensen, Andersen & des Clers, 1994; des Clers &
Prime, 1996), but are also abundant in groundfish
from offshore waters of the U.K., Faroe Islands and
Iceland (Young, 1972; Platt, 1975; Ólafsdóttir,
2001). In the Northwest Atlantic, *P. decipiens* is most
numerous in demersal fish stocks from southern
Newfoundland, the Gulf of St Lawrence, the Breton
and Scotian Shelves and the Gulf of Maine (McClelland, Misra & Martell, 1985, 2000; Brattey, Bishop
& Myers, 1990; Marcogliese, 2001*b*) (Fig. 5). There
is little quantitative information on the distribution
of Arctic-boreal sealworm (*P. bulbosa*) which has
been identified only in the flatfishes, *Hippoglossoides
platessoides* in the Barents Sea, and *H. platessoides*
and *Reinhardtius hippoglossoides* in eastern Canada
(Bristow & Berland, 1992; Brattey & Davidson,
1996; Mattiucci *et al.* 1998).

As infection parameters of larval sealworms often
vary significantly in neighbouring host populations,
these parasites have shown some potential, usually in
combination with other parasite taxa, as biological
indicators of discreteness and movements of host
stocks (Platt, 1976; Hemmingsen, Lombardo &
MacKenzie, 1991; Arthur & Albert, 1993; McClelland & Marcogliese, 1994). The information they
provide could prove particularly vital to management

of migrant fish stocks exploited commercially in
different fisheries, at different times of year.

Sealworm parameters have been documented from
North Atlantic groundfish over the past 60–70 years,
but no clear trends for sealworm prevalence and/or
abundance have emerged from British, Norwegian
and Icelandic fisheries (Bowen, 1990; Desportes &
McClelland, 2001). While industrial data from
Iceland indicate sealworm has become increasingly
abundant in Atlantic cod, scientific data are inconclusive. Both scientific and industrial data from
Germany indicate that larval anisakines have grown
increasingly prevalent or abundant in commercially
exploited fish species since 1982, but no distinction
is made between sealworm and *Anisakis* larvae in
industrial data. In the Northwest Atlantic, the
prevalence and abundance of *P. decipiens* has been
increasing in groundfish from southern Newfoundland, the northern Gulf of St Lawrence and the
Scotian and Breton Shelves since the 1950s, and in
the southern Gulf of St Lawrence groundfish since
1983 (Fig. 5) (McClelland *et al.* 1985, 1990, 2000;
Brattey *et al.* 1990; McClelland & Martell, 2001 *a*, *b*).
Over the last decade, however, sealworm abundances
have declined in an indicator host, Canadian plaice
(*H. platessoides*), from the central Scotian Shelf (Fig.
6), despite the proximity of the large and rapidly
growing Sable Island grey seal colony.

As a rule, sealworm infections in fish do not vary
seasonally. Invasion of tissues and organs is ir-
reversible and the parasites accumulate gradually
over the life span of the host. Normally, nematodes
acquired by a given host cohort over a given season
would not be sufficiently numerous to cause significant change in infection parameters. Samples of
mature plaice taken at given offshore sites at different
times of year, for example, did not vary significantly
with regard to prevalence or abundance of larval
sealworm (McClelland *et al.* 2000). Yearling Scotian
Shelf plaice may be an exception to the rule in that
40 % of the sealworms which they accumulate over
their first year appear to be acquired over a brief
period in late winter or spring (Fig. 3) (McClelland,
2000). Most instances of apparent seasonal variation
in a given host and at a given site, however, would
probably be attributable to sampling of successive
migrant host stocks or mixtures of seasonal migrants
and resident stocks with different infection parameters (Platt, 1976; McClelland & Marcogliese,
1994).

In seal hosts

Grey and harbour seals are the most important hosts
of sealworms throughout the North Atlantic (Desportes & McClelland, 2001). Where the two seal
species coexist, however, *Pseudoterranova* spp. are
usually more abundant in grey seals (Brattey *et al.*

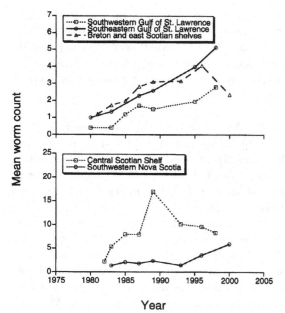

Fig. 6. Variations in abundance of larval sealworm *Pseudoterranova decipiens* (*sensu stricto*) in plaice *Hippoglossoides platessoides* from the sourthern Gulf of St Lawrence, Breton and Scotian shelves, and the Gulf of Maine from 1980 to 2000 (McClelland & Martell, 2001 b).

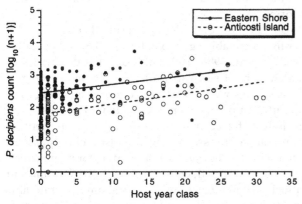

Fig. 7. Relationship of intensity of infection with larval and adult sealworm *Pseudoterranova decipiens* (*sensu stricto*) and age of host grey seals *Halichoerus grypus* from the eastern Nova Scotian mainland (Eastern Shore) and the Gulf of St Lawrence (Anticosti Island) (McClelland, Misra & Martell, unpublished).

1990; Ólafsdóttir, 2001). Harbour seals seem less suitable hosts than grey seals in terms of nematode survival, growth, fecundity and the magnitude of the host response (McClelland, 1980c; Aspholm et al. 1995). Harp (*Phoca groenlandica*), hooded (*Cystophora cristata*) and ring seal (*Phoca hispida*) are generally lightly infected and host few mature worms (Brattey & Ni, 1992; Brattey & Stenson, 1993; Ólafsdóttir, 2001). According to allozyme analyses (Paggi et al. 2000), *P. krabbei* is the dominant sealworm species in grey seals in the Northeast Atlantic, while *P. decipiens* (*sensu stricto*) dominates in Northeast Atlantic harbour seals and is the only species found in both grey and harbour seals in the

Northwest Atlantic. Quantitative data are lacking for Arctic-boreal sealworm, *P. bulbosa*, which has been identified from bearded seals in the Barents Sea and Labrador and northeastern Newfoundland (Bristow & Berland, 1992; Brattey & Stenson, 1993; Mattiucci et al. 1998).

Sealworm abundances in grey and harbour seals vary seasonally, with host age (length) (Fig. 7) and geographical origin (McClelland, 1980b; Brattey et al. 1990; Stobo, Beck & Fanning, 1990; Brattey & Stenson, 1993; Marcogliese, 2001b; Ólafsdóttir, 2001; Stobo & Fowler, 2001). As was the case in fish hosts, worm count distributions are positively skewed. In grey seals, sealworm abundance usually increases with host age or length, but is most closely correlated to the latter. Although male grey seals grow much larger than females, disparities in sealworm abundance between male and female hosts are not statistically significant. The heaviest infections among North Atlantic seals (A = 3368 and maximum I = 21471) were reported in mature grey seals from breeding colonies in western Iceland (Ólafsdóttir, 2001).

In Iceland, sealworm is most numerous in grey seals from the west coast of the main island and least abundant in those found off the northeastern coast (Ólafsdóttir, 2001). In eastern Canada, Scotian Shelf grey seals found along eastern Nova Scotian mainland ('Eastern Shore') are more heavily infected than those from the Gulf of St Lawrence (Fig. 7) (McClelland, Misra & Martell, unpublished). Sealworm abundances in grey seals from Newfoundland waters are, in turn, lower than those found in either Shelf or Gulf grey seals but similar to *P. decipiens* abundances in Newfoundland harbour seals (Brattey & Stenson, 1993).

As *Pseudoterranova* spp. are enteric, with life expectancies of only a few weeks in their final hosts, seasonal fluctuations in sealworm abundance observed in North Atlantic seals are clearly related to host dietary factors (McClelland, 1980c; Stobo et al. 1990; Brattey & Stenson, 1993; Ólafsdóttir, 2001). These factors include periods of fasting during the breeding and moulting seasons, and changes in availability and/or exploitation of infected prey (below). Data on long-term variation of sealworm abundance in North Atlantic seals are either inconclusive or lacking (Brattey et al. 1990; Desportes & McClelland, 2001). Contrasts of 1988 and 1992 samples of adult grey seals from the northern Gulf of St Lawrence indicate that sealworm abundance was significantly lower in the more recent sample (Marcogliese, 2001b) but the sample sizes were small (28 and 31, respectively). Although the overall abundance of sealworm did not differ significantly between 1982 and 1989, samples of grey seals from Sable Island, there were proportionately fewer mature worms in the 1989 sample (Stobo & Fowler, 2001).

Biotic influences

Distributions of definitive hosts. While quantitative data on Arctic-boreal sealworm (*P. bulbosa*) are lacking, the species is confined to demersal fishes within the geographic range of bearded seal (Paggi *et al.* 1991, 2000; Bristow & Berland, 1992; Brattey & Davidson, 1996; Mattiucci *et al.* 1998). In more temperate areas of the North Atlantic, the spatial and temporal distributions of larval *P. decipiens* (and/or *krabbei*) in groundfish seem to be related primarily to the geographical distribution and growth of grey seal populations (Bowen, 1990; Desportes & McClelland, 2001). Grey seals are considered to be the most important final hosts of sealworm in the British Isles (Young, 1972), Faroe Islands (Platt, 1975) and Iceland (Ólafsdóttir, 2001), as well as along the central Norwegian inshore (Jensen *et al.* 1994).

In the Northwest Atlantic, infection parameters are greatest in Canadian plaice and other groundfish species from southern Newfoundland, the Gulf of St Lawrence, the Breton and Scotian Shelves, and the northeastern Gulf of Maine (Fig. 5) (Marcogliese, 2001*b*; McClelland *et al.* 2000; McClelland & Martell, 2001*a, b*). These areas encompass the majority of grey seal breeding and haulout sites in eastern Canada (Stobo, Beck & Horne, 1990). The heaviest infections are found in groundfish from the central Scotian Shelf near Sable Island, site of the largest grey seal colony in the Northwest Atlantic. Known to forage pelagically and migrate over great distances, grey seals seem largely responsible for the dispersal of sealworm in offshore waters far from breeding colonies and haulout sites. A flow chart model, based on recent sealworm abundance data and seal population estimates (Aznar *et al.* 2001), indicates that *P. decipiens* (*sensu stricto*) is transmitted primarily by grey seals in eastern Canada.

In areas where they outnumber grey seals, harbour seals may have considerable influence on larval sealworm abundance in groundfish. This appears to be the case in the Elbe estuary (Möller & Klatt, 1990), southern Norwegian fjords (Aspholm *et al.* 1995; des Clers & Anderson, 1995), and the Wadden Sea, Skagerrak and Kattegat and Scottish firths (des Clers & Prime, 1996). Although they are being rapidly overtaken in numbers by grey seals, harbour seals may still predominate at some inshore sites in eastern Canada, especially off southwestern Nova Scotia and in the lower Bay of Fundy (McClelland *et al.* 2000). ·

Surveys conducted during the whelping season indicate that few *P. decipiens* reach maturity in harp seals (Brattey & Ni, 1992; Marcogliese, Boily & Hammill, 1996). Hence, despite the fact that harp seals greatly outnumber the combined populations of all other seal species in eastern Canada, their role in the transmission of sealworm is believed to be negligible (Aznar *et al.* 2001). The harp seal population, however, is infrequently surveyed during the pre- and postwhelping periods, when, according to earlier records (Scott & Fisher, 1958), transient populations in the southern Gulf of St Lawrence may host significant numbers of mature sealworm.

Host diet. As the early life history of sealworms is benthic, the larvae infect fish that either exploit benthos directly or prey upon smaller benthophagous fish (McClelland, 1990). Heavily infected gadids, cottids and flatfishes in the North Atlantic generally fall into either or both of the above categories. Among gadid hosts, Atlantic cod are benthic consumers as juveniles, but become increasingly piscivorous as they mature. The change to a piscivirous diet in cod and other large demersal fish is usually marked by a dramatic increase in sealworm prevalence and abundance (Desportes & McClelland, 2001). In contrast, natural populations of a second gadid species, pollock (saithe) (*Pollachius virens*), are only lightly infected. Although highly susceptible to infection in laboratory transmissions (McClelland, 1995), pollock are midwater consumers of nekton in their natural environment (Scott & Scott, 1988), and hence, seldom exposed to sealworms.

Surveys of diets and ascaridoid infections of flatfishes inhabiting Sable Island Bank (Martell & McClelland, 1994, 1995) indicate that disparities in parameters of sealworm infection among sympatric host species were largely related to the exploitation of different prey. Juvenile Canadian plaice that were heavily infected with sealworms fed on benthic suprafauna, free swimming organisms closely associated with the bottom, while winter flounder, which was rarely infected, consumed sedentary infauna and attached epifauna. Benthic suprafauna include amphipods and mysids that have proven to be important in the transmission of sealworm (Marcogliese, 2001*a*). While rapid accumulation of smaller nematodes (3–12 mm in length) reflects the frequent consumption of mysids and other suprafauna by 0-group and yearling plaice, the infrequency of infection by smaller sealworm larvae and the decline of the overall rate of re-infection signify exploitation of more sedentary uninfected prey by older plaice (Fig. 3) (Martell & McClelland, 1994; McClelland, 2000; McClelland & Martell, 2001*a*).

It is reasonable to assume that seasonal variations in abundance of *P. decipiens* in seals, as well as variations in sealworm abundance with species, size (age) and geographic origins of seal hosts are related to dietary factors. According to Ólafsdóttir (2001), grey seals are more heavily infected than harbour seals in Icelandic waters because they are larger, consume greater quantities of prey and are thus exposed to greater numbers of larval sealworm. Ólafsdóttir (2001) goes on to attribute the great

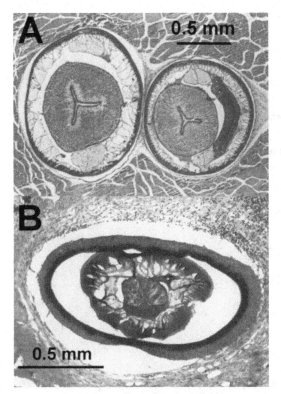

Fig. 8. Cross sections of larval sealworm
Pseudoterranova decipiens (*sensu stricto*) in the
musculature of A, a naturally-infected plaice
Hippoglossoides platessoides from the central Scotian
Shelf, after 68 months in captivity, and B, a sea raven
Hemitripterus americanus from Passamaquoddy Bay,
New Brunswick, Canada.

abundances of the parasite in grey seals from the
west coast of Iceland, during the breeding season, to
more frequent consumption of heavily infected bull
rout (shorthorn sculpin) (*Myoxocephalus scorpius*).
In contrast, Northwest Atlantic grey seals often
reduce food intake or fast during breeding season
leading to declines in abundances of sealworm in
general, and especially mature worms (McClelland,
1980b; Stobo et al. 1990). Unfortunately, relationships drawn between sealworm abundances in seals
and prey exploited by seals are often highly speculative. Surveys of stomach contents or scats do not
provide a comprehensive picture of seal diets (Stobo
& Fowler, 2001). Both approaches reveal prey
consumed over a period of few hours, whereas the
parasite may persist in the seal host for weeks. The
majority of stomachs examined are empty, while
remains from scats permit identification of prey only
in broad taxonomic terms and do not include more
digestible prey items.

Host response. Sealworm transmission is regulated
to varying degrees by the tissue responses of its
intermediate and final hosts. Laboratory studies
(McClelland, 1990) show that while *P. decipiens*
larvae transmitted to mature mysids, cumaceans,
isopods and decapods are encapsulated and de-

stroyed by haemocytes, they seldom encounter a
host haemocytic response in amphipods. Larval
sealworms transmitted to fish are also susceptible to
an encapsulation response that varies with host
species and age, and parasite density. The response
is lacking or benign in European and rainbow smelt
(Möller & Klatt, 1990; McClelland, 1995), and while
encapsulated sealworms are occasionally found in
larger (older), free-living Canadian plaice, there was
no evidence of larval sealworm mortality or a host
encapsulation response in naturally infected specimens held for as long as 68 months in captivity (Fig.
8A) (McClelland, 2000). Sealworms encapsulated in
host fibrous tissue are frequently found in larger,
heavily infected hosts such as mature gadids and
cottids (Fig. 8B) (McClelland, 1995; McClelland,
Misra & Martell, 1990). Encapsulated worms, which
are often necrotic, are usually confined to the body
cavity and surrounding hypaxial musculature. Studies of experimentally infected hosts revealed that
sealworm in rainbow trout (*Oncorhynchus mykiss*)
were encapsulated by fibrous connective tissue by
16–32 d PE (Ramakrishna & Burt, 1991), and that all
encapsulated larvae in mummichog, grubby (*Myoxocephalus aenaeus*, Cottidae) and winter flounder at 56
days PE were necrotic (McClelland, 1995).

Although the encapsulation response to larval
sealworms in fish hosts is typical of a non-specific
granulomatous inflammatory reaction (Ramakrishna
& Burt, 1991), it appears to have an immunological
basis (Silva *et al.* 1999). Moreover, antibodies
specific for *P. decipiens* are found in infected fish
(Coscia & Oreste, 2000). Antibody titres to larval
anisakines increase with the age of fish hosts
indicating, perhaps, that mature fish mount a more
vigorous immune response that interferes with the
migration of nematodes to the musculature (Priebe
et al. 1991) and promotes a stronger cellular reaction.

The poor survival of sealworms from experimental
challenge transmissions to harbour and grey seals
(McClelland, 1980b) is evidence that the parasite is
also vulnerable to a host response in seals. Evidently
nematodes from challenge transmissions failed to
remain anchored in the stomach wall during their
critical maturation moults and were ultimately
passed with host faeces. Possibly, they were extruded
from granulation tissues that developed in response
to sensitizing infections (McClelland, 1980c). Eosi-
nophilic granulomata, grossly apparent as raised
inflammatory areas (in grey seals) and ulcerous
lesions (in harbour seals) in the wall of the fundus
(Fig. 9), however, are seldom observed in naturally
infected seals (Smith, 1999).

Parasite density. In laboratory experiments
(McClelland, 1990), sealworm length was inversely
related to intensity of infection in amphipod hosts,
indicating that growth of the parasite was density
limited. Retarded growth of larval sealworms would

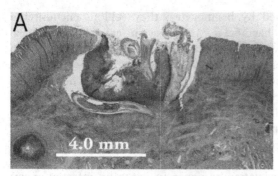

Fig. 9. Transverse sections of A, a lesion associated with a cluster of partially embedded L4 and adult sealworm *Pseudoterranova decipiens* (*sensu stricto*) in the fundic stomach of an experimentally infected juvenile harbour seal *Phoca vitulina*, and B, a raised inflammatory area associated with an aggregation of partially embedded adult sealworm in the fundic stomach of an experimentally infected juvenile grey seal *Halichoerus grypus* (McClelland, 1980c).

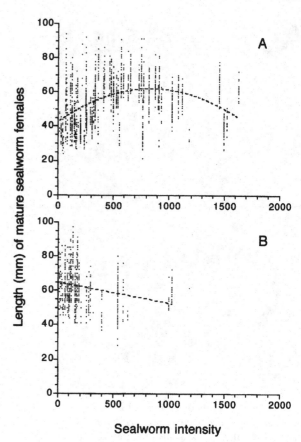

Fig. 10. Relationships of length of adult female sealworm *Pseudoterranova decipiens* (*sensu stricto*) and intensity of larval and adult sealworm infection in A, grey *Halichoerus grypus* and B, harbour seals *Phoca vitulina* from eastern Nova Scotia (McClelland, Martell & Melendy, unpublished).

probably not be a factor, however, in natural invertebrate populations where intensity of infection would seldom exceed one.

An elevated host response to heavy sealworm infection may have a negative impact on the parasite's growth and survival in fish hosts. Retarded growth and mortality of larval sealworms are evident in experimentally infected fish which mount a vigorous response to infection (McClelland, 1995). In natural fish populations, necrotic encapsulated larvae are most frequently found in heavily infected piscivores such as mature Atlantic cod and sea raven (McClelland, Misra & Martell, 1990; McClelland & Martell, 2001a).

Infection in definitive hosts may also be density limited (des Clers, 1990). As *P. decipiens* and other anisakine species grow more numerous in seal stomachs, they may become increasingly prone to the consequences of intra- and interspecific competition ('crowding effects'), and/or a more vigorous host response (McClelland et al. 2000). These effects might include higher mortality rates, retarded growth and maturation rates, and lower uterine egg counts. Marcogliese (1997) found no evidence that sealworm egg counts were density limited, but his analysis was based on counts from only 384 *P. decipiens* sampled from 47 grey seals. Noting that intensities were low (< 600) in all but one of the seals sampled, Marcogliese speculates that the threshold

density (intensity) at which egg counts of the parasite decline may have been reached in a single, heavily infected (I = ~ 1600) seal. More recent analyses show that length of mature female sealworms, which varies directly with uterine egg count but is easier to measure, declines with intensity of infection in harbour and heavily infected grey seals (Fig. 10) (McClelland, Martell & Melendy, unpublished). In grey seals, length/intensity relationships are best described by second-degree polynomial curves showing worm length increasing with sealworm intensity in seals with light to moderately heavy infections, and decreasing in the most heavily infected seals. The threshold intensity varies with size (age) of seals, and may exceed 1500 in mature grey seals.

McClelland et al. (1985) and Burt (1994) speculate that *P. decipiens* abundance in seal stomachs may be limited as a consequence of 'competitive exclusion' by related nematode species, belonging to the *Contracaecum osculatum* species complex. Stobo & Fowler (2001) cited a negative correlation between *P. decipiens* and *C. osculatum* abundances in juvenile grey seals from Sable Island as evidence of competitive exclusion. Marcogliese (1997), remarking on the relatively small size and low uterine egg counts

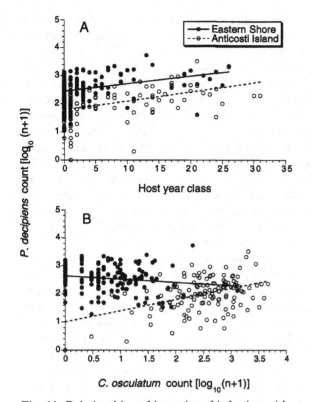

Fig. 11. Relationships of intensity of infection with larval and adult sealworm *Pseudoterranova decipiens* (*sensu stricto*) to intensity of larval and adult *Contracaecum osculatum* in grey seals, *Halichoerus grypus*, from the eastern Nova Scotian mainland (Eastern Shore) and the Gulf of St Lawrence (Anticosti Island) (McClelland, Misra & Martell, unpublished).

of female *P. decipiens* in grey seals from Anticosti Island, further goes on to suggest that sealworm growth and fecundity in grey seals from the Gulf of St Lawrence may be suppressed as a consequence of competitive pressure from *C. osculatum*. The hypothesis is not supported, on the other hand, by survey data from grey seals in Newfoundland (Brattey & Stenson, 1993) and the Gulf of St Lawrence (Marcogliese *et al.* 1996) which indicate that negative correlations between the abundances of the two nematode species are lacking. As shown in Fig. 11, negative *P. decipiens/C. osculatum* correlations may result when samples taken in a given area include not only resident hosts, but migrants with infections acquired in other areas. A negative *P. decipiens/C. osculatum* correlation in grey seals sampled from the Atlantic coast of eastern Nova Scotia (Eastern Shore) is clearly attributable to the influence of data from (suspected) migrant seals with infection parameters similar to those found in seals from the Gulf of St Lawrence (Anticosti Island). The latter are evident in Fig. 11 as Eastern Shore data points lying within the Anticosti Island scatter. According to discriminant function analyses of anisakine infection parameters, the suspected migrants are, in fact, classified as Gulf seals. Notably, the *P. decipiens/C. osculatum* correlation in Anticosti Island grey seals is positive.

Sealworm-induced host mortality. In laboratory experiments, 15 (30 %) of 50 mysids (*Mysis stenolepis*) fed copepods infected with larval *P. decipiens* (*sensu stricto*) died within 5–10 d PE (McClelland, 1990). Infection proved lethal to only the most heavily infected specimens ($I \geqslant 9$), however, and it is unlikely that sealworm is a direct cause of mortality in natural mysid populations where the intensity of infection would rarely exceed one.

Mortalities of experimentally infected mummichog and grubby were attributed to damage to various vital organs (spleen, liver, brain, pericardium and dorsal aorta) caused by migrating and feeding sealworm larvae (McClelland, 1995). Möller *et al.* (1991) report mortalities among juvenile European eel (*Anguilla anguilla*) which were force-fed larval *P. decipiens*. Given that sealworm densities in natural hosts may rival or exceed those achieved in laboratory transmissions, it is possible that the parasite causes mortality by injuring organs and tissues of small, free-living fish. Sealworm densities in individual specimens of *Artediellus atlanticus*, *Triglops murrayi*, *Aspidophoroides monopterygius* and juvenile American plaice from waters surrounding Sable Island often exceed one nematode per gram of body weight (McClelland & Martell, 2001*a*). Recent declines of sealworm infection parameters in mature plaice from the Sable Island Bank complex may be attributable to parasite-induced mortality of heavily infected juveniles (McClelland & Martell, 2001*b*). While prevalence of infection remains at or near 100 %, sealworm abundance in plaice in the vicinity of Sable Island has fallen dramatically over the last decade (Fig. 6) and the tails of intensity distributions have become truncated.

Larval sealworms may also cause mortality of invertebrate and fish hosts through chemical impairment of the their ability of forage and avoid predators (McClelland, 1995). This mechanism could be considered a form of parasite-induced behavioural modification which, by rendering the host more vulnerable to predators, promotes transmission of sealworm to a subsequent intermediate of final host. In support of this hypothesis, infected amphipods and mysids exhibited sluggish and/or erratic swimming behaviour in laboratory experiments (McClelland, 1990), while field studies reveal that sealworm prevalence in mysids found in the stomachs of Scotian Shelf plaice (Martell & McClelland, 1995) greatly exceed the prevalence of the parasite in local mysid populations (Marcogliese, 2001*a*).

Volatile ketones which larval sealworms produce as metabolic by-products may act as local anaesthetics in surrounding host musculature (Ackman & Gjelstad, 1975), and may have been a factor in experiments conducted by Sprengel & Lüchtenberg (1991) wherein the maximum swimming speed of European smelt declined with increasing intensity of larval *P. decipiens* infection. Evidently, infection with even a single worm reduces swimming speed and stamina (Rowling, Palm & Rosenthal, 1998).

Abiotic Influences

Temperature. Since their final hosts are homeotherms, temperature-dependent aspects of the life cycles of marine mammal ascaridoids may influence their geographic (and presumably, temporal) distributions (Cheng, 1976). Near bottom temperature has perhaps the greatest impact on sealworm distribution, transmission and developmental rates. It directly regulates prehatch developmental rate, posthatch survival of the free-living ensheathed L2 (L3?), efficiency of transmission of larvae to and between poikilothermic intermediate hosts, and developmental rates of larvae in these hosts (McClelland, 1982, 1990, 1995; Measures, 1996). Temperature may also affect sealworm transmission and survival indirectly by influencing growth rates, abundance and distribution (hence, availability) of important intermediate hosts (Campana *et al.* 1995; Swain, 1999).

Ólafsdóttir (2001) attributes greater abundance of sealworms in fish and seals on the west coast of Iceland to, among other factors (above), more temperate waters which favour the development and transmission of the parasite. Conversely, larval sealworms (*P. decipiens* and/or *krabbei*) are not frequently found in fish from the Norwegian Arctic, Greenland, Labrador, and northeastern Newfoundland (Platt, 1975; Brattey *et al.* 1990; McClelland *et al.* 2000). This is probably attributable to retarded or arrested development of the parasite's eggs and larval stages in the cold waters found in these areas, rather than a scarcity of suitable definitive hosts (McClelland, 1982).

Pseudoterranova bulbosa is indigenous to Arctic-Boreal waters, but, being confined to the body cavity of fish hosts (Bristow & Berland, 1992), has been largely overlooked in major surveys (above) where only the musculature of fish hosts was examined.

Among other factors, relatively high near-bottom temperatures prevalent in eastern Canadian waters during the late 1970s and early 1980s (Campana *et al.* 1995) may have promoted marked increases in *P. decipiens* parameters in Canadian groundfish during the 1980s (Fig. 5) (McClelland *et al.* 1990, 2000). Subsequent declines of infection parameters in eastern Canadian grey seals and groundfish during the late 1980s and early 1990s may have been attributable, in turn, to low near-bottom temperatures prevalent in Atlantic Canadian waters since the mid 1980s (Marcogliese, 2001*b*; McClelland & Martell, 2001*b*).

Ocean currents. Although the spatial distribution of sealworm is largely attributable to the movements of its hosts, and especially its final hosts (above), ocean currents undoubtedly play an important role in the dispersal of sealworm eggs passed near seal haulouts and other areas frequented by foraging seals. With an estimated settling rate of $1 \cdot 01 \times 10^{-4}$ m sec^{-1} in seawater, and current speeds similar to those encountered on the Scotian Shelf, sealworm eggs passed near the surface of waters 100 m deep would be transported 50 km before reaching bottom 12 days later (McConnell, Marcogliese & Stacey, 1997). Eggs passed in shallow inshore waters near seal haulouts or those passed near-bottom by seals foraging for demersal prey would reach the substrate in a matter of hours and be carried only a few hundred metres. At temperatures of 5–9 °C, warm by North Atlantic standards, sealworm eggs would hatch in 20–50 days, long after settling to bottom at depths found in shallow coastal zones. Beyond the continental shelf, where depths exceed 200 m and upwelling activity prevails, it is possible that the eggs may be suspended in the water column indefinitely, dispersed over vast areas and hatch before reaching bottom where the newly hatched larvae can be ingested by suitable benthic hosts. The latter scenario, perhaps, explains why sealworm (*P. decipiens* E?) parameters found in fish along the Weddell Sea ice shelf are so low, despite the presence of large numbers of definitive hosts (Palm, 1999).

CONTROLS

Industrial

Commercial fishing practices. Selective fishing for lightly infected populations or age (size) classes of fish has often been suggested as a method for reducing production costs, but this approach offers only temporary relief for processors (Young, 1972; Platt, 1975; McClelland, Misra & Marcogliese, 1983; Hafsteinsson & Rizvi, 1987). Moreover, concentration of the fishing effort on particular age groups or stocks may be neither economically feasible nor healthy for the stocks in question. Selective fishing can put excessive pressure on lightly infected stocks (age classes) and, ultimately, it would be necessary to exploit wormy fish again. Moreover, the wormiest fish are usually found in close association with seals and probably serve to re-infect them. Failure to exploit fish serving as reservoirs for larval sealworms may lead to increasing abundance of the parasite in the surviving fish stocks.

Viscera of gadids and other large groundfish are often discarded at sea to prevent degradation of the flesh during storage on ice for prolonged periods, prior to reaching the processing plant. While this practice may carry the added benefit of precluding migrations of nematodes from the alimentary canal (*Hysterothylacium* sp.) (Cheng, 1976) and body cavity (*Anisakis* sp.) (Smith & Wootten, 1975) to the

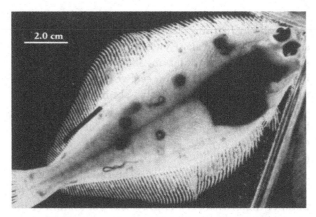

Fig. 12. Larval sealworm *Pseudoterranova decipiens* (*sensu stricto*) in the flesh of a live, naturally infected juvenile plaice *Hippoglossoides platessoides* from the central Scotian Shelf: the fish was photographed in a Plexiglas container of seawater, between crossed polarized filters, on a light table.

napes and fillets, it could also result, unfortunately, in heavier infections in fish which feed on the discarded viscera (McClelland, Misra & Martell, 1990).

Detection and removal. At the processing plant, fillets and napes of groundfish are examined for parasites by 'candling' on a light table (see Hafsteinsson & Rizvi, 1987 for a detailed account). Forceps, hooks or vacuum tubes are used to remove the worms from the flesh, and heavily infected napes and flesh trimmed from fillets are often discarded. The candling procedure is notoriously inefficient. Candling efficiency at eastern Canadian plants has been estimated to vary from 33 to 93 % for heavily infected (> 3 worms/kg). Overall candling efficiency for heavily infected fillets increases to 95 % when fillets are cut longitudinally into slices 13 mm thick beforehand, but this approach has been deemed unacceptable in that it would increase processing costs while yielding a product which is less marketable than whole fillets.

Intensities and wavelengths of both transmitted and incident light in the candling procedure have been investigated in an effort to enhance the contrast between the nematodes and the surrounding flesh (reviewed by Hafsteinsson & Rizvi, 1987). While light in the visible range is optimal in terms of the light absorption disparity between worm and flesh, the visibility of the worm within this wavelength range is obscured as a consequence of light scattering by the flesh. Although wavelengths approaching the infrared range reduce 'scattering', they have an undesirable 'heating' effect. Exposure to ultraviolet light, on the other hand, causes sealworms to fluoresce, but worms embedded > 0·5 mm deep in fish tissues are not visible. Odense (1978) suggests the use of polarised transmitted light to control

scattering and improve contrast between worm and flesh. This approach has the added advantage that, when the fillet is then viewed through a second polarised filter at right angles to the first, background light is eliminated, while the fillet, because of residual scattering effect, remains brightly illuminated (see Fig. 12).

As a consequence of the high cost and low efficiency of detecting parasites in fish fillets by candling, various alternative approaches have been investigated (Choudhury & Bublitz, 1994). Laser candling proved unsuitable for detecting parasites due to beam scattering by the flesh. Methods employing X-rays, the scanning laser acoustic microscope (SLAM) and pulse-echo technology have also failed because of similarities in the physical properties of the parasites and the surrounding flesh. Choudhury & Bublitz (1994) appear to have struck upon the most promising approach, an electromagnetic detection technique based on the observation that the electrical conductivity of cod flesh is ~ 200 times greater than the conductivity of anisakid nematodes. An industrial application of this latter approach, however, remains to be developed.

As previously surmised by Hafsteinsson & Rizvi (1987), the basic light table, employing white, non-collimated transmitted light is still used at most processing plants. Although fillets range up to 4 cm in thickness, it unlikely that sealworms embedded > 0·6 cm below the surface are detected. Also of some concern is the fact that discarded parasites and heavily infected flesh, often find their way back into the ocean in fish plant effluents, and may cause elevated infection parameters in local fish and marine mammals.

Seafood preparation. Fillets containing larval sealworms which have escaped detection by candling procedures are not only objectionable to the consumer from an aesthetic standpoint, but also represent a health risk. Thorough cooking of fresh fish, however, will kill anisakid larvae, thereby removing the health hazard (Bier *et al.* 1987). Worms in fillets 3 cm thick survive < 10 min at a temperature of 60 °C, and < 7 min at 70 °C. Bier *et al.* suggest that fresh fish be cooked for 10–12 min for each inch (~ 2·5 cm) of thickness. This practice is, perhaps, more critical in the preparation of fish served with bones in and skin on, than in the cooking of fillets.

Freezing is recommended for killing anisakid nematodes and other pathogens in fish destined to be smoked, marinated, undercooked or served raw (US Food and Drug Administration/Center for Food Safety and Applied Nutrition, 1992). Fish should be held for a minimum of 15 h at −30 °C in commercial blast freezers, and for at least 7 d at −20 °C in a domestic freezer.

Biological

Seal population control. As sealworms mature and reproduce in the stomachs of seals, management of seal populations has been the primary biological approach to controlling parasite abundances in groundfish (Hafsteinsson & Rizvi, 1987). A two part workshop was held in Halifax, Nova Scotia (Canada) in 1987 and 1988 in order to integrate current information on the biology and population dynamics of sealworms, clarify relationships between numbers of seals and abundances of sealworm in commercially important fish and, ultimately, develop predictive models for sealworm control (Bowen, 1990). Models developed from data available at the time looked into the effects of various seal management scenarios on sealworm abundances in fish (Mohn, 1990; des Clers, 1990), but the consensus was that data were lacking for many important model parameters (des Clers *et al.* 1990). Although many of the gaps in data had been addressed by the time sealworm researchers convened for a workshop in Tromsø, Norway in 1997 (Desportes & McClelland, 2001), working predictive models for sealworm control are still pending. In eastern Canada, interest in the sealworm problem has subsided in recent years as groundfish have declined in size and abundance, and fisheries have closed. There is growing concern that the failure of these stocks to rebuild may be attributable, at least in part, to predation by seals (Mohn & Bowen, 1996).

In the North Atlantic, grey and harbour seals appear to be the most important definitive hosts of the parasite(s) with correlations between the distributions of grey seals and heavy sealworm infestations in groundfish stocks being particularly strong (Young, 1972; Platt, 1975; McClelland *et al.* 2000; Ólafsdóttir, 2001). Consequently, there have been numerous efforts to control population growth of both species through commercial hunts, culls and bounty programmes in the British Isles, Norway, Iceland and eastern Canada (Hafsteinsson & Rizvi, 1987). Recently, investigators at Dalhousie University (Halifax, Nova Scotia) (Brown *et al.* 1997 *a*, *b*) developed an immuno-contraceptive approach for controlling seal populations as a more humane alternative to hunts and pup culls employing firearms and clubs.

There is little conclusive evidence that declines in seal numbers resulting from hunts, culls and natural causes have led to lower sealworm infection parameters in fish. For example, although the harbour seal population in the outer Oslofjord, Norway fell from 350 in the early 1980s to ~ 100 after a phocine distemper epizootic in 1988, there were no perceptible changes in abundances of larval sealworm in local demersal fishes (Aspholm *et al.* 1995). Lunneryd, Ugland & Aspholm (2001) subsequently developed a model illustrating that, in the confines of a Norwegian fjord, heavy larval sealworm infections in fish may be maintained by a relatively small number of seals. Evidently, fluctuations in the size of the seal population above this 'threshold' number have little effect on sealworm abundances in intermediate hosts. Similarly, abundances of sealworm in Icelandic cod have not declined significantly although the size of the grey seal population decreased by ~ 50% between 1986 and 1998 (Ólafsdóttir, 2001).

On the other hand, increasing abundances of larval sealworm in eastern Canadian groundfish, seem to parallel the growth of the grey seal population (Fig. 4) (McClelland *et al.* 1985, 2000; McClelland & Martell, 2001 *a*). In the southern Gulf of St Lawrence, infection parameters in Canada plaice had apparently remained stable since the 1950s, but increased dramatically after an annual commercial hunt/pup cull, which was undertaken in the southeastern Gulf beginning in 1967, was discontinued in 1982. Increases in abundance of the parasite in the 1980s seem disproportionately high, however, in comparison to the modest growth of the seal population (Zwanenburg & Bowen, 1990), and may have been promoted by a warming trend in seawater temperatures. The recent collapse of sealworm abundances in plaice from the central Scotian Shelf (Fig. 5), despite their proximity to the large and rapidly growing Sable Island grey seal colony, is further evidence that the influence of definitive host populations on infection parameters in groundfish may be mitigated by other factors (McClelland, 2001 *b*).

'Worm medicine'. Odense (1978) suggests that, as an alternative to population controls, seals might be treated with anthelmintics or vaccines which would render them unsuitable hosts for sealworms. Such treatments are used routinely for control of parasitic diseases in domestic or zoo animals where the patients are available for follow-up treatment and re-exposure to the parasites can be prevented, but unfortunately they seem impractical for controlling sealworms in free-living marine mammals (McClelland *et al.* 1983). Although pups and mature seals in large island or ice-breeding colonies are accessible for treatment, pups do not become infected with sealworms until they are weaned and foraging for themselves. Moreover, both pups and adult seals are subsequently re-infected throughout the year. In many parasitic diseases, the host develops a resistance to re-infection but seals appear to adapt to the presence of the parasite and support progressively heavier infections as they mature (McClelland, 1980*b*). As seals do not seem to develop immunity to *Pseudoterranova* spp. in nature, production of vaccines that promote effective immunity to these parasites may prove extremely difficult.

In light of recent use of liposome, sustained-release vehicles in the delivery of immuno-contraceptives to free-living seals (Brown *et al.* 1997 *a*, *b*), however, the possibility of controlling *Pseudoterranova* infections in seals with anthelmintics or vaccines merits further investigation. Ivermectin, for example, has proven lethal to L3, L4 and adult sealworms in an *in vitro* system (Manley & Embil, 1989) and, when injected into seals in encapsulated form, might prove efficacious, not only in eliminating existing infections, but also in preventing re-infection for periods of weeks or even months. Similarly, a single administration of a vaccine incorporating liposome-encapsulated antigenic material and an appropriate adjuvent could promote long lasting immunity to sealworm infection in wild seals.

CONCLUSIONS

Sealworms, larval ascaridoid nematodes belonging to the *Pseudoterranova decipiens* species complex, infect the flesh of numerous species of marine and euryhaline fish and have proven a chronic and costly cosmetic problem for seafood processors. Over the past two decades there have been significant advances in our knowledge of the phylogeny, life cycles and distributions of sealworms, and our understanding of factors which influence the distribution of these parasites in their intermediate and final hosts. Much research has also been conducted on methods of mitigating the problem through innovations in seafood processing technology and biological controls. While better solutions to the problem may be close at hand, inefficient and costly candling procedures are still employed for the detection and removal of the nematodes in fish fillets at most processing plants. Moreover, new biological and quantitative information on sealworms remains to be incorporated into predictive models describing the impact of seal management strategies on sealworm dynamics. The relationship between seal abundances and sealworm infection parameters in groundfish is complex, and declines in seal populations through control programs or natural mortalities do not necessarily result in decreasing sealworm abundances in fish. Ironically, the problem has become almost irrelevant in some areas in recent years not as a result of falling infection parameters in fish stocks or improvements in processing technology, but as a consequence of closure of fisheries or drastically reduced quotas due to the collapse of the groundfish stocks.

ACKNOWLEDGEMENTS

The author thanks Jason Melendy of the Gulf fisheries Institute, Moncton, New Brunswick, Canada for preparing the map (Fig. 1) showing distributions of sealworm species and John Smith of Aberdeen, Scotland for providing a facsimile copy for his recent (1999) review article. Mark Tupper and Ione von Herbing photographed the infected plaice (Fig. 12).

REFERENCES

ACKMAN, R. G. & GJELSTAD, R. T. (1975). Gas-chromatographic resolution of isomeric pentanols and pentanones in the identification of volatile alcohols and ketones in the codworm *Terranova decipiens*. *Analytical Biochemistry* **67**, 684–687.

AMIN, O. M., EIDELMAN, W. S., DOMKE, W., BAILEY, J. & PFEIFER, G. (2000). An unusual case of Anisakiasis in California, U.S.A. *Comparative Parasitology* **67**, 71–75.

APPLETON, T. E. & BURT, M. D. B. (1991). Biochemical characterization of third-stage larval sealworm, *Pseudoterranova decipiens* (Nematoda: Anisakidae), in eastern Canada using isoelectric focusing of soluble proteins. *Canadian Journal of Fisheries and Aquatic Sciences* **48**, 1800–1803.

ARTHUR, J. R. & ALBERT, E. (1993). Use of parasites for separating stocks of Greenland halibut (*Reinhardtius hippoglossoides*) in the Canadian northwest Atlantic. *Canadian Journal of Fisheries and Aquatic Sciences* **50**, 2175–2181.

ARYEE, E. B. & POEHLMAN, W. F. S. (1991). A neural-network-based system to recognize parasites/seal worms on cod fish images. *Engineering Applications of Artificial Intelligence* **4**, 341–350.

ASPHOLM, P. E., UGLAND, K. E., JØDESTØL, K. A. & BERLAND, B. (1995). Sealworm (*Pseudoterranova decipiens*) infection in common seals (*Phoca vitulina*) and potential intermediate hosts from the outer Oslofjord. *International Journal for Parasitology* **25**, 367–373.

AZNAR, F. J., BALBUENA, J. A., FERNÁNDEZ, M. & RAGA, J. A. (2001). Establishing the relative importance of sympatric definitive hosts in the transmission of the sealworm, *Pseudoterranova decipiens*: a host-community approach. In *Sealworms in the North Atlantic: Ecology and Population Dynamics*. (ed. Desportes, G. & McClelland, G.), *NAMMCO Scientific Publications* **3**, 161–171. Tromsø Norway, The North Atlantic Marine Mammal Commission.

BIER, J. W., DEARDORFF, G. F., JACKSON, G. J. & RAYBOURNE, R. B. (1987). Human anisakiasis. In *Baillier's Clinical Tropical Medicine and Communicable Diseases* **2**, pp. 723–733. London, Harcourt Brace Jovanavich.

BOUREE, P., PAUGAM, A. & PETITHORY, J.-C. (1995). Anisakidosis: report of 25 cases and review of the literature. *Comparative Immunology and Infectious Disease* **18**, 75–84.

BOWEN, W. D. (ed.). (1990). *Population biology of sealworm (Pseudoterranova decipiens) in relation to its intermediate and seal hosts. Canadian Bulletin of Fisheries and Aquatic Sciences* **222**, 306 pp.

BOWEN, W. D. & HARRISON, G. D. (1996). Comparison of harbour seal diets in two inshore habitats of Atlantic Canada. *Canadian Journal of Zoology* **74**, 125–135.

BOWEN, W. D., LAWSON, J. W. & BECK, B. (1993). Seasonal and geographic variation in the species composition and size of prey consumed by grey seals (*Halichoerus* grypus) on the Scotian Shelf. *Canadian Journal of Fisheries and Aquatic Sciences* **50**, 168–178.

BRATTEY, J., BISHOP, C. A. & MYERS, R. A. (1990). Geographic distribution and abundance of *Pseudoterranova decipiens* (Nematoda: Ascaridoidea) in the musculature of Atlantic cod, *Gadus morhua*, from Newfoundland and Labrador. In *Population Biology of Sealworm (Pseudoterranova decipiens) in Relation to its Intermediate and Seal Hosts*. (ed. Bowen, W. D.) *Canadian Bulletin of Fisheries and Aquatic Sciences* **222**, 67–82.

BRATTEY, J. & DAVIDSON, W. S. (1996). Genetic variation within *Pseudoterranova decipiens* (Nematoda: Ascaridoidea) from Canadian Atlantic marine fishes and seals: characterization by RFLP analysis of genomic DNA. *Canadian Journal of Fisheries and Aquatic Sciences* **53**, 333–341.

BRATTEY, J. & NI, I.-H. (1992). Ascaridoid nematodes from the stomach of harp seals, *Phoca groenlandicus*, from Newfoundland and Labrador. *Canadian Journal of Fisheries and Aquatic Sciences* **49**, 956–966.

BRATTEY, J. & STENSON, G. B. (1993). Host specificity and abundance of parasitic nematodes (Ascaridoidea) from the stomachs of five phocid species from Newfoundland and Labrador. *Canadian Journal of Zoology* **71**, 2156–2166.

BRATTEY, J. & STOBO, W. T. (*Rapporteurs*), BJØRGE, A., BURT, M. D. B., DES CLERS, S., FANNING, L. P., HAUKSSON, E., JARECKA, L., LANDRY, T., MANSFIELD, A., MCCLELLAND, G., MOHN, R., MÖLLER, H., MYERS, R. A., NI, I. H., PÁLSSON, J., SMITH, J. W., THOMPSON, D. & WOOTTEN, R. (1990). Group report 2: Infection of definitive hosts. In *Population Biology of Sealworm (Pseudoterranova decipiens) in Relation to its Intermediate and Seal Hosts*. (ed. Bowen, W. D.) *Canadian Bulletin of Fisheries and Aquatic Sciences* **222**, 139–145.

BRISTOW, G. A. & BERLAND, B. (1992). On the ecology and distribution of *Pseudoterranova decipiens* (Nematoda: Anisakidae) in an intermediate host, *Hippoglossoides platessoides*, in northern Norwegian waters. *International Journal for Parasitology* **22**, 203–208.

BROWN, R. G., BOWEN, W. D., EDDINGTON, J. D., KIMMINS, W. C., MEZEI, M., PARSONS, J. L. & POHAJDAK, B. (1997a). Evidence for a long-lasting single administration contraceptive vaccine in wild grey seals. *Journal of Reproductive Immunology* **35**, 43–51.

BROWN, R. G., BOWEN, W. D., EDDINGTON, J. D., KIMMINS, W. C., MEZEI, M., PARSONS, J. L. & POHAJDAK, B. (1997b). Temporal trends in antibody production in captive grey, harp and hooded seals to a single administration immunocontraceptive vaccine. *Journal of Reproductive Immunology* **35**, 53–64.

BUBLITZ, C. G. & CHOUDHURY, G. S. (1992). Effect of light intensity and color on worker productivity and parasite detection efficiency during candling of cod fillets. *Journal of Aquatic Food Product Technology* **1**, 75–89.

BULLINI, L., ARDUINO, P., CIANCHI, R., NASCETTI, G., D'AMELIO, S., MATTIUCHI, S., PAGGI, L., ORECCHIA, P., PLÖTZ, J., BERLAND, B., SMITH, J. W. & BRATTEY, J. (1997). Genetic and ecological research on Anisakid endoparasites of fish and marine mammals in the Antarctic and Artic-Boreal regions. In *Antarctic*

Communitites: Species, Structure and Survival. (ed. Battaglia, B., Valencia, J. & Walton, D. W. H.), pp. 362–383. Cambridge, Cambridge University Press.

BURT, M. D. B. (1994). The sealworm situation. In *Parasitic and infectious diseases: epidemiology and ecology* (ed. Scott, M.E. & Smith G.), pp. 347–362. Academic Press, San Diego.

BURT, M. D. B., SMITH, J. W., CAMPBELL, J. D. & LIKELY, C. G. (1990). Serial passage of larval *Pseudoterranova decipiens* (Nematoda: Ascaridoidea) in fish. *Canadian Journal of Fisheries and Aquatic Sciences* **47**, 693–695.

CAMPANA, S. E., MOHN, R. K., SMITH, S. J. & CHOUINARD, G. A. (1995). Spatial implications of a temperature-based growth model for Atlantic cod (*Gadus morhua*) off the eastern coast of Canada. *Canadian Journal of Fisheries and Aquatic Sciences* **52**, 2445–2456.

CARAVAJAL, J. & CATTAN, P. E. (1985). A study of the anisakid infection in the Chilean hake, *Merluccius gayi* (Guichenot), 1848. *Parasitology Research* **3**, 245–250.

CHENG, T. C. (1976). The natural history of anisakiasis in animals. *Journal of Food and Milk Technology* **39**, 32–46.

CHOUDHURY, G. S. & BUBLITZ, C. G. (1994). Electromagnetic method for detection of parasites in fish. *Journal of Aquatic Food Product Technology* **3**, 49–63.

COSCIA, M. R. & ORESTE, U. (2000). Plasma and bile antibodies of the teleost *Trematomus bernacchii* specific for the nematode *Pseudoterranova decipiens*. *Diseases of Aquatic Organisms* **4**, 37–42.

DES CLERS, S. (1989). Modelling regional differences in 'sealworm' *Pseudoterranova decipiens* (Nematoda, Ascaridodidea), infections in some North Atlantic cod, *Gadus morhua*, stocks. *Journal of Fish Biology* **35** (Supplement A), 187–192.

DES CLERS, S. (1990). Modeling the life cycle of the sealworm (*Pseudoterranova decipiens*) in Scottish waters. In *Population Biology of Sealworm (Pseudoterranova decipiens) in Relation to its Intermediate and Seal Hosts*. (ed. Bowen, W. D.) *Canadian Bulletin of Fisheries and Aquatic Sciences* **222**, 273–288.

DES CLERS, S. & ANDERSON, K. (1995). Sealworm (*Pseudoterranova decipiens*) transmission to fish trawled from Hvaler, Oslofjord, Norway. *Journal of Fish Biology* **46**, 8–17.

DES CLERS, S. & MOHN, R. (*Rapporteurs*), BOWEN, D., FANNING, P., LANDRY, T., MIZRA, R., MYERS, R., NI, I., SMITH, J. & ZWANENBURG, K. (1990). Group report 4: Models. In *Population Biology of Sealworm (Pseudoterranova decipiens) in Relation to its Intermediate and Seal Hosts*. (ed. Bowen, W. D.) *Canadian Bulletin of Fisheries and Aquatic Sciences* **222**, 39–145.

DES CLERS, S. & PRIME, J. (1996). Chapter 17. Seals and fisheries interactions: models in the Firth of Clyde, Scotland. In *Aquatic Predators and their Prey*. (ed. Greenstreet, S. P. R. & Tasker, M. L.) *Fishing News Books*, pp. 124–132. Oxford, Blackwell Science.

DESPORTES, G. & MCCLELLAND, G. (ed.) (2001). *Sealworms in the North Atlantic: Ecology and Population Dynamics, NAMMCO Scientific Publications* **3**, Tromsø Norway, The North Atlantic Marine Mammal Commission.

FAIR, P. A. (2000). Health-related parasites in seafoods. In *Marine and Freshwater Products Handbook*, pp. 761–776. Lancaster PA, Technomic Publishing Co., Inc.

GEORGE-NASCIMENTO, M. & LLANOS, A. (1995). Micro-evolutionary implications of allozymic and morphometric variations in sealworms *Pseudoterranova* sp. (Ascaridoidea: Anisakidae) among sympatric hosts from the southeastern Pacific Ocean. *International Journal for Parasitology* **25**, 1163–1171.

GEORGE-NASCIMENTO, M. & URRUTIA, X. (2000). *Pseudoterranova cattani* sp. nov. (Ascaridoidea: Anisakidae), a parasite of the South American sea lion *Otaria byronia* De Blainville from Chile. *Revista Chilena de Historia Natural* **73**, 93–98.

GIBSON, D. I. (1983). The systematics of ascaridoid nematodes. A current assessment. In *Concepts in Nematode Systematics*. (ed. Stone, A. R., Platt, H. M. & Khalil, H. F.) *Systematics Association Special Volume No. 22*, pp. 321–338. New York & London, Academic Press Inc.

HAFSTEINSSON, H. & RIZVI, S. S. H. (1987). A review of the sealworm problem: Biology, implications and solutions. *Journal of Food Protection* **50**, 70–84.

HEMMINGSEN, W., LOMBARDO, I. & MacKENZIE, K. (1991). Parasites as biological tags, *Gadus morhua* L., in northern Norway: a pilot study. *Fisheries Research* **12**, 365–373.

HERRERAS, M. V., AZNAR, F. J., BALBUENA, J. A. & RAGA, J. A. (2000). Anisakid larvae in the musculature of the Argentian hake, *Merluccius hubbsi*. *Journal of Food Protection* **63**, 141–143.

HOLMES, J. C. (1990). Helminth communities in marine fishes. In *Parasite Communities: Patterns and Processes*. (ed. Esch, G. W., Bush, A. O. & Aho, J. M.), pp. 101–130. New York, Chapman and Hall.

ISHIKURA, H., KIKUCHI, K., NAGASAWA, K., OOIWA, T., TAKAMIYA, H., SATO, N. & SUGANE, K. (1993). Anisakidae and anisakidosis. *Progress in Clinical Parasitology* **3**, 43–102.

JENSEN, T. (1997). Experimental infection/transmission of sculpins (*Myoxocephalus scorpius*) and cod (*Gadus morhua*) by sealworm (*Pseudoterranova decipiens*) larvae. *Parasitology Research* **83**, 380–382.

JENSEN, T., ANDERSEN, K. & DES CLERS, S. (1994). Sealworm (*Pseudoterranova decipiens*) infections in demersal fish from two areas of Norway. *Canadian Journal of Zoology* **72**, 598–608.

KØIE, M., BERLAND, B. & BURT, M. D. B. (1995). Development to third-stage larvae occurs in the eggs of *Anisakis simplex* and *Pseudoterranova decipiens* (Nematoda, Ascaridoidea, Anisakidae). *Canadian Journal of Fisheries and Aquatic Sciences* **52** (Suppl 1), 134–139.

LUNNERYD, S.-G., UGLAND, K. I. & ASPHOLM, P. E. (2001). Sealworm (*Pseudoterranova decipiens*) infection in the benthic cottid (*Taurulus bubalis*) in relation to population increase of harbour seal (*Phoca vitulina*) in Skagerrak, Sweden. In *Sealworms in the North Atlantic: Ecology and Population Dynamics*. (ed. Desportes, G. & McClelland, G.) *NAMMCO Scientific Publications* **3**, 47–55. Tromsø Norway, The North Atlantic Marine Mammal Commission.

MALOUF, A. H. (1986). *Report of the Royal Commission on Seals and Sealing in Canada*. Vol. 3. Pt. 5. *Biological issues*. Ottawa, Canada.

MANLEY, K. M. & EMBIL, J. A. (1989). *In vitro* effect of ivermectin on *Pseudoterranova decipiens* survival. *Journal of Helminthology* **63**, 72–74.

MARCOGLIESE, D. J. (1997). Fecundity of sealworm (*Pseudoterranova decipiens*) infecting grey seals (*Halichoerus grypus*) in the Gulf of St. Lawrence, Canada: lack of density-dependent effects. *International Journal for Parasitology* **27**, 1401–1409.

MARCOGLIESE, D. J. (2001a). Review of experimental and natural invertebrate hosts of sealworm (*Pseudoterranova decipiens*) and its distribution and abundance in macroinvertebrates in eastern Canada. In *Sealworms in the North Atlantic: Ecology and Population Dynamics*. (ed. Desportes, G. & McClelland, G.) *NAMMCO Scientific Publications* **3**, 27–37. Tromsø Norway, The North Atlantic Marine Mammal Commission.

MARCOGLIESE, D. J. (2001b). Distribution and abundance of sealworm (*Pseudoterranova decipiens*) and other anisakid nematodes in fish and seals in the Gulf of St. Lawrence: potential importance of climatic conditions. In *Sealworms in the North Atlantic: Ecology and Population Dynamics*. (eds. Desportes, G. & McClelland, G.) *NAMMCO Scientific Publications* **3**, 113–128. Tromsø Norway, The North Atlantic Marine Mammal Commission.

MARCOGLIESE, D. J., BOILY, F. & HAMMILL, M. O. (1996). Distribution and abundance of stomach nematodes (Anisakidae) among grey seals (*Halichoerus grypus*) and harp seals (*Phoca groenlandica*) in the Gulf of St. Lawrence. *Canadian Journal of Fisheries and Aquatic Sciences* **53**, 2829–2836.

MARGOLIS, L. (1977). Public health aspects of "codworm infections". A review. *Journal of the Fisheries Research Board of Canada* **34**, 887–898.

MARTELL, D. J. & McCLELLAND, G. (1994). Diets of sympatric flatfishes, *Hippoglossoides platessoides* (Fabricius), *Pleuronectes ferrugineus* (Storer), *Pleuronectes americanus* (Walbaum), from Sable Island Bank, Canada. *Journal of Fish Biology* **44**, 821–848.

MARTELL, D. J. & McCLELLAND, G. (1995). Transmission of *Pseudoterranova decipiens* (Nematoda: Ascaridoidea) via benthic macrofauna to sympatric flatfishes (*Hippoglossoides platessoides, Pleuronectes ferrugineus, Pleuronectes americanus*) on Sable Island Bank, Canada. *Marine Biology* **122**, 129–135.

MATTIUCCI, S., PAGGI, L., NASCETTI, G., ISHIKURA, H., KIKUCHI, K., SATO, N., CIANCHI, R. & BULLINI, L. (1998). Allozyme and morphological identification of *Anisakis, Contracaecum* and *Pseudoterranova* (Nematoda: Ascaridoidea) from Japanese waters. *Systematic Parasitology* **40**, 81–92.

McCLELLAND, G. (1980a). *Phocanema decipiens*: molting in seals. *Experimental Parasitology* **49**, 128–136.

McCLELLAND, G. (1980b). *Phocanema decipiens*: growth, reproduction, and survival in seals. *Experimental Parasitology* **49**, 175–187.

McCLELLAND, G. (1980c). *Phocanema decipiens*: pathology in seals. *Experimental Parasitology* **49**, 405–419.

McCLELLAND, G. (1982). *Phocanema decipiens* (Nematoda: Anisakinae): experimental infections in marine copepods. *Canadian Journal of Zoology* **60**, 502–509.

McCLELLAND, G. (1990). Larval sealworm (*Pseudoterranova decipiens*) infections in benthic macrofauna. In *Population Biology of Sealworm (Pseudoterranova decipiens) in Relation to its Intermediate and Seal Hosts*. (ed. Bowen, W. D.) *Canadian Bulletin of Fisheries and Aquatic Sciences* **222**, 47–65.

McCLELLAND, G. (1995). Experimental Infection of fish with larval sealworm, *Pseudoterranova decipiens* (Nematoda, Anisakinae), transmitted by amphipods. *Canadian Journal of Fisheries and Aquatic Sciences* **52** (Suppl 1), 140–155.

McCLELLAND, G. (2000). Natural transmission of larval sealworm *Pseudoterranova decipiens* to juvenile Canadian plaice *Hippoglossoides platessoides* and other small benthic consumers. *Bulletin of the Canadian Society of Zoologists* **31**, 83.

McCLELLAND, G. (*Rapporteur*), BJØRGE, A., BRATTEY, J., BURT, M., DES CLERS, S., FANNING, P., HARE, G., JARECKA, L., LANDRY, T., MARGOLIS, L., McGLADDERY, S., MISRA, R., MOHN, R., MÖLLER, H., PÁLSSON, J., SMITH, J., STOBO, W. & WOOTTEN, R. (1990). Group report 1: Hatching and infection of intermediate hosts. In *Population Biology of Sealworm (Pseudoterranova decipiens) in Relation to its Intermediate and Seal Hosts*. (ed. Bowen, W. D.) *Canadian Bulletin of Fisheries and Aquatic Sciences* **222**, 17–25.

McCLELLAND, G. & MARCOGLIESE, D. J. (1994). Larval anisakine nematodes as biological indicators of cod (*Gadus morhua*) populations in the southern Gulf of St. Lawrence and on the Breton Shelf, Canada. *Bulletin of the Scandinavian Society for Parasitology* **4**, 97–116.

McCLELLAND, G. & MARTELL, D. J. (2001*a*). Surveys of larval sealworm (*Pseudoterranova decipiens*) infection in various fish species sampled from Nova Scotian waters between 1988 and 1996, with an assessment of examination procedures. In *Sealworms in the North Atlantic : Ecololgy and Population Dynamics*. (ed. Desportes, G. & McClelland, G.) *NAMMCO Scientific Publications* **3**, 57–76. Tromsø Norway, The North Atlantic Marine Mammal Commission.

McCLELLAND, G. & MARTELL, D. J. (2001*b*). Spatial and temporal distributions of larval sealworm, *Pseudoterranova decipiens* (Nematoda: Anisakinae), in *Hippoglossoides platessoides* (Pleuronectidae) in the Canadian Maritime Region from 1993 to 1999. In *Sealworms in the North Atlantic : Ecololgy and Population Dynamics*. (ed. Desportes, G. & McClelland, G.) *NAMMCO Scientific Publications* **3**, 77–94. Tromsø Norway, The North Atlantic Marine Mammal Commission.

McCLELLAND, G., MISRA, R. K. & MARCOGLIESE, D. J. (1983). Variations in abundance of larval anisakines, sealworm (*Phocanema decipiens*) and related species in cod and flatfish from the southern Gulf of St. Lawrence (4T) and the Breton Shelf (4Vn). *Canadian Technical Report of Fisheries and Aquatic Sciences* **1201**, ix + 51 pp.

McCLELLAND, G., MISRA, R. K. & MARTELL, D. J. (1985). Variations in abundance of larval anisakines, sealworm (*Pseudoterranova decipiens*) and related species, in eastern Canadian flatfish. *Canadian Technical Report of Fisheries and Aquatic Sciences* **1392**, xi + 57 pp.

McCLELLAND, G., MISRA, R. K. & MARTELL, D. J. (1990). Larval anisakine nematodes in various fish species from Sable Island Bank and vicinity. In *Population Biology of Sealworm (Pseudoterranova decipiens) in Relation to its Intermediate and Seal Hosts*. (ed. Bowen, W. D.) *Canadian Bulletin of Fisheries and Aquatic Sciences* **222**, 83–118.

McCLELLAND, G., MISRA, R. K. & MARTELL, D. J. (2000). Spatial and temporal distributions of larval sealworm (*Pseudoterranova decipiens*, Nematoda: Anisakinae), in *Hippoglossoides platessoides* (Pleuronectidae) in eastern Canada from 1980 to 1990. *ICES Journal of Marine Science* **57**, 69–88.

McCONNELL, C. J., MARCOGLIESE, D. J. & STACEY, M. W. (1997). Settling rate and dispersal of sealworm eggs (Nematoda) determined using a revised protocol for myxozoan spores. *Journal of Parasitology* **83**, 203–206.

MEASURES, L. N. (1996). Effect of temperature and salinity on development and survival of eggs and free-living larvae of sealworm (*Pseudoterranova decipiens*). *Canadian Journal of Fisheries and Aquatic Sciences* **53**, 2804–2807.

MEASURES, L. N. & HONG, H. (1995). The number of moults in the egg of sealworm, *Pseudoterranova decipiens* (Nematoda: Ascaridoidea): an ultrastructural study. *Canadian Journal of Fisheries and Aquatic Sciences* **52** (Suppl 1), 156–160.

MOHN, R. K. (1990). A synthesis to explore internal consistency and sensitivity of sealworm dynamics. In *Population Biology of Sealworm (Pseudoterranova decipiens) in Relation to its Intermediate and Seal Hosts*. (ed. Bowen, W. D.) *Canadian Bulletin of Fisheries and Aquatic Sciences* **222**, 261–272.

MOHN, R. & BOWEN, W. D. (1996). Grey seal predation on the eastern Scotian Shelf: modeling the impact on Atlantic cod. *Canadian Journal of Fisheries and Aquatic Sciences* **53**, 2722–2738.

MÖLLER, H. & KLATT, S. (1990). Smelt as host of the sealworm (*Pseudoterranova decipiens*) in the Elbe estuary. In *Population Biology of Sealworm (Pseudoterranova decipiens) in Relation to its Intermediate and Seal Hosts*. (ed. Bowen, W. D.) *Canadian Bulletin of Fisheries and Aquatic Sciences* **222**, 129–138.

MÖLLER, H., HOLST, S., LÜCHTENBERG, H. & PETERSEN, F. (1991). Infection of eel *Anguilla anguilla* from the River Elbe estuary with two nematodes, *Anguillicola crassus* and *Pseudoterranova decipiens*. *Diseases of Aquatic Organisms* **11**, 193–199.

MYERS, R. A. & BRATTEY, J. (1990). Statistical models for the age-specific and length-specific aggregation of *Pseudoterranova decipiens* (Nematoda: Ascaridoidea) in Atlantic cod, *Gadus morhua*. In *Population Biology of Sealworm (Pseudoterranova decipiens) in Relation to its Intermediate and Seal Hosts*. (ed. Bowen, W. D.) *Canadian Bulletin of Fisheries and Aquatic Sciences* **222**, 289–301.

ODENSE, P. H. (1978). Some aspects of the codworm problem. *Fisheries and Environment Canada Fisheries and Marine Service Industry Report* **106**. Halifax, Canada.

ÓLAFSDÓTTIR, D. (2001). A review of the ecology of sealworm, *Pseudoterranova decipiens* (Nematoda: Ascaridoidea) in Icelandic waters. In *Sealworms in the North Atlantic : Ecology and Population Dynamics*. (ed.

Desportes, G. & McClelland, G.) *NAMMCO Scientific Publications* **3**, 95–111. Tromsø Norway, The North Atlantic Marine Mammal Commission.

OSHIMA, T. (1987). Anisakiasis – is the sushi bar guilty? *Parasitology Today* **3**, 44–48.

PAGGI, L., NASCETTI, G., CIANCHI, R., ORECCHIA, P., MATIUCCI, S., D'AMELIO, S., BERLAND, B., BRATTEY, J., SMITH, J. W. & BULLINI, L. (1991). Genetic evidence for three species within *Pseudoterranova decipiens* (Nematoda, Ascaridida, Ascaridoidea) in the North Atlantic and Norwegian and Barents Seas. *International Journal for Parasitology* **21**, 195–212.

PAGGI, L., MATTIUCCI, S., GIBSON, D. I., BERLAND, B., NASCETTI, G., CIANCHI, R. & BULLINI, L. (2000). *Pseudoterranova decipiens* species A and B (Nematoda: Ascaridoidea): nomenclatural designation, morphological diagnostic characters and genetic markers. *Systematic Parasitology* **45**, 185–197.

PALM, H. W. (1999). Ecology of *Pseudoterranova decipiens* (Krabbe, 1878) (Nematoda: Anisakidae) from Antarctic waters. *Parasitology Research* **85**, 638–646.

PALM, H., ANDERSEN, K. & KLÖSER, H. (1994). Occurrence of *Pseudoterranova decipiens* (Nematoda) in fish from the southeastern Weddell Sea (Antarctic). *Polar Biology* **14**, 539–544.

PLATT, N. E. (1975). Infestation of cod (*Gadus morhua* L.) with the larvae of codworm (*Terranova decipiens*) and herringworm, *Anisakis* sp. (Nematoda: Ascaridata) in North Atlantic and Arctic waters. *Journal of Applied Ecology* **12**, 437–450.

PLATT, N. E. (1976). Codworm – a possible biological indicator of the degree of mixing of Greenland and Iceland cod stocks. *Jounal du Conseil* **37**, 41–45.

PRIEBE, K., HUBER, C., MÄRTLBAUER, E. & TERPLAN, G. (1991). Detection of antibodies against larvae of *Anisakis simplex* by ELISA. (in German). *Journal of Veterinary Medicine, Series B* **38**, 209–214.

QUARRAR, P. (1999). *The Life Cycle, Morphology, Ultrastructure, Biochemical Characterization, and Sibling Status of the Sealworm (Pseudoterranova decipiens) in the North West Atlantic.* PhD Thesis. University of New Brunswick, Fredericton, NB, Canada.

RAMAKRISHNA, N. R. & BURT, M. D. B. (1991). Tissue response to invasion by *Pseudoterranova decipiens* (Nematoda: Ascaridoidea). *Canadian Journal of Fisheries and Aquatic Sciences* **48**, 1623–1628.

ROWLING, T. M., PALM, H. W. & ROSENTHAL, H. (1998). Parasitisation with *Pseudoterranova decipiens* (Nematoda) influences the survival rate of the European smelt *Osmerus eperlanus* retained by a screen wall of a nuclear power plant. *Diseases of Aquatic Organisms* **32**, 233–236.

SCOTT, D. M. & FISHER, H. D. (1958). Incidence of the ascarid *Porrocaecum decipiens* in the stomachs of three species of seals along the southern Canadian Atlantic mainland. *Journal of the Fisheries Research Board of Canada* **15**, 495–516.

SCOTT, W. B. & SCOTT, M. G. (1988). Atlantic fishes of Canada. *Canadian Bulletin of Fisheries and Aquatic Sciences* **219**, 731 pp.

SILVA, J. R. M. C., STAINES, N. A., PARRA, O. M. & HERNANDEZ-BLAZQUEZ, F. J. (1999). Experimental studies on the response of the fish (*Notothenia*

coriiceps Richardson, 1844) to parasite (*Pseudoterranova decipiens* Krabbe, 1878) and other irritant stimuli at Antarctic temperatures. *Polar Biology* **22**, 417–424.

SMITH, J. W. (1999). Ascaridoid nematodes and pathology of the alimentary tract and its associated organs in vertebrates. *Helminthological Abstracts* **68**, 49–96.

SMITH, J. W. & WOOTTEN, R. (1975). Experimental studies on the migration of *Anisakis* sp. larvae (Nematoda: Ascaridata) into the flesh of herring *Clupea harengus* L. *International Journal for Parasitology* **5**, 133–136.

SPRENGEL, G. & LÜCHTENBERG, H. (1991). Infection by endoparasites reduces swimming speed of European smelt *Osmerus operlanus* and European eel *Anguilla anguilla*. *Diseases of Aquatic Organisms* **11**, 31–35.

STOBO, W. T. & FOWLER, G. M. (2001). Sealworm (*Pseudoterranova decipiens*) dynamics in Sable Island grey seals (*Halichoerus grypus*): seasonal fluctuations and other changes in worm infections during the 1980s. In *Sealworms in the North Atlantic: Ecology and Population Dynamics.* (ed. Desportes, G. & McClelland, G.) *NAMMCO Scientific Publications* **3**, 129–147. Tromsø Norway, The North Atlantic Marine Mammal Commission.

STOBO, W. T., BECK, B. & FANNING, L. P. (1990). Seasonal sealworm (*Pseudoterranova decipiens*) abundance in grey seals (*Halichoerus grypus*). In *Population Biology of Sealworm (Pseudoterranova decipiens) in Relation to its Intermediate and Seal Hosts.* (ed. Bowen, W. D.) *Canadian Bulletin of Fisheries and Aquatic Sciences* **222**, 147–162.

STOBO, W. T., BECK, B. & HORNE, J. K. (1990). Seasonal movements of grey seals (*Halichoerus grypus*) in the Northwest Atlantic. In *Population Biology of Sealworm (Pseudoterranova decipiens) in Relation to its Intermediate and Seal Hosts.* (ed. Bowen, W. D.) *Canadian Bulletin of Fisheries and Aquatic Sciences* **222**, 199–213.

SWAIN, D. P. (1999). Changes in the distribution of Atlantic cod (*Gadus morhua*) in the southern Gulf of St. Lawrence – effects of environmental change or change in environmental preferences? *Fisheries Oceanography* **8**, 1–17.

U.S. FOOD AND DRUG ADMINISTRATION/CENTER FOR FOOD SAFETY AND APPLIED NUTRITION. (1992). Bad bug book. *Anisakis simplex* and related worms. In *Foodborne Pathogenic Microorganisms and Natural Toxins Handbook*, pp. 1–3. World Wide Web http://vm.cfsan.fda.gov/-mow/chap25.html

WOOTTEN, R. & WADDELL, J. F. (1977). Studies on the biology of larval nematodes from the musculature of cod and whiting in Scottish waters. *Journal du Conseil* **37**, 266–273.

YOUNG, P. C. (1972). The relationship between the presence of larval Anisakine nematodes in cod and marine mammals in British home waters. *Journal of Applied Ecology* **9**, 459–485.

ZWANENBURG, K. T. C. & BOWEN, W. D. (1990). Population trends of the grey seal (*Halichoerus grypus*) in eastern Canada. In *Population Biology of Sealworm (Pseudoterranova decipiens) in Relation to its Intermediate and Seal Hosts.* (ed. Bowen, W. D.) *Canadian Bulletin of Fisheries and Aquatic Sciences* **222**, 185–197.

Subject Index

Index compiled by Dr Laurence Errington

Page numbers are for the first page number only of each article. The reader may find many mentions of the topic throughout the article.